Genetic Algorithms in Engineering and Computer Science

Genetic Algorithms in Engineering and Computer Science

Edited by

G. Winter

University of Las Palmas, Canary Islands, Spain

J. Périaux

Dassault Aviation, Saint Cloud, France

M. Galán
P. Cuesta

University of Las Palmas, Canary Islands, Spain

John Wiley & Sons
Chichester · New York · Brisbane · Toronto · Singapore

Copyright © 1995 by   John Wiley & Sons Ltd.
                         Baffins Lane, Chichester,
                         West Sussex PO19 1UD, England

                         National   01243 779777
                         International (+44) 1243 779777

Chapter 2 © English language only

Other Wiley Editorial Offices

John Wiley & Sons, Inc., 605 Third Avenue,
New York, NY 10158-0012, USA

Jacaranda Wiley Ltd, 33 Park Road, Milton,
Queensland 4064, Australia

John Wiley & Sons (Canada) Ltd, 22 Worcester Road,
Rexdale, Ontario M9W 1L1, Canada

John Wiley & Sons (SEA) Pte Ltd, 37 Jalan Pemimpin #05-04,
Block B, Union Industrial Building, Singapore 2057

British Library Cataloguing in Publication Data

A catalogue record for this book is available from the British Library

ISBN 0 471 95859 X

Produced from camera-ready copy supplied by the authors.
Printed and bound in Great Britain by Bookcraft (Bath) Ltd.

This book is printed on acid-free paper responsibly manufactured from sustainable forestation, for
which at least two trees are planted for each one used for paper production.

Contents

Preface

This book entitled *Genetic Algorithms in Engineering and Computer Sciences* has been motivated by the EUROGEN95 Short-Course (Dec. 1995, Las Palmas de Gran Canaria, Spain).

The aim of the book is to present, to an audience of pure mathematicians, computer scientists, physicists, practising engineers and postgraduate students, several "state of the art" lectures on an emergent optimization approach based on Artificial Intelligence coupled with Darwin's rules of evolution.

Genetic Algorithms were initially proposed by John Holland (University of Michigan) as a brand new approach to the old "global optimization" problem, and, after its maturation in several Universities and Research Institutions at both sides of the Atlantic, have moved on the field of applied technology (Automation, Control, Design, Economics, Electronics, etc.) with considerable success.

Genetic Algorithms and Evolution Strategies provide innovative stochastic non-deterministic improvement tools, very well suited to a parallel computational environment, able to deal with a wide range of complex, real-life problems that are not feasible with the classical deterministic methods.

On the other side, the spring of Genetic Algorithms and Evolution Strategies, branch of the river of Artificial Intelligence, seems promising and abundant. The results obtained in these first stages of its development forecast a fruitful flourishing in the next future.

This volume is divided into two sections which reflect the focus of the Short-Course. The main areas discussed include the theoretical and computational aspects in Genetic Algorithms and Evolution Strategies in one side, coupled with several applications and its computational implementation.

A big share in the success of this volume should be given to the following people, because of their kind cooperation, support and contribution:

J. L. Diaz (Univ. Complutense de Madrid and Sociedad Española de Matematica Aplicada, SEMA.)

E. Oñate (Univ. Politecnica de Cataluña and Sociedad Española de Metodos Numericos en la Ingenieria, SEMNI.)

F. Rubio (Excmo. Mgfco. Rector de la Universidad de Las Palmas de G.C.)

M. Hermoso (Excmo. Presidente del Gobierno de Canarias)

At the same time, the members of the Scientific Organizing Committee would like to address their warmest thanks to those institutions and associations for their support, in particular the European Union's CEC (Dir. Gen. XII), ECCOMAS, GAMNI/SMAI, SEMA, SEMNI, the University of Las Palmas de Gran Canaria, the Center for Numerical Methods in Engineering (CEANI), the Government of the Canary Islands and Dassault Aviation.

We would like to express our particular thanks to the lecturers whose kind cooperation has proven indispensable.

And, to conclude, we hope that the present volume be useful and interesting not only to the one who deals with theoretical aspects but to the practical engineer; after all, Nature has been quite successful during 3.5 billion years using Genetics.

Dedication

To the memory of Miguel Galante for whom life and work were so strongly motivated by nature.

Contributors

T. Bäck
Informatik Centrum Dortmund e.V.
Joseph-von-Fraunhofer Str. 20
D-44227 Dortmund
Germany

L. Bull
Faculty of Computer Studies and
Mathematics
University of the West of England
Frenchay Campus
Coldharbour Lane
Bristol BS16 1QY
UK

B. Carse
Faculty of Computer Studies and
Mathematics
University of the West of England
Frenchay Campus
Coldharbour Lane
Bristol BS16 1QY
UK

O. Cordón
Escuela Tecnica Superior de
Ingenieros de Informatica
Departamento de Ciencia de la
Computacion e Inteligencia Artificial
Universidad de Granada
18071 Granada
Spain

P. Cuesta
Centro de Aplicaciones Numéricas
en Ingeniería de la U.L.P.G.C.
(CEANI)
Edificio de Ingenierías I
Campus Universitario de Tafira
Tafira Baja
E-35017 Las Palmas de Gran Canaria
Spain

D. Doorly
Imperial College
Aeronautics Department
Prince Consort Road
London SW7 2BY
UK

T. C. Fogarty
Faculty of Computer Studies and
Mathematics
University of the West of England
Frenchay Campus
Coldharbour Lane
Bristol BS16 1QY
UK

M. Galán
Centro de Aplicaciones Numéricas
en Ingeniería de la U.L.P.G.C.
(CEANI)
Edificio de Infenierías I
Campus Universitario de Tafira
Tafira Baja
E-35017 Las Palmas de Gran Canaria
Spain

S. Garg
Advanced Controls Technology
Branch
NASA Lewis Research Center
21000 Brookpark Road
MS 77-1
Cleveland
OH 44135
USA

D. E. Goldberg
Department of General Engineering
University of Illinois at Urbana-
Champaign
117 Transportation Bldg.
104 South Matheus Av.
Urbana
Illinois 61801
USA

D. Greiner
Centro de Aplicaciones Numéricas
en Ingeniería de la U.L.P.G.C.
(CEANI)
Edificio de Infenierías I
Campus Universitario de Tafira
Tafira Baja
E-35017 Las Palmas de Gran Canaria
Spain

F. Herrera
Escuela Tecnica Superior de
Ingenieros de Informatica
Departamento de Ciencia de la
Computacion e Inteligencia Artificial
Universidad de Granada
18071 Granada
Spain

K. Krishnakumar
Department of Aerospace
Engineering
University of Alabama
Box 870280
Tuscaloosa
AL 35487-0280
USA

E. Laporte
Dassault Aviation
78 Quai Marcel Dassault
Cedex 300
F-92552 Saint Cloud Cedex
France

M. Lozano
Escuela Tecnica Superior de
Ingenieros de Informatica
Departamento de Ciencia de la
Computacion e Inteligencia Artificial
Universidad de Granada
18071 Granada
Spain

L. Magdalena
Dept. Matemática Aplicada a lad
Tenologías de la Información
E.T.S.I. Telecomunicación
Universidad Politécnica de Madrid
Ciudad Universitaria
E-28040 Madrid
Spain

B. Mantel
Dassault Aviation
78 Quai Marcel Dassault
Cedex 300
F-92552 Saint Cloud Cedex
France

F. Michielssen
University of Illinois at Urbana-
Champaign
Department of Electrical and
Computer Engineering
Electromagnetic Communication
Laboratory
1406 West Green Street
Urbana
IL 61801-2991
USA

R. Moreno-Diaz
CIICC-ULPGC
Facultad de Informatica
Campus Universitario de Tafira Baja
Edificio de Informatica y
Matematicas
35017 Las Palmas de Gran Canaria
Spain

H. W. Mühlenbein
GMD (Gesellschaft für Mathematik
und Datenverarbeitung MBH)
Schloß Birlinghoven
D-53754 Sankt Augustin
Germany

S. Narayanaswamy
Department of Aerospace
Engineering
University of Alabama
Box 870280
Tuscaloosa
AL 35487-0280
USA

J. Périaux
DGT / DEA
Dassault Aviation
78 Quai Marcel Dassault
Cedex 300
F-92552 Saint Cloud Cedex
France

C. Poloni
Universita degli Studi di Triesto
Dipartimento di Energetica
Via A. Valerio 10
I-34127 Trieste
Italy

D. Qualiarella
C.I.R.A.
Centro Italiano Ricerche
Aerospazialli
Via Maiorise
I-81043 Capua (CE)
Italy

A. Satyadas
Flexible Intelligence Group
LLC
P.O. Box 1477
Tuscaloosa
AL 35487-0280
USA

M. Schoenauer
Centre de Mathématiques Appliquées
CNRS - URA 756
Ecole Polytechnique
Route Saclay
F-91128 Palaiseau
France

H. P. Schwefel
University of Dortmund
Department of Computer Science
D-44221 Dortmund
Germany

M. Sefrioui
Dassault Aviation
78 Quai Marcel Dassault
Cedex 300
F-92552 Saint Cloud Cedex
France
and
Laforia-IBP-CNRS
Université Paris VI
4 place Jussicu
F-75252 Paris Cedex 05
France

B. Stoufflet
Dassault Aviation
78 Quai Marcel Dassault
Cedex 300
F-92552 Saint Cloud Cedex
France

F. Vavak
Faculty of Computer Studies and
Mathematics
University of the West of England
Frenchay Campus
Coldharbour Lane
Bristol BS16 1QY
UK

J. R. Velasco
Dept. Ingeniera de Sistemas
Telemáticos
E.T.S.I. Telecomunicación
Universidad Politécnica de Madrid
Ciudad Universitaria
E-28040 Madrid
Spain

J. L. Verdegay
Escuela Tecnica Superior de
Ingenieros de Informatica
Departamento de Ciencia de la
Computacion e Inteligencia Artificial
Universidad de Granada
18071 Granada
Spain

D. S. Weile
University of Illinois at Urbana-
Champaign
Department of Electrical and
Computer Engineering
Electromagnetic Communication
Laboratory
1406 West Green Street
Urbana
IL 61801-2991
USA

D. Whitley
Colorado State University
Department of Computer Science
Fort Collins
CO 80524
USA

G. Winter
Centro de Aplicaciones Numéricas
en Ingeniería de la U.L.P.G.C.
(CEANI)
Edificio de Infenierías I
Campus Universitario de Tafira
Tafira Baja
E-35017 Las Palmas de Gran Canaria
Spain

Part 1
Theoretical and Computational Aspects in GAs and ESs

1

Evolving Multi-Agent Systems

TERENCE C. FOGARTY, LAWRENCE BULL AND
BRIAN CARSE[1]

1.1 USING MANY INDIVIDUALS TO SOLVE A PROBLEM

Evolutionary computing can be distinguished from other machine learning, search and optimization techniques by the fact that it is a process which uses a population of many individuals, rather than a single individual, to solve a problem. The result of the process, however, is usually of the same order as that of any of the other techniques - a single individual as the solution to the problem. In life complex problems cannot be solved by single individuals but require the co-operation of many individuals. In engineering and computer science the same is true - solutions to complex problems often consist of many interacting systems.

1.1.1 Evolving rule-based systems in multi-agent environments

Evolutionary computing techniques have been used to automate the construction of rule-based systems in multi-agent environments. Rule-bases are used as the representation for the agents because they are more comprehensible to human inspection than other possible representations. Genetic algorithms have been used to produce rule-based systems for handling communication [Fogarty and Bull, 1995], network routing [Carse, Fogarty and Munro, 1995a; Carse, Fogarty and Munro, 1995b] and co-operative traffic control [Mikami, Kakazu and Fogarty, 1995].

In designing a number of interacting agents, evolutionary computing has been extended and developed by the invention of appropriate techniques to solve the problems involved. Allowing a rule-base to post messages onto any other active rule-base's message list provides a powerful formalism for the evolution of communication between agents in multi-agent environments. Using fuzzy as well as

1. University of the West of England, Bristol, BS16 1QY, UK
{tc-fogar, l-bull, b-carse}@uwe.ac.uk

discrete matching of rules extends the range of domains in which rule-based systems can be applied. Having more than one agent active in an environment gives evolutionary computing the opportunity to show its advantages over the hand building of agents. Automating the production of rule-based agents in the context of multi-agent environments allows for the creation of a number of agents, each of which is suited to its particular niche in the system. Given that these agents have the added functionality of learning and of communicating with each other and with us it is possible to evolve highly complex multi-agent systems.

1.1.2 Heterogeneous and Homogeneous Systems

Many multi-agent environments may require a mixture of agents, some of which are the same and some of which are different, to solve the problem.

Work done so far has been on either completely heterogeneous or completely homogenous problems and this chapter will concentrate on these two extremes. The application area from which exemplars have been chosen to demonstrate the approach is distributed control. First we look at the evolution of multiple communicating classifier systems in the heterogeneous environment of a distributed control system for a walking robot and second we look at the evolution of multiple fuzzy controllers in the homogeneous environment of a distributed control system for a communication network.

1.2   EVOLVING HETEROGENEOUS AGENTS

Symbiosis is the term used to describe the close association of heterogeneous species. The popular definition of symbiosis refers to associations between cooperative species, i.e. species which mutually benefit from the relationship. We suggest that this natural phenomenon provides a good analogy for the evolution of heterogeneous populations within multi-agent environments. That is, the evolution of heterogeneous agents, each operating from within their own niche, toward a common optimum (this is in contrast to Game Theory [Maynard-Smith 1982], in which cheating is possible, e.g. [Axlerod 1987]). The most "intimate" [Ehrman 1987] of symbioses is termed endosymbiosis, in which one of the partners, the "host", incorporates the other(s) internally. This intimacy can range from being intracellular to them being stored in the gut of the host. The evolutionary significance of such symbioses has long been noted in that it results in an "interspecies supraorganism about as well integrated as parts of an individual organism... with selection operating on the system as a functional whole" [Allee et al. 1949]. The formation of such supraorganisms has been described as a macro-level evolutionary phenomenon, or "megamutation" [Haynes 1991], part of a process termed "symbiogenesis" - "the origin of organisms through combination and unification of two or many beings, entering into symbiosis" [Merezhovsky 1920]. We also therefore extend the symbiosis analogy to introduce a macro-level operator to the evolution of heterogeneous species in multi-agent environments which allows for this closer integration of interacting agents.

We begin by presenting an extension to the classifier system [Holland & Reitman 1978], enabling them to communicate, to enhance its use within such systems.

1.2.1 Evolving Communicating Classifier Systems

To allow communication we assert that all action strings of the classifier system's rules contain an extra "address" tag and that each classifier system has its own address. Then whenever a rule's conditions are satisfied in the usual way, resulting in its action being posted onto the message list, a check is made on the contents of the tag. If the address is set to one of the other classifier system's, a copy of the string is posted onto that classifier system's message list, with the tag altered to the address of the sender. This allows the most suitable organisation of the classifiers, be it a hierarchy or a democracy, to evolve along with the classifier systems themselves. The architecture will also allow the relationships to alter if or when the system's world changes. Whether a particular communicating classifier system ends up influencing others is determined by the message passing between them, specified by evolution, rather than prescribed by a designer. We define an effective communication to have taken place when a passed message is used by its receiver in the satisfaction of a rule that is subsequently used - either as an effector, internally, or in further communication (this is similar to Burghardt's [1970] definition, as used by MacLennan in his detailed work on communication [e.g. 1989]). The aim is that in this way local models of the system will be built up between the classifier systems' rules, while a global model of the whole system will be built up by the communication between the rule-bases.

1.2.2 Symbiotic Classifier Systems: an architecture

Typically systems are of the form $F_{sys}(x) = \{F_1(x) + ... + F_n(x)\}$ or $F_{sys}(x) = F(x)$. We now outline an approach to using the GA within such systems and further suggest a macro-level genetic operator, or "megamutation", which can be applied to systems of either form.

What we suggest is to take a system with its identified fitness $F_{sys}(x) = F(x)$ and divide it into a number of smaller sub-systems (i.e. create a multi-agent system), each evolved by a separate genetic algorithm (GA) [Holland 1975] and represented by a communicating classifier system (CCS); view the system as a supraorganism and artificially identify a number of dependent endosymbiotic agents that make up the organism over which selection operates. Any structure, indicating epistasis between the identified sub-problems, would be allowed to evolve along with the solutions to the smaller sub-problems via communication.

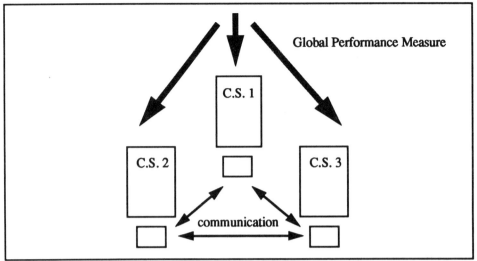

Figure 1.1: Cooperative Communicating Classifier Systems as an Endosymbiotic
Supraorganism.

In the case where $F_{sys}(x) = \{F_1(x) + \ldots + F_n(x)\}$ the same approach can be taken.
Here the divisions can be made such that each sub-system has access to its own
fitness function $F_i(x)$; view the system as an ecology [Hubermann 1988] and artificially identify a number of symbionts living together in close association. Here
again they would be represented by CCSs, with any structure allowed to emerge
through communication. If the identified symbiont sub-system for $F_i(x)$ is still
complex, it can be further treated as a supraorganism in the way described
above.

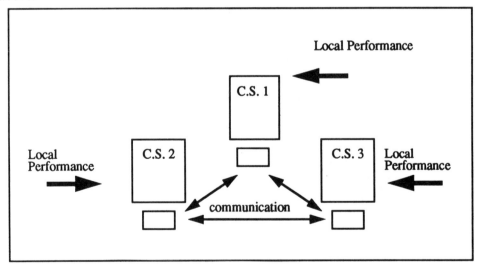

Figure 1.2: Cooperative Communicating Classifier Systems as an Ecology of
Symbionts.

We further suggest a megamutation operator to add to cooperative multi-agent systems - the formation of hereditary endosymbiotic associations. The operator considers the amount of interdependence between the identified subsystems in that it allows for when inappropriate divisions are made in a given system's search space (perhaps due to the complexity of the task). By adding the ability for agents to become genetically linked, the adverse effects of evolving in a highly interdependent search space (i.e. landscape oscillation - see [Kauffman 1993]) may be potentially avoided without relying on the evolution of complex communication protocols; incorrect a priori divisions in the global search space are remedied by allowing sub-systems to become joined again.

1.2.3   Evolving Communicating Classifier Systems for Gait in a Quadrupedal Robot

To demonstrate our approach we describe its application to the development of the control system for a wall climbing robot used in the inspection and maintenance of buildings. The control of walking in natural systems is known to be distributed among a number of parallel sub-systems - "flexible control appears to arise from the cooperation of several autonomous control centres... each of which controls the movement of one leg" [Cruse et al. 1992]. This type of system therefore represents an ideal platform on which to implement our approach.

The robot is built on a framework of steel tube and measures approximately 400mm x 400mm x 150mm. It consists of four vacuum feet, one on each corner of its chassis, and a central assembly, also consisting of four vacuum feet. The corner feet are each connected to a leg consisting of two pneumatic cylinders allowing vertical and lateral movement, i.e. each foot is connected to a leg consisting of two pistons which enables overall backwards/forwards movement by the robot. The central assembly of four feet is able to move vertically and rotate enabling the robot to alter its path. The robot also carries a distance sensor which can return the robot's distance to some fixed point (e.g. the ground); forward/backward movement is detectable. For our purposes we are concerned with the evolution of gait in the qaudruped - the evolution of a complete control system represents possible future work.

The control of a walking system has two aspects: the control of each leg, to create the power and return strokes; and the coordination of all such legs, thus enabling the system to move. To implement our approach on this system we use Pittsburgh-style classifier systems [Smith 1980] to control the movement of each leg, with coordination being achievable via emergent communication between the classifier systems. Such an approach is closely analogous to the biological control mechanism - "Coordination of the legs is not determined by a hierarchically superior control system... the gait pattern emerges from the cooperation of the separate control systems... [this] cooperation is based on different types of signals which convey information on the actual state of the sender to the control systems of the neighbouring legs" [Cruse et al. 1992].

The robot's pneumatic leg joints lend themselves nicely to digital control, as they represent an almost discrete system. Therefore in the simulation each leg controller can return one of three actions for each of the two pistons - move forward or backward, or do nothing - which are coded for by two two-bit pairs (figure 1.3). Actions such as move forward when a piston is already in the forward position are simply said to have no effect. For this work we assume a stable gait to consist of at least two opposite corner feet being down. Feet in contact with a surface are said to have enough contact for a vacuum to have (automatically) formed and hence the task is equivalent to learning to walk on the floor. Failure to satisfy the stability constraint on a given time step means that all outputs are ignored and the system remains in its previous state. The legs also fail to execute their outputs if any two or more legs try to give a power stroke in opposing directions, or if any legs would effectively be dragged, since they are attached to the surface by their vacuum feet.

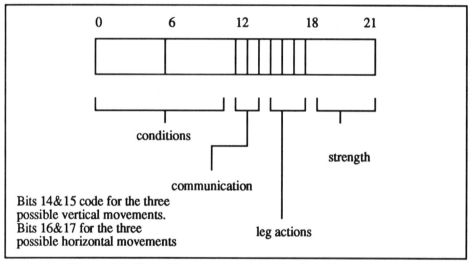

Figure 1.3: Showing rule structure for classifier systems.

Each leg receives its own environmental input message (detector). This detector string contains information as to whether or not the robot moved on the previous time step, which would be available from the distance sensor.

The robot scores on every one of the twenty discrete time step it moves forward, that is every time at least two legs give the correct power stroke (front to back) in a stable gait. All agents receive this (global) fitness. We have also found it necessary to reward individual legs for executing either a power or return stroke (local fitness) to obtain walking systems from experiments with populations of reasonable size. The maximum achievable fitness for each leg is therefore (2x20) 40. This system is a mixture of the two previously defined, as shown in figure 1.4.

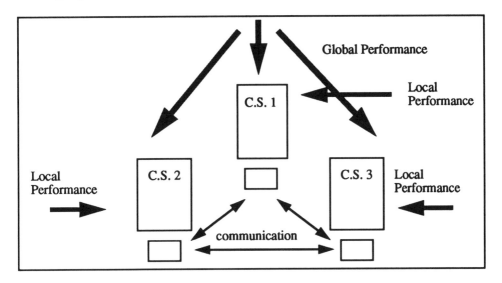

Figure 1.4: Cooperative Classifier Systems (in a hybrid form of those previously presented).

We use roulette wheel selection, allele mutation (rate 0.001 per bit), and multi-point crossover (rate 0.01 per bit) with the GA(s) operating over the genomes of concatenated rules. The classifier systems are a simplified version of the Pittsburgh-style in that they have fixed length genomes and we don't make any of Smith's macro/micro level distinctions over the rules when using the crossover operator. Effectively we turn a normal fixed length GA string into a classifier system for evaluation. The internal matching/firing is the same, with all satisfied rules posting actions onto the message list and the highest bidder taken as the effector (conflict resolution is arbitrary when strengths match). All classifier systems contain six bi-conditional rules.

The simulation starts with all four feet being on the ground and with diagonally opposite legs in the same horizontal position, as Cruse et al. [1992] point out "the start is particularly difficult when [opposite] legs begin from the same x-position".

To implement an heterogeneous multi-agent system on the quadruped we create one separate population of evolving agents for each leg, where the members of each population are responsible only for the control of their corresponding leg. Therefore the genomes of the classifier systems are treated as those of four distinct cooperative "species"; they are stored in four separate populations of size P, evolving via separate generational GAs. Genomes are again matched for evaluation simply by taking the corresponding members from each population, i.e. leg1[x] is matched with leg2[x], leg3[x] and leg4[x] (where 0<x<P). The populations are never ordered in any way.

With four sub-systems making up the overall system two bits of the classifier systems' action strings are used for communication (figure 1.3), where the legs are numbered clockwise from the top-left of the robot; each population (for each leg) has a unique communication address. Genomes are (6x22) 132 bits long here.

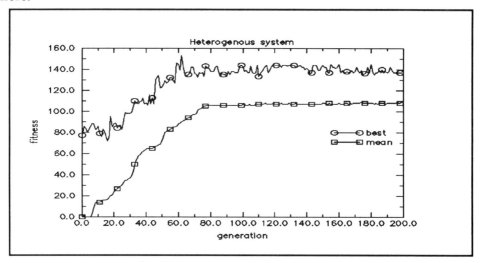

Figure 1.5: Graph showing heterogeneous system performance.

Figure 1.5 shows the typical evolutionary progress (averaged over 20 runs) of the four heterogeneous populations, each containing 2000 members, over 200 generations. The results show that on average the GAs come close to generating a group of agents capable of walking forward on each discrete time step and heterogeneous multi-agent systems have been evolved capable of walking on every step, that is scoring (2x20x4) 160.

When examining the interactions between the classifier systems we have found effective communication strategies to have evolved. The amount of communication varies but we have found legs at the same end of the robot, both the front and the back, passing messages to each other. These strings are then used by the receiver in the satisfaction of the conditions of its effector on the next time step. That is "contralateral" connections have been seen to emerge in the quadruped, which matches the biologists observations of natural quadrupedal systems [Cruse 1991]. However "ipsilateral" communications between legs on the same side of the robot have also been seen to emerge, which biologists suggest is of less importance to a quadruped than say a stick insect [ibid.]. The evolved gait of robots capable of walking on every time step is the "opposite stance gait" whereby the legs on opposing corners are synchronous.

To implement the macro-level operator within this multi-agent system we modify the above heterogeneous agent system such that there is the possibility of a sub-population of agents emerging where each agent carries all (four) genomes,. Here we arbitrarily assign the leg1 members to be the hosts and say that each member has endosymbiotic genomes for legs 2-4. Consequently in the evaluation of each individual of this sub-population all four genomes are expressed as communicating classifier systems just as they would if they had come from separate populations; the hereditary endosymbiotic association keeps the separate genomes together during evolution. During the application of the

GA all genomes of the chosen parents are treated as if they were separate, so that a given child's leg1 genome and its legs 2-4 genomes are a mix of its parents'. Genomes therefore consist of four 132 bit strings here. Therefore there are five possible sub-populations, four for the legs as cooperative separate species and one for the four in an hereditary endosymbiosis. Initially each of the leg sub-populations evolve separately, each receiving their own separate fitness measure and the global fitness, with these populations being initially set to size P (P=2000). The other sub-population of hereditary endosymbionts members are treated as a supraorganism for selection by receiving a combined fitness of all legs. The size of this sub-population is initially set to zero, but during the course of a run individuals can move from the other sub-populations to this, and back, via the coupling megamutation operator. Members of the hereditary endosymbiotic sub-population compete against each other and those of the other four, such that over evolutionary time the most efficient association will come to dominate the overall population space.

The communication mechanism is the same as that above.

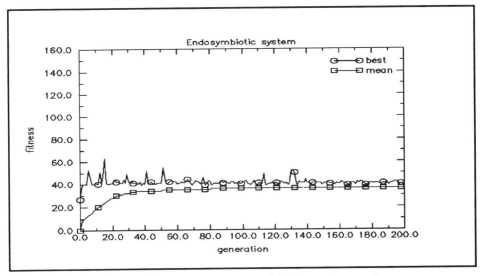

Figure 1.6 Graph showing endosymbiotic system performance.

We tried various rates for the operator but were unable to get the sub-population of hereditary endosymbionts to establish itself. Figure 1.6 shows the typical progress of a fixed population of 2000 hereditary endosymbionts at the walking task over 200 generations. It can be seen that on average (20 runs) the configuration has been able to evolve controllers which can individually make power or return strokes but the evolution of the necessary coordination has yet to emerge (correspondingly when we examine the communication between the classifiers no coherent strategy is found). This explains why we were unable to successfully evolve a supraorganism sub-population; it is more efficient in this system for the symbionts to stay separated.

With this multi-agent task we are able to create a single controller function-
ally equivalent to the multi-agent systems, i.e. one responsible for all four legs of
the robot.

The equivalent single agent model consists of one population, evolving via a
single GA. Each member of this population is obviously more complex than those
of the multi-agent systems in that each must give output for all four legs. As
stated above, each leg requires four bits to code for its possible actions, so each
single agent classifier system gives effectors of 4x4 bits (no bits are needed for
external communication of course); genomes are 312 bits long here. The genomes
still code for six rules, as before, and the GA has the same parameters as those
used above (all rates are per bit and are therefore equivalent).

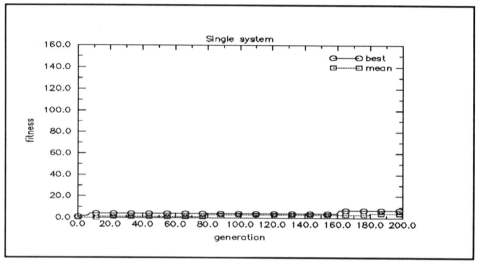

Figure 1.7: Graph showing single agent system performance.

Figure 1.7 shows the typical evolutionary progress of the population (2000
members) of single agent systems over 200 generations. It can be seen that this
implementation of a controller has performed very poorly.

1.2.4 Discussion

It has been shown that the heterogeneous multi-agent system of figure 1.5 performed better than the equivalent single agent system for this task. In terms of
search space size this is not unexpected; the species operate within their own
niches, where those niches are as small as possible for the task. Indeed the biological control system for walking appears to have evolved a similar approach
[Cruse et al. 1992]. The emergent communication strategies seen were also predicted by the natural phenomenon of quadrupedal walking [Cruse 1991], in that
consideration of contralateral connections was more likely than ipsilateral. The
endosymbiotic systems, that is the application of a coupling megamutation operator to the heterogeneous system, whilst sharing many features of the heteroge-

neous model, performed less well, being unable to generate a coherent strategy for gait. We suggest this is due to the fact that population mixing within the niches of the individual legs is reduced as all legs recombine with the corresponding individuals of the same single parent, and not with any individual also operating within their niche, as in the heterogeneous case (the conditions under which endosymbiosis represents an evolutionarily advanced multi-agent configuration examined in [Bull et al. 1995] indeed shows them to be a small sub-set of the total space of possibilities). However we maintain that such an operator be available to these types of systems since the process has proved so significant in natural evolution [Margulis 1992]. The single system performed least well with the GA parameters used here, this being due mainly to the much larger search space it had to operate within. This result demonstrates our reasons for examining the performance of the GA in heterogeneous multi-agent environments (the reader is referred to [Bull 1995] for a full review of this and related work).

1.3   EVOLVING HOMOGENEOUS AGENTS

In this section we describe how a collection of identical and geographically dispersed agents may be evolved to perform distributed control of a distributed system. The specific control problem addressed is that of adaptive distributed routing in communication networks. This problem is particularly interesting since it provides a good example of the classic dichotomy between individual and social optimization, whereby agents are required to locally adapt their actions using incomplete and delayed information in order to optimise global criteria. In addition, such a system displays complex dynamics due to non-linear interactions between agents' actions.

The approach proposed may be summarised as follows. Each routing control agent is implemented as a fuzzy rule-set, which maps measured state inputs to control outputs. A population of such encoded rule-sets is maintained and evolved under application of the GA. Fitness evaluation of each population member is achieved by deploying copies of the encoded rule-set at distributed nodes in a simulated network. During simulation, the global performance of the collection of identical agents is measured and used to derive the fitness evaluation for that individual. Therefore, although from the GA point of view this is basically straight forward optimization, fitness evaluation implicitly takes into account interactions between agents and the degree to which they cooperate to achieve cohesion in actions.

We begin by first describing the adaptive distributed routing problem in more detail. Following that we outline the particular representation employed to encode each individual fuzzy rule-set, and give details of the simulation test-bed used in experiments. Finally, in this section, we present experimental results that demonstrate that a collection of homogeneous agents can be evolved for competent routing control in a simulated network.

1.3.1 *Adaptive Distributed Routing (ADR) in Communication Networks*

Two commonly used switching techniques in communication networks are circuit-switching and packet-switching [Stallings 1994]. The differences in philosophy of the two methods mainly derive from the way that network resources are utilised. In circuit- switched networks, a route and a fixed amount of bandwidth are allocated and dedicated to a connection during a circuit set- up phase. Transmitted information appears at the destination with a more-or-less fixed time delay. Existing packet-switched networks, by contrast, do not dedicate bandwidth to connections, but rather employ a form of distributed dynamic time-division multiplexing, in which discrete packets of information offered to the network are routed through a network of packet-switching exchanges (PSEs) on a store-and- forward basis. The end-to-end delay of packets between a particular source-destination pair is variable and is influenced by the amount of traffic offered to the network. Clearly in both types of network, routing policies which determine the route that circuits or packets take through the network affect the quality of service perceived by users of the network. In the case of circuit-switching, poor routing decisions result in the denial of a requested connection, whereas in packet-switched networks poor routing decisions can lead to unacceptable packet delays or reduced network throughput due to congestion. The discussion and experiments presented in this paper pertain specifically to packet- switched networks, although there is no reason why the general approach could not be applied to circuit-switched networks.

Routing policies in communication networks may be static or adaptive, centralised or distributed. Static policies fix network routes (or fix the probability of selecting different particular routes) between each source-destination pair. Adaptive policies alter routes (or probabilities of selecting routes) in response to perceptions of changes in network traffic conditions or topology. Adaptive, centralised routing employs a central routing control centre which gathers information regarding changing network traffic conditions and topology, calculates routing parameters for individual switching nodes, then informs switching nodes how to make routing decisions. On the other hand, distributed routing has network switching nodes reach their own routing decisions based upon information available to them. Here we focus on the problem of adaptive, distributed routing. For a more detailed discussion of routing policies, consult [Stallings 1994].

Virtually all packet-switched networks base their routing decisions using some form of least-cost criterion. This criterion may be, for example, to minimise the number of hops, or to minimise packet delay. Two elegant algorithms in widespread use in both centralised and distributed form, are those of Dijkstra [1959] and Ford and Fulkerson [1962], both of which translate to shortest-path routing algorithms in the communication network context. However, a pure shortest path strategy at each routing node suffers from known deficiencies in terms of stability. Routes measured as heavily used are simultaneously avoided by all routing nodes and routes measured as lightly used are simultaneously selected, thus causing unwanted oscillations in routing decisions and inefficient network usage. The conclusion to be reached from these observations is that every routing node is "selfishly" attempting to obtain the best route for all destinations and that there is no use of explicit cooperation in actions between routing controllers.

The field of distributed artificial intelligence (DAI) [Mandiau et al. 1992] lends many insights to new methods of approaching the adaptive, distributed control problem. In particular, the "social abilities" of individual agents, their organisation and the dynamic behaviour of the assembly of distributed agents addressed by the DAI paradigm are highly relevant to many aspects of control in communication networks. Work by Huberman and Hogg [1988] on the dynamics of computational ecologies is also germane to the discussion of distributed network control. We now briefly discuss these paradigms with reference to the ADR problem.

In terms of social abilities, the capacity of routing controllers to reason about the beliefs, actions and behaviour of others, and to influence these, is important. This could apply to both selfish and cooperative controllers, although in this paper we concentrate on cooperative behaviour. As stated above, a shortest-path routing algorithm running on each routing node does not lead to social behaviour in any way (excepting the communication of measured link delays) and does not take into account the actions and behaviours of other agents. Organisation in a multi-agent system reflects the control relationships and information passing mechanisms. Organisation also decides the degree of cooperation or antagonism between agents, and it is argued in [Mandiau et al. 1992] that the presence of incomplete or imprecise information (a characteristic of the ADR problem) that total cooperation is crucial for success. Information passing mechanisms range from no communication at all to exchange of local state information and, at a higher level, communication of plans and intentions. The dynamics of the system of agents is determined by the individual behaviours of the agents, and the degree to which they are coherent in the global sense, which in turn may depend on agents coordination and negotiation abilities.

Huberman and Hogg [1988] present a dynamical theory of computational ecologies in which distributed agents select actions based on incomplete and delayed information. In particular, they address the problem of distributed resource allocation using game dynamics. Agents are able to choose from a number of strategies according to perceived pay-offs. Delays in the evaluation of payoff are introduced by delaying state information measured by the agents. A mathematical model is developed which demonstrates how the stability of agents' choices of strategies depends on measurement delays. Three operational regimes are identified as measurement delay is increased: direct relaxation to a fixed set of strategies for small delay, then damped oscillations in strategy selection for moderate delay, and finally at large delay, undamped oscillations with a sharp, periodic and non-linear switch-over in strategies. This latter situation is highly reminiscent of the instabilities observed using the shortest-path routing algorithm in communication networks. Indeed, the similarities in behaviour of the real network of routing controllers and the theoretical computational ecology should not come as a surprise. In Huberman and Hogg's treatment, agents evaluate perceived payoff for different strategies and switch to the one with highest payoff. In a distributed shortest-path routing algorithm, routing nodes switch their routes along the shortest path. Both employ "pure" strategies in the sense that an agent employs a single strategy at each decision instant.

Next, we describe a routing agent representation and details of experiments conducted to determine whether or not it is possible to evolve a homogeneous, "self-social" assembly of fuzzy routing controllers which are capable, collectively, of controlling routing in a stable manner.

1.3.2 Fuzzy Agent Representation and Simulation Details

In the approach proposed, each fuzzy routing controller is represented as a variable sized fuzzy rule-set. A population of encoded fuzzy controllers is maintained and operated upon by the GA. Since the unit of selection and replacement is the complete rule-set, this method is effectively a fuzzy implementation of the Pittsburgh classifier system. The fuzzy classifier system employed is a version of P-FCS1 (Pittsburgh Fuzzy Classifier System #1), described and evaluated in [6,7]. This classifier system employs a representation and genetic operators particularly suited to the evolution of fuzzy rule bases.

In PFCS-1, each rule, R_k, for an n-input, m-output system, is represented as:

$$R_k: (x_{c1k}, x_{w1k}); .. (x_{cnk}, x_{wnk}) \Rightarrow (y_{c1k}, y_{w1k}); .. (y_{cmk}, y_{wmk})$$

The bracketed terms represent the centres and widths of fuzzy set membership functions over the range of input and output variables. The genome representing a complete rule-set is a variable length concatenated string of such fuzzy rules. The GA employs a specially designed crossover operator (described in detail in [Carse, Fogarty and Munro 1995]) which enhances the probability that good fuzzy "building blocks" (i.e. high-fitness assemblies of fuzzy rules with overlapping input membership functions) survive and proliferate in future generations.

In experiments performed, an assembly of fuzzy controllers are required to perform adaptive, distributed routing control in a simulated 3-node datagram packet switched network (see Figure 1.8). The network is fully connected with bidirectional full duplex links between each switching node pair. Packets requiring transmission over a particular link are queued using a first-come first-served discipline. Packets arrive from outside the network at network source node i, (i = A,B,C), to be delivered to destination node j, (j = A,B,C), j≠ i, at an average rate of λ_{ij}.

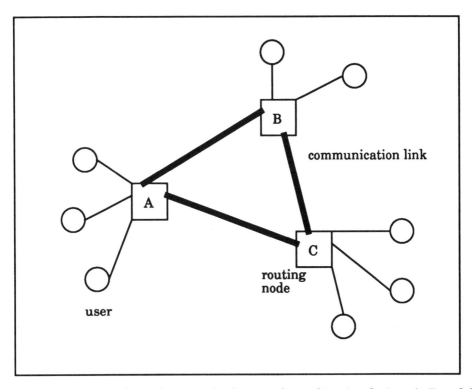

Figure 1.8: Three node packet switched network used in simulation. A, B and C
are routing nodes connected by bidirectional links

A controller situated at each node decides whether to route each packet
directly to its destination or via an intermediate node. Controller decisions are
based on packet delay measurements over different paths. The goal is to minimise average global packet delay (i.e. the mean delay between packet arrival at
the source node and packet delivery to the destination node for all packets which
arrive during the period of simulation).

Each routing controller is implemented as a variable size fuzzy classifier system with four inputs and two outputs. At each node the controller inputs are:

• DelayLeftDirect: The measured packet delay from the source node for packets
destined for the node to the left of the source node and which are routed directly.

• DelayLeftIndirect: The measured packet delay from the source node for packets destined for the node to the left of the source node and which are routed indirectly.

• DelayRightDirect: The measured packet delay from the source node for packets destined for the node to the right of the source node and which are routed
directly.

- DelayRightIndirect: The measured packet delay from the source node for packets destined for the node to the right of the source node and which are routed indirectly.

Packet delays are measured at the destination node (each packet is time-stamped on arrival in the system) and averaged over the last $N_{Measure}$ packets for each route taken for each source node. In the simulation, we assume this information is transmitted without delay to source nodes once the averages have been taken and transmission of control information does not consume network bandwidth. In a real network such information would be sent as control packets which would incur a finite delay and utilise network bandwidth. $N_{Measure}$ is a parameter we vary by hand in the experiments described later and determines the granularity of measurements. Also, in a real network, a trade-off would have to be made in choosing the value of $N_{Measure}$. If too small a value is chosen, the network becomes swamped with control packets which compete with user data packets for use of the shared bandwidth. If too large a value is chosen, measurements become out of date and meaningless.

At each node, the controller outputs are:

- PLeftDirect: The probability that a packet arriving at the source node which is destined for the node to the left of the source node is routed directly.

- PRightDirect: The probability that a packet arriving at the source node which is destined for the node to the right of the source node is routed directly.

By dynamically adjusting local PLeftDirect and PRightDirect control outputs based on network delay measurements, the distributed assembly of controllers should attempt, in a cooperative fashion, to spread the network load to minimise global mean packet delay in response to changing traffic conditions in the network.

Each network simulation is run for a simulation time of 500 seconds. The data rates of all network links are set to 10,000 bits per second. Mean packet arrival rates used in the simulation, and their variation in time, are given by:

$\lambda_{BC}, \lambda_{CA}, \lambda_{BC}, \lambda_{CA} = 3$ packets/sec, $0<t<500$

$\lambda_{AB} = 3$ packets/sec $0<t<125$, $t>250$ 15 packets/sec $125<t<250$

$\lambda_{CB} = 3$ packets/sec $0<t250$, $t>375$ 15 packets/sec $250<t<375$

These traffic patterns were chosen to exercise the dynamic capabilities of routing controllers in moving from relatively light network load, when direct routing is optimal, to heavy load when controllers must balance the offered load between direct and indirect network paths. In the simulation, packets arriving at an intermediate node are always forwarded to the destination node to avoid a "ping-pong" effect. The evaluation function for each rule-set returns the inverse of the mean measured packet delay for all packets delivered during the simulation. Experiments were carried out using deterministic and Poisson packet arrival processes. In the first case, fixed size packets of 1000 bits were used. In the second case, packet sizes were exponentially distributed with a mean size of 1000 bits.

1.3.3 Results and Interpretation

To evaluate the controllers evolved by the fuzzy classifier system, we compared their performance with a shortest-path routing algorithm which routes all packets along the route whose measured delay is least between a particular source/destination pair. A range of measurement intervals, $N_{Measure}$ from 2 packets to 100 packets were used. Ten independent runs of P-FCS1 were conducted with different random seeds for population initialisation in each case. In addition, different initial random seeds were also used for each of the network simulations used in evaluating a particular individual. The latter introduces noise in the evaluation function. Each of the 10 learned fuzzy controllers using P-FCS1 were evaluated in 20 subsequent simulations and the results are presented in Figure 1.9 where they are compared with the shortest path routing algorithm. This figure plots the mean packet delay (in seconds) over a complete simulation against the measurement interval defined above.

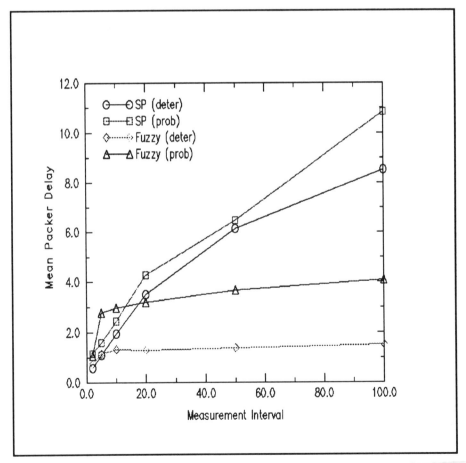

Figure 1.9: Mean packet delay versus measurement interval using evolved fuzzy controllers and shortest path (SP) routers

The results depicted in Figure 1.9 indicate that, when the measurement interval is small, the shortest-path algorithm outperforms the learned fuzzy controllers, although not by that large a margin. As the measurement interval increases, the learned fuzzy controllers begin to outperform the shortest path algorithm significantly. As mentioned earlier, an important characteristic of a routing algorithm is that routing control information exchange should not consume excessive network bandwidth. A value of $N_{Measure}$ greater than (at least) 20 is realistic for a real network and the results using a homogeneous assembly of GA-derived fuzzy controllers appear to be better than an assembly of shortest-path algorithm controllers in this region of rate of feedback. When measured delays and control actions were observed, it was seen that the shortest-path controllers showed pronounced instabilities in switching from one route to another. This appeared to a much lesser extent with evolved fuzzy controllers.

We propose two major reasons why, when feedback information is realistically delayed, the learned fuzzy controllers appear superior in performance to the shortest-path routers in the experiments presented. Firstly, fuzzy controllers adapt routing decisions in a continuous fashion i.e. they does not choose a single route for all traffic, even if that route looks attractive at the time. In terms of Huberman and Hogg's [1988] model this can be interpreted as an agent adopting an adaptive "meta-strategy", which determines the probability of selecting lower level strategies. Secondly, the learning environment used in evolving fuzzy controllers creates evolutionary pressure which encourages the replication of genotypes whose phenotypes are "self-social" and able to cooperate in their actions to optimise global network performance (since, after all, global network performance determines the fitness of a particular individual).

1.4   FURTHER WORK

1.4.1 Machine Learning

While evolutionary computing techniques are aimed at designing agents that are adaptive over the long term, machine learning techniques are aimed at allowing agents to be responsive to short term changes. Furthermore, in multi-agent systems agents can learn from one another as well as from the environment. The interaction between evolution and learning using both Baldwin and Lamarckian effects will be investigated [e.g.Bull and Fogarty 1994] as will ways of allowing agents to learn from each other [e.g. Tan 1993].

1.4.2 Communication

Both systemic and stigmergic communication between rule-based agents in large multi-agent systems will now be investigated more fully. Results from the field of semiotics and knowledge acquisition will also be utilised in facilitating productive communication both with and between the rule-based agents.

1.5 ACKNOWLEDGEMENT

This work was supported partially by the EPSRC under Grant No: GR/J7168

1.6 REFERENCES

Allee, W. C., Emerson A. E., Schmidt K P, Park T & Park O (1949), "Principles of Animal Ecology", Saunders Company.

Axlerod, R. (1987), "The Evolution of Strategies in the Iterated Prisoners Dilemma", Genetic Algorithms and Simulated Annealing, Pittman, London.

Burghardt, G. M. (1970), "Defining Communication by Chemical Signals", Appleton-Century-Crofts, New York.

Bull, L. (1995), "Artificial Symbiology: Evolution in Cooperative Multi-agent Environments", PhD Dissertation, UWE Bristol.

Bull L. and Fogarty T. C. (1994), "An Evolution Strategy and Genetic Algorithm Hybrid: An initial implementation and first results", Evolutionary Computing, Springer Verlag.

Bull, L., Fogarty, T. C., and Pipe, A. G. (1995), "Artificial Endosymbiosis", Third European Conference on Artificial Life, Springer-Verlag

Carse,B., Fogarty,T.C. and Munro,A. (1995a) "Adaptive distributed routing using evolutionary fuzzy control", Proceedings of the Sixth Int. Conf. on Genetic Algorithms, Morgan Kaufmann.

Carse B., Fogarty T.C. and Munro A. (1995b) "Evolving fuzzy rule based controllers using genetic algorithms". To appear in the International Journal on Fuzzy Sets and Systems

Carse B. and Fogarty T.C. (1994) "A fuzzy classifier system using the Pittsburgh approach". In: Davidor Y., Schwefel H-P., Manner R. (Eds) (1994), Parallel Problem Solving from Nature - PPSN III, Lecture Notes in Computer Science vol. 866. pp260-269.

Cruse, H, (1991), "Coordination of Leg Movement in Walking Animals", Animals to Animats, MIT Press.

Cruse H, Muller-Wilmm U, and Dean, J. (1992), "Artificial Neural Nets for Controlling a 6-legged Walking System", Animals to Animats 2, MIT Press.

Dijkstra E.W. (1959) "A note on two problems in connection with graphs". Numerical Mathematics 1: pp269-271.

Ehrman, L. (1983), "Endosymbiosis", Coevolution, Sinauer Associates.

Fogarty, T. C. and Bull, L (1995) "Optimising individual control rules and multiple communicating rule-based control systems with parallel distributed genetic algorithms", IEE Proc. on Control Theory and Applications, 142 (3):211-5.

Ford, L.R. Jnr. and Fulkerson, D.R. (1962) "Flows in Networks", Princeton University Press, Princeton,N.J.

Haynes, R. H. (1991), "Modes of Mutation and Repair in Evolutionary Rhythms", Symbiosis as a Source of Evolutionary Innovation, MIT Press.

Holland, J. H. (1975), "Adaption in Natural and Artificial Systems", Univ. of Michigan Press, Ann Arbor.

Holland, J. H. and Reitman, J. S. (1978), "Cognitive Systems Based on Adaptive Algorithms", Pattern-directed Inference Systems, New York.

Hubermann, B. A. (1988), "The Ecology of Computation", North-Holland.

Huberman, B.A. and Hogg, T. (1988) "The behaviour of computational ecologies". In: Huberman, B.A. (Ed.) (1988) The Ecology of Computation pp77-115. North Holland.

Kauffman, S. A. (1993), "The Origins of Order: Self-organisation and Selection in Evolution", Oxford University Press.

MacLennan, B. (1989), "Synthetic Ethology: An Approach to the Study of Communication", Artificial Life II, Addison-Wesley.

Margulis L (1992), "Symbiosis in Cell Evolution", W.H Freeman and Company.

Maynard-Smith, J. (1982), "Evolution and the Theory of Games", Cambridge University Press.

Mandiau, R., Chaib-Draa, B., Moulin, B. and Millot, P. (1992) "Trends in distributed artificial intelligence". In: Artificial Intelligence Review 6, pp35-66.

Mikami,S., Kakazu,Y. and Fogarty,T.C. (1995) Co-operative reinforcement learning by payoff filters. Proc. of the European Conference on Machine Learning, 1995: 319-322.

Smith, S. F. (1980), "A Learning System Based on Genetic Adaptive Algorithms", PhD Dissertation, University of Pittsburgh.

Stallings ,W. (1994) "Data and computer communications". 4th edition. Macmillan Publishing Company.

Tan M. (1993), "Multi-Agent Reinforcement Learning: Independent vs. Cooperative Agents", Tenth International Conference on Machine Learning, Morgan Kaufmann.

2

The Existential Pleasures of Genetic Algorithms

DAVID E. GOLDBERG

2.1 INTRODUCTION

Some readers will recognize that I owe Samuel Florman an apology for first borrowing the title of his well-known book (Florman, 1976) and then for twisting it to my evolutionary purposes. In that book, *The Existential Pleasures of Engineering*, Florman discusses, among other things, two rarely articulated facts about engineering: (1) engineering is a multifacetted profession, and (2) engineering is fun. In this article, I make the same two points about genetic algorithms (GAs)—that GAs are multifacetted and fun—and I believe the analogy I am drawing is a fairly tight one if it is viewed in the following way. The world employs both engineering and genetic algorithms for what they can *do*—the external interest in the two subjects is almost always *utilitarian*—and both engineers and genetic algorithmists themselves take great pride in what they can *accomplish*. But inside their heart of hearts both engineers and genetic algorithmists know that there is a good deal of intrinsic satisfaction that comes from working in these fields. Moreover, both sets of individuals know that there are many facets within their respective fields to find such *engagement*.

In this essay, I try to identify some of the fun and many facets one can find in the study and application of genetic algorithms. The list is far from complete—it is unclear whether such a list can ever be complete—and the categories I choose are not orthogonal. To make matters worse the coverage in this essay is not uniform. Nonetheless, I believe the costs in ambiguity, inprecision, and non-uniformity of a coarse exploration are offset by the value of a fuller articulation of the intrinsic merit of our computations, their uses, and their higher meaning.

I start by listing some of the different ways to view genetic algorithms. From problem solver to technical puzzle to a way of understanding more about life itself, GAs are

---

[1] The author's current address is Department of General Engineering, University of Illinois at Urbana-Champaign, 117 Transportation Building, 104 S. Mathews Avenue, Urbana, Illinois 61801, USA, e-mail: deg@uiuc.edu.

[2] This article was originally published in Spanish as *Los Placeres de los Algoritmos Genéticos* in the January 1995 edition of *Soluciones Avanzadas*, *3*(17), 44–51.

much more than the problems they help us solve, much more than the individuals involved in the solving, and much more important than most genetic algorithmists realize. It is this largeness that this essay seeks to explore.

2.2 GENETIC ALGORITHMS AS X

There are many ways to view genetic algorithms, and perhaps most users come to GAs looking for a problem solver, but this is an overly restrictive view (De Jong, 1993). Herein, we will examine GAs *as* a number of different things:

- GAs as problem solvers
- GAs as challenging technical puzzle
- GAs as basis for competent machine learning
- GAs as computational model of innovation and creativity
- GAs as computational model of other innovating systems
- GAs as guiding philosophy

In the remainder of the section we examine each of these viewpoints in turn.

GAs as problem solvers. Many genetic algorithm users initially come to the field in search of computationally tractable solutions to difficult problems, and indeed the list of successful solutions has grown considerably in recent years as I've noted elsewhere (Goldberg, 1994). From the design of gas turbines to that of fiber-optic communications networks, from the scheduling of complex manufacturing facilities to the creation of complex musical compositions, GAs are playing an increasing role in difficult problems once thought to beyond the realm of automated or computer-assisted solution. As a GA user, the primary value of the enterprise is *utility*. How *well* does the GA work? Does it *reliably* solve the *range of problems* of interest with sufficient *accuracy* and *speed*? For many years, the answers to these questions were mixed. Sometimes things worked well and at others they did not, and there was little or no framework for making sense of these mixed signals. It has required a different viewpoint to put these signals into proper perspective, a viewpoint that is being better grasped by a larger segment of the community.

GAs as challenging technical puzzle. On the face of it, what we attempt to do with GAs may be viewed as patently absurd. After all, here we are, using a stochastic, large-memory, dynamical system to operate on possibly stochastic, non-stationary problems of infinite variety and high dimension and complexity. This doesn't sound like a sensible way to earn a living unless one realizes that (1) the effort is supported by natural example and (2) some basic understanding of underlying mechanism is in hand. Turning this around, we recognize that under this perspective the fundamental value of the GA enterprise shifts from one that is utilitarian to one that is *prescriptive* or *normative*. In other words, the goal shifts to *designing* effective GAs from simply *using* them, and this dictates the adoption of an effective methodology of invention or engineering. Elsewhere (Goldberg, 1993) I have drawn a connection between the methodology of invention that led the Wright Brothers to success in the skies above Kitty Hawk and the methodology that has led to successful GA flight.

Recapping that argument briefly, the Wright Brothers succeeded where so many others failed because they (1) decomposed the difficult problem of powered flight into quasi-separate subproblems, (2) solved the subproblems using empirical study, *facetwise* modeling, and dimensional analysis (Ipsen, 1960), and (3) integrated the subsolutions into a full problem solution with much regard for results and little regard for elegance, proof, or other mathematical niceties (Bradshaw & Lienert, 1991).

A decomposition of the problem of designing a selectorecombinative genetic algorithm that reflects the current state of affairs has been presented elsewhere (Goldberg, Deb, & Clark, 1992):

1. Know what you're processing: building blocks (BBs)
2. Ensure there is an adequate initial supply of BBs
3. Ensure that necessary BBs are expected to grow
4. Ensure that BB-decisions are well made
5. Solve problem of bounded BB-difficulty
6. Ensure that BBs can be properly mixed (exchanged)

The paper cited above and some of the others cited in this section should be consulted for more details on how this decomposition has taken us beyond Holland's pioneering theories (1975), but the situation has progressed to the point where it is now possible to demonstrate the limitations of simple GAs analytically through the construction of *control maps*, and the same methods make it possible for the creation of *fast, messy genetic algorithm* that solve hard problems quickly.

Recently, simple models of each of the portions of the GA design decomposition were assembled to predict the region of effective convergence for a simple GA operating on an easy problem in terms of the selection pressure s and crossover probability p_c (Goldberg, Deb, & Thierens, 1993). Figure 2.1 shows the shape of the theoretically predicted control map, which shows how to set up a GA over a large range of GA parameters. Figure 2.2 displays a contour of constant solution quality from empirical results on a one-max problem, and the empirical results follow the model quite well. Moreover, in these experiments good results were obtained in a subquadratic number of function evaluations: since the number of generations is $O(\log n)$, n the population size, and since the population size to achieve good decision making probabilistically is $O(\ell)$ (Goldberg, Deb, & Clark, 1992), with ℓ the problem size, good convergence is achieved with high probability in a number of function evaluations that grows as $O(\ell \log \ell)$. This is a tantalizing result, but unfortunately theory and experiments (Thierens & Goldberg, 1993) suggest that the result does not generalize to problems of bounded difficulty. Work has continued to see if adding elitism, niching, mating restriction, or other relatively straightforward mechanisms speed traditional GAs sufficiently; recent computations and experiments (Thierens, 1995) have not been encouraging.

These other mechanisms may help improve convergence marginally, but none is likely to achieve the subquadratic potential GAs have been dangling before us as of late. A different approach has recently led to a practical, if nontraditional, GA that appears to achieve subquadratic results to problems of bounded difficulty. Work began on such *messy genetic algorithms* (mGAs) at the University of Alabama in 1988 (Goldberg, Deb, & Korb, 1990; Goldberg, Korb, & Deb, 1989), and work here at the University of Illinois since 1990 has been directed at overcoming the *initialization bottleneck* of first-generation mGAs. In those early mGAs a partial enumeration was

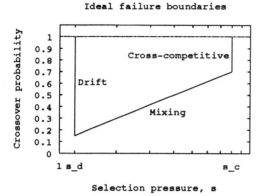

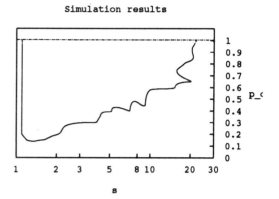

Figure 2.1   A control map of a simple GA operating on an easy problem shows the drift, mixing, and cross-competitive limits of GA success. Within those boundaries, the GA should be expected to work well.

Figure 2.2   Simulation results of a simple GA operating on an easy problem outline the border of the region of effective solution. The shape of the predicted control map is confirmed.

performed to ensure the presence of tight building blocks of a given length. Accurate convergence was obtained, but these results were purchased at a cost of $O(\ell^k)$ function evaluations, where k is a number that goes up with increased problem difficulty. This computational barrier was recently smashed via a fast mGA that replaced the enumerative initialization with one based on *probabilistically complete enumeration* and *building-block filtering* (Goldberg, Deb, Kargupta, & Harik, 1993). Despite the need for additional work, figure 2.3 shows that order-5 deceptive problems have been solved to problem sizes $\ell = 150$. As the figure shows, the fast mGA solves the problem faster than the old mGA and the old mGA is known to be faster than the usual simple GA. Although some have criticized the choice of these test functions, the 150-bit problem is very difficult, containing over 10^{45} points with over a billion local optima! The ability to solve this kind of problem in a few hundred thousand function evaluations gives us hope that our stiffest applications challenges may soon be tackled quickly and reliably.

GAs as basis for competent machine learning. Even before there were computers there were dreams of intelligent machines, and even before there were competent GAs, there were dreams of using evolutionary computations in machine learning systems. Holland's early (1962) vision in "An Outline for a Logical Theory of Adaptive Systems" painted a swashbuckling computational landscape of bands of computer programs roaming around a cellular machine space, fighting, mating, and improving their lot. Disappointment in early efforts to select and mutate one's way toward finite-state machines appropriate to a given machine-learning task have given way to a string of empirical successes in using GAs in classification, sequential learning, and automated programming (Goldberg, 1989). Much work remains, and the design challenges that these systems represents should not be underestimated. Nonetheless, it is clear that

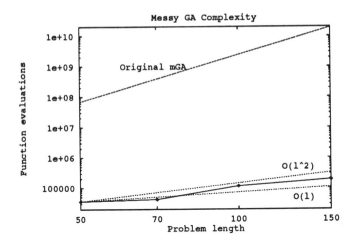

Figure 2.3   A fast messy GA finds solutions to hard problems quickly and reliably. The original mGA for the order-five problem is $O(\ell^5)$, but the modified procedure is apparently subquadratic.

we have moved beyond the stage where researchers blithely assume that learning is some "chrome cognitive hubcap" that we add in the showroom to improve our systems once the "heavy horsepower" of knowledge representation and acquisition are chosen. Selectionists and connectionists have been on the forefront of the movement to put learning first, and others appear to be joining us as disappointment with traditional artificial intelligence (AI) mounts.

GAs as computational model of innovation and creativity. From my first writings about genetic algorithms, I have likened the power of GAs to the power of human creativity and innovation. What is it we are doing when we are being innovative or creative. Often we take a set of solution features that worked well in one context and a set of solution of features that worked well in a different context and bring them together—possibly for the first time—to try to solve the problem at hand. This *emphasis* and *juxtaposition* of human creativity is similar to the *selection* and *recombination* of genetic algorithms. Of course, the willy-nilly juxtaposition of GAs seems much less directed than that of our own creative efforts—or so we would like to believe; nonetheless, the alignment of the fundamental processes in the two situations is appealing, and in a different setting we could survey some of the writers who have identified similar processes in their explorations of human creativity.

Here, it seems more important that I address a fundamental objection to this line of reasoning. In response to this argument I sometimes get what I call the *innovation-not-creativity* objection. In its usual form the objector will accept that selectorecombinative GAs sound *innovative* enough, but they don't seem to qualify for the loftier term *creative*. I believe there is merit to this objection, but metaphors are as interesting for the connections that do not exist between two situations as for those that do. If we take the objection seriously we must ask the question, "What is missing?" No doubt there is quite a lot, but in my opinion the single most important lacuna is the lack of an analog to *analogy* or *metaphor* in garden-variety GAs.

When we appraise great art, music, architecture, literature, or other human outputs that are commonly associated with the term "creativity," some features that commonly distinguish great from not-so-great work are the number, quality, and type of interconnections that the work makes with other works and other aspects of life. For example, we look at a piece of art or literature for what it *represents* (the direct connections it makes with the object or situation it portrays), for its *allusions* (the direct interconnections it makes to other works of art or literature), and for its *symbolism* (the meaning we infer by analogy through conceptually constructing interelations between the work of art and some situation not directly represented in the work).

In things creative, analogy *transfers* a set of fit features to the subject work, thereby infusing it with a well-adapted richness, a complexity that would have been difficult to obtain by other means. In genetic algorithms a method of *analogizing* or *metaphoric transfer* would enable *deep, difficult building blocks* to be transferred from a well-understood situation to a poorly understood situation without the cost of explicit search. Mechanisms for allusion—direct transfer of knowledge from a different domain would not be all that different than simple recombination and would simply involve taking genetic material from one context and moving it to another. It is tempting to say the vectoring of genetic material via a virus from one species to another looks a lot like allusion as is.

The richer type of transfer associated with metaphor would appear to require more explicit alignment of context in the two situations, and the usual AI mechanisms to achieve this require explicit mapping of model abstractions one to the other to achieve the desired transfer. One route that should be explored before taking this obvious pathway is to examine a more sophisticated role for *development* in the genetic process and to consider whether analogic transfer might occur as a combination of genetic operators acting on genetic structures or sets of structures that were themselves a combined structural-developmental blueprint.

GAs as computational models of other innovating systems. After one plays with genetic algorithms for even a fairly short time and marvels at the many strange and counterintuitive things the populations discover, it is more than a little tempting to try to transfer these lessons to other population-dominated situations. Economies, organizations, social systems, political systems, whole societies, seem a little less mysterious after one has spent some time watching GA runs and learning from them. The reasons for this are twofold. First, much of the mystery of such systems emanates from their innovative nature, and GAs replace the mystery shrouding innovation with a healthy dosage of mechanism. Many of the difficulties in the social sciences come from the lack of a computational theory of actor innovation (Sargent, 1993), and genetic algorithms increasingly fill that bill quite nicely. Second, population-oriented systems are dominated by what economists call *the law of unintended consequences* (which is itself largely the result of the innovative capability of the actors) and interacting with GAs provides hands-on experience in understanding what for most people is counterintuitive behavior.

In addition, the shift to *constructing* models that match population behavior *qualitatively* is a departure from traditional modeling that requires adherence to fairly rigorous notions of *modeling fidelity*. The *engineering* of *qualitatively interesting* models is a fairly radical departure from the the science of developing high fidelity models, and although much of the attention paid to work under the rubric *artificial life* comes from novelty and hype, the remaining portion that is scientifically interesting is the result of doing *good science through good engineering*. The reverse situation—doing good engineering through good science—is commonly thought to be the more fruitful path, but relaxation of almost any perceived constraint in a thoughtful manner can at one time or another be a useful methodological device. The combination of relaxing modeling fidelity and engineering models that match a range of qualitative behavior is proving to be a useful approach in many disciplines whose objects are themselves complex systems.

GAs as guiding philosophy. After seeing GAs as powerful problem solvers, as a technical puzzle, as a model of creativity, and as models of other innovating systems, it is difficult not to start using GAs as a metaphor for *everything*. The telltale signs that a genetic algorithmist has reached this point of development are (1) seeing selection and genetics everywhere, and (2) telling selectionist "jokes." When I reached this stage, I was somewhat suspicious of it, because in a previous life I did computations of transients in flow systems, and at that time I developed a tendency to (1) see wave propagation in all human events, and (2) tell waterhammer jokes. I was therefore quite leery of reading too much into my musings about the cosmic significance of GAs until I was introduced to a serious philosophical literature that takes evolution as its basis in Csikszentmihalyi's (1990) *Flow: The Psychology of Optimal Experience*. There he argues for a need, a set of criteria, and a candidate for a new belief system. The need is straightforward (p. 238):

> But it seems clear that an increasing majority are not being helped by traditional religions and belief systems. Many are unable to separate the truth in old doctrines from the distortions and degradations that time has added, and since they cannot accept error, they reject the truth as well. Others are so desperate for some order that they cling rigidly to whatever belief happens to be at hand—warts and all—and become fundamentalist Christians, or Muslims, or communists.

The criteria for a new belief system are sweeping (pp. 238–239):

> If a new faith is to capture our imagination, it must be one that will account rationally for the things we know, the things we feel, the things we hope for, and the ones we dread. It must be a system of beliefs that will marshal our psychic energy toward meaningful goals, a system that provides rules for a way of life that can provide flow.

This sounds like a tall order to fill, and where should we turn to fill it? In a word, Csikszentmihalyi says we can do worse than to turn to *evolution* (p. 239):

One way to accomplish this is through the concept of evolution.
Everything that matters most to us—such questions as: Where
did we come from? Where are we going? What powers shape our
lives? What is good and bad? How are we related to one another,
and to the rest of the universe? What are the consequences of
our actions?—could be discussed in a systematic way in terms
of what we now know about evolution and even more in terms
of what we are going to know about it in the future.

This argument for replacing religion with a faith in evolutionary mechanism has some
appeal on its face, but I am bothered by Csikszentmihalyi's suggestion in the same way
I've been bothered by one aspect of working on GAs. For as long as I've been a genetic
algorithmist I've had the nagging suspicion that I was something of a party wrecker,
because by trying to make rational the creative or innovative process, in a sense, I
was threatening to unmask what has always been an important mystical or spiritual
experience for human beings. On the other hand, I've not dwelled on this thought
because I've imagined that a rational understanding of the *juxtapositional calculus of
innovation* should have little or no effect on the magic of the eureka moment in much
the same way that human understanding of the neurobiology of endorphins does not
lessen the feeling of the "runner's high."

When the locus of concern shifts away from the individual and toward the
population, however, I believe my intuitions are better grounded. If the ethical
and moral principles taught by the world's religions end up having their basis in
evolutionary principles, and their "purpose" is to promote social cohesion and group
survival, will evolutionary faith be enough to encourage individuals to take the "right"
action in those instances where they can benefit as individuals by defecting from the
social good? Religions encourage compliance with moral precepts through unverifiable
promises of life in the hereafter, favorable treatment by a deity, punishment for
wrongdoing, or the like. If the evolutionary faithful seek mechanism in all things,
finding it opens the door to ruthless behavior by self-interested individuals who might
otherwise be deterred by the uncertainty of the consequences of such behavior under
a more traditional religion.

Thus, while I agree with Csikszentmihalyi's focus on the importance of evolutionary
thinking, and I believe that GAmists are in a special position because of their hands-
on experience with these systems in miniature, perhaps calls for the substitution of an
evolutionary faith are extreme. Evolutionary thinking will no doubt chip away at the
mystical kernel, attaching mechanism to more of what was once considered beyond
human comprehension. But perhaps it is neither possible nor desirable to cause that
set to vanish, and maybe a better aim is to work in this way to approach a better
understanding of the essence of our spiritual nature.

2.3 CONCLUSION

Most of us come to genetic algorithms to solve problems, but some of us leave the
encounter with much more. Although this essay has considered how GAs have solved
problems and how they are becoming more competent as we solve the design challenges

they pose, perhaps the greatest lessons are still ahead of us as we better understand what genetic algorithms are trying to tell us about innovation, creativity, and the complex systems that depend on those processes. Moreover, understanding evolution in a detailed, mechanistic way provides insights into many walks of life, thereby making the study of these computational entities a matter of greater importance than the solution of any given single problem.

ACKNOWLEDGMENTS

Support for this work was provided by the U.S. Air Force Office of Scientific Research under Grant No. F49620-94-1-0103.

REFERENCES

Bradshaw, G. L., & Lienert, M. (1991). The invention of the airplane. *Proceedings of the Thirteenth Annual Conference of the Cognitive Science Society*, 605–610.

Csikszentmihalyi, M. (1990). *Flow: The psychology of optimal experience.* New York: Harper & Row.

De Jong, K. A. (1993). Genetic algorithms are not function optimizers. *Foundations of Genetic Algorithms*, 5–17.

Florman, S. C. (1976). *The existential pleasures of engineering.* New York: St. Martin's Press.

Goldberg, D. E. (1989). *Genetic algorithms in search, optimization, and machine learning.* Reading, MA: Addison-Wesley.

Goldberg, D. E. (1993). Making genetic algorithms fly: A lesson from the Wright Brothers. *Advanced Technology for Developers*, *2*, 1–8.

Goldberg, D. E. (1994). Genetic and evolutionary algorithms come of age. *Communications of the Assocation for Computing Machinery*, *37*(3), 113–119.

Goldberg, D. E., Deb, K., & Clark, J. H. (1992). Genetic algorithms, noise, and the sizing of populations. *Complex Systems*, *6*, 333–362.

Goldberg, D. E., Deb, K., Kargupta, H., & Harik, G. (1993). Rapid, accurate optimization of difficult problems using fast messy genetic algorithms. *Proceedings of the Fifth International Conference on Genetic Algorithms*, 56–64.

Goldberg, D. E., Deb, K., & Korb, B. (1990). Messy genetic algorithms revisited: Studies in mixed size and scale. *Complex Systems*, *4*(4), 415–44.

Goldberg, D. E., Deb, K., & Thierens, D. (1993). Toward a better understanding of mixing in genetic algorithms. *Society of Instrument and Control Engineers Journal*, *32*(1), 10–16.

Goldberg, D. E., Korb, B., & Deb, K. (1989). Messy genetic algorithms: Motivation, analysis, and first results. *Complex Systems*, *3*(5), 493–530.

Holland, J. H. (1962). Outline for a logical theory of adaptive systems. *Journal of the Association for Computing Machinery*, *3*, 297–314.

Holland, J. H. (1975). *Adaptation in natural and artificial systems.* Ann Arbor: University of Michigan Press.

Ipsen, D. C. (1960). *Units, dimensions, and dimensionless numbers.* New York: McGraw-Hill.

Sargent, T. J. (1993). *Bounded rationality in macroeconomics.* Oxford: Clarendon Press.

Thierens, D., & Goldberg, D. E. (1993). Mixing in genetic algorithms. *Proceedings of the Fifth International Conference on Genetic Algorithms*, 38–45.

Thierens, D. (1995). *Analysis and design of genetic algorithms.* Unpublished doctoral dissertation, Catholic University of Leuven, Leuven.

3

A General Study on Genetic Fuzzy Systems

OSCAR CORDÓN, FRANCISCO HERRERA

3.1 INTRODUCTION

As it is known, a rule based system (production rule system) has been successfully used to model human problem-solving activity and adaptive behavior, where a classic way to represent the human knowledge is the use of IF/THEN rules. The satisfaction of the rule antecedents gives rise to the execution of the consequent, one action is performed. The conventional approaches to knowledge representation are based on bivalent logic. A serious shortcoming of such approaches is their inability to come to grips with the issue of uncertainty and imprecision. As a consequence, the conventional approaches do not provide an adequate model for modes of reasoning and all commonsense reasoning fall into this category.

Fuzzy Logic (FL) may be viewed as an extension of classical logical systems, provides an effective conceptual framework for dealing with the problem of knowledge representation in an environment of uncertainty and imprecision. FL, as its name suggests, is the logic underlying modes of reasoning which are approximate rather than exact. The importance of FL derives from the fact that most modes of human reasoning -and especially commonsense reasoning- are approximate in nature. FL is concerned in the main with imprecision and approximate reasoning.

The applications of FL to rule based systems have been widely developed. From a very broad point of view a Fuzzy System (FS) is any Fuzzy Logic Based Sytems, where FL can be used either as the basis for the representation of different forms of knowledge systems, or to model the interactions and relationships among the system variables. FS have been shown to be an important tool for modelling complex systems, in which, due to the complexity or the imprecision, classical tools are unsuccessful.

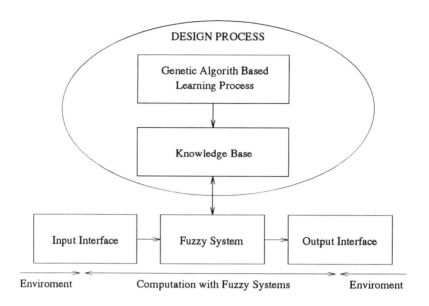

Figure 3.1 Genetic Fuzzy Sysems

Among the most successful applications of this systems has been the area of Fuzzy logic controllers (FLCs). FLCs are rule based systems useful in the context of complex ill-defined processes, especially those which can be controlled by a skilled human operator without knowledge of their underlying dynamics. Recently fuzzy control techniques have been applied to many industrial processes, FLCs have been widthly used in automation and engineering. The experience of skilled operators and the knowledge of control engineers are expressed qualitetively by a set of fuzzy control rules. In fact, one of the features of the FLCs is that the *IF-THEN* rules are described on the base of the conventional control strategy and the experts' knowledge. Each fuzzy rule has an antecedent, or *IF*, part containing several preconditions, and a consequent, or *THEN*, part which prescribes the value.

Recentely, numerous papers and applications combining fuzzy concepts and genetic algorithms (GAs) have become known, and there is an increasing concern in the integration of these two topics. In particular, there are a great number of publications exploring the use of GAs for developing fuzzy systems, the called genetic fuzzy systems (GFSs). Figure 1 shows this idea.

This paper presents an overview of the GFSs, showing the use of the GAs in the construction of the fuzzy logic controllers knowledge bases comprising the known knowledge about the controlled system.

To achieve that, this paper is divided into 4 sections the first being this introduction. The section 2 introduces the fuzzy systems with a special attention to FLCs, while section 3 presents the GFSs. Some final remarks are made in section 4.

3.2 FUZZY SYSTEMS

Fuzzy logic and fuzzy sets in a wide interpretation of FL (in terms of which fuzzy logic is coextensive with the theory of fuzzy sets, that is, classes of objects in which the transition from membership to nonmembership is gradual rather than abrupt) have placed modeling into a new and broader perspective by providing innovative tools to cope with complex and ill-defined systems. The area of fuzzy sets has emerged following some pioneering works of Zadeh [Zad65, Zad73] where the first fundamentals of fuzzy systems were established.

As we aforesaid, a rule based system has been successfully used to model human problem-solving activity and adaptive behavior. The conventional approaches to knowledge representation, are based on bivalent logic. A serious shortcoming of such approaches is their inability to come to grips with the issue of uncertainty and imprecision. As a consequence, the conventional approaches do not provide an adequate model for modes of reasoning. Unfortunatelly, all commonsense reasoning fall into this category.

The application of FL to rule based systems leads us to the fuzzy systems. The main role of fuzzy sets is representing knowledge about the problem, or to model the interactions and relationships among the system variables. There are two essential advantages for the design of rule-based systems with fuzzy sets and logic:

- the key features of knowledge captured by fuzzy sets involve handling uncertainty, and
- inference methods become more robust and flexible with approximate reasoning methods of fuzzy logic.

Knowledge representation is enhanced with the use of linguistic variables and their linguistic values that are defined by context-dependent fuzzy sets whose meanings are specified by graded membership functions. On other hand, inference methods such as generalized modus ponens, tollens, etc., which are based on fuzzy logic form the bases of approximate reasoning with pattern matching scores of similarity. Fuzzy logic provides an unique computational base for inference in rule based systems. Unlike traditional logical systems, fuzzy logic is aimed at providing modes of reasoning which are approximate and analogical rather than exact.

Abording the fuzzy system modeling issue, it is essentially developed into two different types of system models identified as acquisition of rules and their parameters: i) fuzzy expert system models, and ii) fuzzy logic controllers.

Fuzzy expert system models are designed, developed and implemented with a direct participation of a system's expert who is throughly familiar with the characteristic behaviour of the system under investigation. The knowledge of the expert is extracted from the expert through experimental methods of questionnaires, protocols and interviews which may be conducted by people or by computers for the purpose of identifying the form and the structure of the rules, i.e., structure identification as well as the membership functions of the linguistic values of linguistic variables, i.e., parameter identification.

On the other hand, fuzzy control model design, development and implementation are dependent on the availability of input-output data sets. The system structure identification in terms of rules and specification of membership functions that define

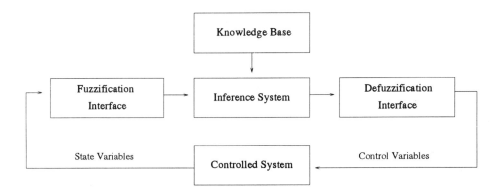

Figure 3.2   Generic structure of a fuzzy logic controller

the meaning representation of linguistic values of linguistic variables are determined by learning techniques. Also there is a third kind that is the combination of the others two that may be called *fuzzy expert-control*.

Here, we will center in the second fuzzy system model, the FLCs, where the GAs have been used for design the structure identification of the system.

FLCs, initiated by Mamdani and Assilian in the work [MA75], are now considered as one of the most important applications of the fuzzy set theory. FLCs are knowledge based controllers that make use of the known knowledge of the process, expressed in form of fuzzy linguistic control rules collected in a knowledge base (KB), to control it. The advantage of this approach with respect to the classical *Control Theory* is that it has not necessity of expressing the relationships existing in the system by means of a mathematical model, what constitutes a very difficult task in many real situations presenting nonlinear characteristics or complex dynamic. In the following two subsections we present an introduction to FLCs and to the FLC KBs.

3.2.1   Fuzzy Logic Controllers

An FLC is composed by a *Knowledge Base*, that comprises the information given by the process operator in form of linguistic control rules, a *Fuzzification Interface*, which has the effect of transforming crisp data into fuzzy sets, an *Inference System*, that uses them joined with the Knowledge Base to make inference by means of a reasoning method, and a *Defuzzification Interface*, that translates the fuzzy control action so obtained to a real control action using a defuzzification method. The generic structure of an FLC is shown in figure 2 [Lee90].

Several factors with a significant influence have to be analyzed in order to design an FLC for a concrete process. Concretely, there are two main decisions to make in order to design a FLC, to derive a KB for the system and to decide the reasoning method to use. As can be viewed, only the first one depends directly on the concrete application although several reasoning methods will perform better with some kind of systems than others.

The **Knowledge Base** is the FLC component comprising the expert knowledge known about the controlled system. So it is the only component of the FLC depending on the concrete application and it makes the accuracy of the FLC depends directly on its composition. It is composed by a set of fuzzy control rules with the form:

If X_1 is A_{i1} and X_2 is A_{i2} and ... and X_n is A_{in} then Y is B_i

being the X_i and Y linguistic system variables and the A_i and the B_i linguistic labels associated with fuzzy sets specifying their meaning.

The **Fuzzification Interface** defines a mapping from an observed input space to fuzzy sets in certain input universes of discourse, obtaining the membership function associated to each one of the crisp system inputs.

The **Inference System** is based on the application of the Generalized Modus Ponens, extension of the classical logic Modus Ponens proposed by Zadeh. It is done by means of the Compositional Rule of Inference (CRI) given by the following expression:

$$\mu_{B'}(y) = Sup_{x \in X} \left\{ T'\left(\mu_{A'}(x), I(\mu_{A_i}(x), \mu_B(y))\right) \right\} \tag{3.1}$$

being T, T' connectives, $\mu_{A_i}(x) = T(\mu_{A_{i1}}(x), \ldots, \mu_{A_{in}}(x))$ and I an implication operator.

Since the input x corresponding to the state variables of the controlled system is crisp, $x = x_0$, the fuzzy set A' is a singleton, that is, $\mu_{A'}(x) = 1$ if $x = x_0$ and $\mu_{A'}(x) = 0$ if $x \neq x_0$. Thus the CRI is reduced to:

$$\mu_{B'}(y) = I(\mu_{A_i}(x_0), \mu_B(y)) \tag{3.2}$$

Since from each rule R_i is obtained a fuzzy set B'_i from the inference process, the **Defuzzification Interface** uses an aggregation operator G which composes them and applies a defuzzification method D to translate the fuzzy sets obtained in this way into values corresponding to the control variables of the system. So, calling S to the FLC, x_0 to the inputs value and y_0 to the crisp value obtained from the defuzzification, we have:

$$\mu_{B'}(y) = G\left\{ \mu_{B'_1}(y), \mu_{B'_2}(y), \ldots, \mu_{B'_n}(y) \right\} \tag{3.3}$$

$$y_0 = S(x_0) = D(\mu_{B'}(y)) \tag{3.4}$$

At present, the commonly used defuzzification methods may be described as the *Max Criterion*, the *Mean of Maximum (MOM)* and the *Center of Area (COA)* [Lee90].

The design tasks that have to be developed in order to decide the FLC reasoning method are the selection of the fuzzy operators I, T and G and the defuzzification operator D [KKS85]. The problem of selection them have been analyzed in several works such us [CCC+94, CCC+95, CHP95b, CHP95a, KKS85].

For more information about FLCs see [CHP95a, DHR93, HMB93, Lee90].

3.2.2 The Fuzzy Logic Controller Knowledge Base

The Knowledge Base is comprised of two components, a *Data Base* (DB), containing the definitions of the fuzzy control rules linguistic labels, that is, the membership

functions of the fuzzy sets specifying the meaning of the linguistic terms, and a
Rule Base (RB), constituted by the collection of fuzzy control rules representing the
expert knowledge. We are going to analize more concretely this two components in
the following sections.

The Data Base

The concepts associated with a DB are used to characterize fuzzy control rules and
fuzzy data manipulation in an FLC [Lee90]. In this way, the main task to be done
in order to design an FLC DB is to associate a membership function to every one of
the linguistic terms that the system input and output variables can take as possible
values. There are two modes of DB definition:

- By means of a quantization or normalization process.
- By means of a tuning or learning process.

Every linguistic variable involved in the FLC KB forms a fuzzy space with respect to
a certain universe of discourse and have associated a label set containing the possible
linguistic values that it can take. A *fuzzy partition* determines how many terms should
exist in the label set. The choice of the term set is equivalent to finding the primary
fuzzy sets or linguistic labels (terms). This is a previous task for the first mode and
for some of the methods included in the second one.

The first definition mode makes use only of a little part of the a priori known
knowledge of the system. In order to define the meaning of the linguistic term set,
a discretization of the process input and output variables continuous universes of
discourse have to be performed. This process is usually called *quantization* and is
done in a number of steps [DHR93, HMB93, Lee90]:

1. The continuous domain is discretized (quantized) into a certain finite
   number of segments called quantization levels. Each segment is labeled
   as a generic element and the set of all generic elements forms a discrete
   universe of discourse.
2. Given a linguistic value from a certain term set, the fuzzy set defining
   the meaning of this linguistic value is built by assigning a degree of
   membership to each generic element. This is done for every linguistic
   value in a term set.

The use of quantized or normalized domains requires a scale transformation which
maps the physical values of the process variables into the discretized universe. In both
cases, the mapping can be uniform (linear), non-uniform (nonlinear) or both. Either a
numerical or functional definition may be used to assign the degrees of membership to
the primary fuzzy sets [DHR93, HMB93, Lee90]. This choice is based on the subjective
criteria and it should be more convenient that the human processes operators could
represent the meaning of their usually employed linguistic terms in form of fuzzy sets.
Unfortunately, in many real cases it is not possible for him to make that, and it is
very common the use of an uniform fuzzy partition as the one proposed by Liaw and
Wang [LW91] shown in figure 3.

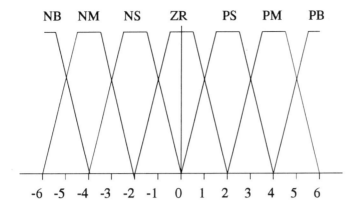

Figure 3.3   Fuzzy partition proposed by Liaw

The second definition mode employs many of the known knowledge and leads to a more application specific DB requiring less intervention of the controller designer and presenting more automatically developed tasks. Belonging to this group, we can distinguish two different subgroups:

- Definition by means of a tuning process
- Definition by means of a learning process

Methods included in the first one make use of a primary DB definition developed by means of a quantization or normalization process and then apply a process that modifies the meaning of the linguistic labels, that is, the definitions of the fuzzy sets associated to them. On the other hand, it would be a learning process when there is not an initial DB definition. Usually, this last process is carried out joined to a Rule Base learning process.

Moreover, an important decision in order to define the FLC is to determine the nature of its DB. There are two different approximations for it depending on the scope in what there is assigned the meaning to the linguistic labels belonging to the different term sets. On one hand, an usual DB definition process in what the fuzzy sets giving meaning to the linguistic labels are uniformly defined for all rules included in the RB constitutes a *descriptive* approach since the linguistic labels represents a real world semantics. On the other hand, it can be considered a KB whose rules present different meaning for the same linguistic terms. The meaning associated to a concrete label will depend on the concrete rule in what this label appears. In this case, the KB and the FLC using it present a different philosofy. The approach is *approximative* and the system is in the line of an *Universal Approximator* [Cas95].

The Rule Base

There are different kinds of fuzzy control rules proposed in the specialized literature regarding to the expression of the consequent. Mamdani and Assilian employ rules in which the consequent is another fuzzy variable [MA75], while Sugeno et al. use rules whose conclusion is a polynomial function of the inputs (TSK rules) [TS85]. Another kind of rules present also the consequent being a function of the input parameters. The following three rules show respectively the generical expressions of the three types commented:

$If\ X_1\ is\ A_1\ and\ ...\ and\ X_n\ is\ A_n\ then\ Y\ is\ B$

$If\ X_1\ is\ A_1\ and\ ...\ and\ X_n\ is\ A_n\ then\ Y = p_0 + p_1 X_1 + ... + p_n X_n$

$If\ X_1\ is\ A_1\ and\ ...\ and\ X_n\ is\ A_n\ then\ Y = f(X_1, ..., X_n)$

being the X_i and Y, linguistic system variables and the A_i and B, linguistic labels associated with fuzzy sets specifying their meaning.

There are four modes of derivation of fuzzy control rules that are not mutually exclusive [HMB93, Lee90]. These modes are the following:

1. Expert Experience and Control Engineering Knowledge.
2. Modeling of the Operator's Control Actions.
3. Based on the Fuzzy Model of a Process.
4. Based on Learning and Self-Organization.

The first method has been widely used. This method is effective when expert human operators can express what they use to control the system in terms of control rules. The rules more usually obtained by means of this process are Mamdani type because they present an adequate form to represent the expert knowledge. The second method directly models the control actions of the process operator. Instead of interviewing the operator, the types of control actions taken by him are modeled. The third approach is based on developing a model of the plant and construct an FLC to control the fuzzy model generating the fuzzy control rules of the RB by means of the fuzzy model of the system. It makes this approach similar to that traditionally used in Control Theory. Hence, structure and parameter identification are needed. Finally, the fourth method is focused on learning. In this case, the ability to create fuzzy control rules and to modify them based on experience in order to improve the controller performance is considered.

3.3   DESIGNING GENETIC FUZZY SYSTEMS

As it has been shown in the above section, there are many tasks that have to be performed in order to design an FLC to control a concrete system. We have commented yet that the derivation of the KB is the only one directly depending on the controlled system and it presents a significative importance in the design process. It is known that the more used method in order to perform this task is based directly on extracting the expert experience from the human processes operators. The problem arises when these are not able to express their knowledge in terms of fuzzy control rules. In order to avoid this drawback, researches have been investigating automatic learning methods for designing FLCs by deriving automatically an appropiate KB for the controlled system without necessity of its human operator.

The genetic algorithms (GAs) have demostrated to be a powerful tool for automating the definition of the KB since adaptative control, learning and self-organization can be considered in a lot of cases as optimization or search processes. The fuzzy systems making use of a GA in their design process are called generically GFSs.

These advantages have extended the use of the GAs in the development of a wide range of approaches for designing FLCs in the last years. Some of these approaches of *genetic FLC design* will be shown in the present chapter. It is possible to distinguish three different groups of *genetic FLC design processes* according to the KB components included in the learning process. These ones are the following:

1. Genetic definition of the Fuzzy Logic Controller Data Base.
2. Genetic derivation of the Fuzzy Logic Controller Rule Base.
3. Genetic learning of the Fuzzy Logic Controller Knowledge Base.

In the following subsections we are going to analyze each one of the approaches for genetic design of FLCs.

3.3.1 *Defining the Fuzzy Logic Controller Data Base using Genetic Algorithms*

As we have commented already, one of the modes of definition of the FLC DB is based on learning or tuning this FLC component. The difference between these two approaches depends on the existence of a previously primary DB definition. While learning processes do not need this previous definition, tuning processes works over it obtaining a more accurated one.

Several methods have proposed in order to define the FLC DB using GAs [BN95, BMU95, FTH94, HLV95b, HTS93, Kar91b]. All of them are based on the existence of a previously defined RB, usually extracted from the process operator. Each chromosome involved in the evolution process will represent different DB definitions, that is, each one of the chromosomes will contain a coding of the whole membership functions giving meaning to the linguistic terms. The degree of adaptation of an individual is measured using a fitness function that usually is based on the aplication of the FLC to the controlled system, using a KB formed by the RB and the DB encoded by the chromosome.

There are two different approaches for the genetic definition of FLC DBs depending on the scope of the association of membership functions to linguistic labels in the KB, either all fuzzy control rules using the same meaning for the system variables linguistic terms or a different approximation in what each rule presents its own meaning for the labels involved by it. In this subsection we analyze an example of each one of both groups. The method proposed by Karr [Kar91b], belonging to the first one, and the method of Herrera et al. [HLV95b], belonging to the second one.

The DB definition method proposed by Karr

The approach of Karr [Kar91b] is based on the existence of primary fuzzy partitions of the different system variables input and output spaces. The GA is applied for defining the meaning of the different linguistic term sets, that is, for learning the fuzzy sets associated to each one of the linguistic labels belonging to the fuzzy partitions. As it can be viewed, this approach is a descriptive one because all the fuzzy terms appearing in the fuzzy control rules will use the same meaning (that defined by means of the GA learning process).

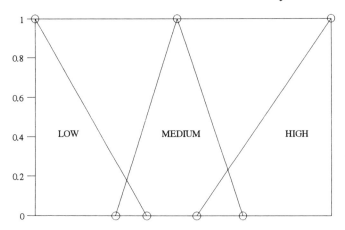

Figure 3.4 An example of fuzzy partition in the Karr's DB definition method

In order to develop this task, Karr uses a Simple GA with binary coding, proportional selection mechanism, simple crossover and random mutation. The binary coding outline will represent all the membership functions associated to the different linguistic labels belonging to each one of the linguistic term sets into a single chromosome.

The membership functions selected by Karr to define the meaning of the linguistic labels are triangular-shaped. He consider only two points, extremes of the fuzzy sets support, for defining every triangle, fixing its central point which presents value 1 in the membership function. In this way, the evolution performed by the GA can make the triangles to be distorted (when the base width is altered) and translated (when these two points are shifted along the x-axis) freely. The process is different for the extreme triangles, which only requires a single point with value of the membership functions equal to 0 to be defined. Their modal point is clearly placed in the correspondent extreme of the concrete system variable universe of discurse. Only one operation can be performed with these triangles, altering their base width in one of the two x-axis directions, that is, making it bigger or smaller. Figure 4 shows an example.

The constraint placed over the membership functions is that the ones associated to the extreme labels must remain right triangles while those associated to the interior terms must remain isosceles triangles. It is clear that this constraint avoid the GA to obtain incorrectlly defined membership functions.

A chromosome coding all the membership functions is built finally by joining all the individual coding of these ones into a single string. As each point is represented by a binary number with a fixed number of bits, the chromosomes are of fixed-length, that is, all individuals in the population will present the same length.

Finally, Karr do not present a concrete fitness function but introduces several considerations in order to define it. His idea for measuring the accuracy of a concrete DB in the optimal control of the system is based on an application-dependent measure, that is, any error or convergence measure (see [CCC⁺94, CHP95a]).

The DB definition method proposed by Herrera et al.

In [HLV95b] it was presented a DB definition process used to tune the DB parameters. The process is based on the existence of a previous complete KB, that is, an initial DB definition and a RB constituted by m control rules. The chromosomes will encode a complete KB since each one of them contains the RB with a different DB associated.

The GA designed for the process present real coding issue and use the stochastic universal sampling as selection procedure and the Michaelewicz's non-uniform mutation operator. Regarding to the crossover, two different operators are employed: the simple and the Max-Min-Arithmetical crossover. This last operator have been proposed by the authors and makes use of fuzzy operators in order to improve the behavior of the GA crossover operator.

The membership functions selected in order to define the DB are trapezoidal-shaped. They have associated a parametric representation based on a 4-tupla of real values. Let the following rule be the ith rule of the previous RB:

$$\text{If } X_1 \text{ is } A_{i1} \text{ and } X_2 \text{ is } A_{i2} \text{ and } \ldots \text{ and } X_n \text{ is } A_{in} \text{ then } Y \text{ is } B_i$$

Then the fuzzy sets giving meaning to the linguistic labels A_{ij} associated to the input variables X_i will be represented by the 4-tuple $(c_{ij}, a_{ij}, b_{ij}, d_{ij})$ and the ones associated to the output variable linguistic labels B_i by (c_i', a_i', b_i', d_i'). Thus each one of the rules will be encoded in pieces of chromosome C_{ri}, $i = 1, \ldots, m$, in the following way:

$$C_{ri} = (c_{i1}, a_{i1}, b_{i1}, d_{i1}, ..., c_{in}, a_{in}, b_{in}, d_{in}, c_i', a_i', b_i', d_i') \qquad (3.5)$$

Therefore the complete RB with an associated DB is represented by a complete chromosome C_r:

$$C_r = C_{r1} \, C_{r2} \, ... \, C_{rm} \qquad (3.6)$$

As it can be viewed, each individual of the population represents a complete KB. More concretelly, all of them encode the derived system RB and the difference between them is the meaning associated to the linguistic variables taking part in the fuzzy control rules, that is, the DB definition. As each rule is coded in a piece of chromosome, the fuzzy set giving meaning to a linguistic term can be changed in one rule and not in the other ones in which it appears or, in a more extreme case, it can be changed in many rules presenting different forms in several ones. The meaning of the linguistic terms will depend then on the rule in which they are involved. The KB so obtained will present an approximative behavior.

The initial gene pool is created from the initial KB. This KB is encoded directly in a chromosome, denoted as C_1. The remaining individuals are generated by associating an interval of performance, $[c_h^l, c_h^r]$ for every gene c_h of C_1, $h = 1 \ldots (n + 1) \times m \times 4$. Each interval of performance will be the interval of adjustment for the correspondent variable, $c_h \in [c_h^l, c_h^r]$.

If $(t \bmod 4) = 1$ then c_t is the left value of the support of a fuzzy number. The fuzzy number is defined by the four parameters $(c_t, c_{t+1}, c_{t+2}, c_{t+3})$ and the intervals of performance are the following:

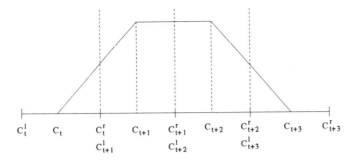

Figure 3.5 Intervals of performance

$$c_t \in [c_t^l, c_t^r] = [c_t - \tfrac{c_{t+1} - c_t}{2}, c_t + \tfrac{c_{t+1} - c_t}{2}]$$
$$c_{t+1} \in [c_{t+1}^l, c_{t+1}^r] = [c_{t+1} - \tfrac{c_{t+1} - c_t}{2}, c_{t+1} + \tfrac{c_{t+2} - c_{t+1}}{2}]$$
$$c_{t+2} \in [c_{t+2}^l, c_{t+2}^r] = [c_{t+2} - \tfrac{c_{t+2} - c_{t+1}}{2}, c_{t+2} + \tfrac{c_{t+3} - c_{t+2}}{2}]$$
$$c_{t+3} \in [c_{t+3}^l, c_{t+3}^r] = [c_{t+3} - \tfrac{c_{t+3} - c_{t+2}}{2}, c_{t+3} + \tfrac{c_{t+3} - c_{t+2}}{2}]$$

Figure 5 shows these intervals. Therefore, we create a population of chromosomes containing C_1 as its first individual and the remaining ones initiated randomly, with each gene being in its respective interval of performance.

By using a training input-output data set, E_{TDS}, and a concrete error measure, the square medium error, the fitness function of a chromosome is defined. In this way, the adaptation value associated to an individual is obtained by computing the error between the outputs given by the FLC using the KB encoded in the chromosome and those contained in the training data set. The fitness function is represented by the following expression:

$$E(C) = \frac{1}{2|E_{TDS}|} \sum_{e_k \in E_{TDS}} (ey^k - S(ex^k))^2 \qquad (3.7)$$

3.3.2 Deriving the Fuzzy Logic Controller Rule Base using Genetic Algorithms

The great majority of the approaches belonging to this group are based on learning the consequents of the fuzzy control rules included in the FLC RB. In this way, many of these genetic processes encode the complete system decision table in the chromosomes. There are different methods developing this task [Bon93, HT94, Kar91a, KB93, Thr91] but in many cases the only difference existing among them is the cappability of learning the number of rules forming the RB. This characteristic is presented when there exist a possible alelle representing the absence of consequent for a rule with a concrete antecedent, that is, the absence of the correspondent rule in the RB.

All methods belonging to this family suppose the existence of a defined DB. Thus, a primary partition of the fuzzy spaces, that is, a term set for each one of them, and a collection of fuzzy sets giving meaning to these primary labels are considered. Other common characteristic for a majority of them is that they consider the FLC

RB represented in form of *decision table* (called too *look-up table*). As it is known, an usual FLC RB constituted by control rules presenting n input variables and a single output variable can be represented using an n-dimensional decision table, each dimension corresponding to each one of the input variables. Every dimension will have associated an array containing the labels of the concrete variable term set and the cells of the decision table will contain the linguistic label that the output variable take for the combination of antecedents represented in this cell. Therefore each table cell represents a fuzzy control rule that can belong to the FLC RB.

This structure is encoded in the individuals forming the GA population. If there not exist a value for the alleles representing the absence of value for the rule consequent, there is not possible to derive a FLC RB with an optimal number of rules but all the possible rules have to be considered belonging to it. This is the case of the methods proposed by Karr [Kar91a], and Kropp and Baitinger [KB93] altough the first of them do not encode the complete decision table as we are going to see in the following. The remaining ones commented are able to learn the number of fuzzy rules.

In this subsection we are going to study three different approaches. The methods selected were proposed by Thrift [Thr91], Karr [Kar91a] and Bonarini [Bon93]. This last one constitutes an original approach for learning FLC RB and differs a lot from the others belonging to this family as we are going to see in the following.

The RB derivation method proposed by Thrift

This method, as many others belonging to these category, is based on encoding all the cells of the complete decision table in the chromosomes. In this way, Thrift [Thr91] establishes a mapping between the label set associated to the system output variable and an increasing integer set (containing one element more and taking 0 as first element) representing the alelle set. An example is shown to clarify the concept. Being $\{NB, NS, ZR, PS, PB\}$ the term set associated to the output variable, it can be noticed the absence of value for the output variable by the symbol -. The complete set formed joining this symbol to the term set is mapped into the set $\{0, 1, 2, 3, 4, 5\}$. Hence the label NB is associated with the value 0, NS with 1, ..., PB with 4 and the blank symbol - with 5.

Therefore the GA emploies an integer coding. Each one of the chromosomes is constituted by joining the partial coding associated to each one of the linguistic labels contained in the decision table cells. A gen presenting the alelle - will represent the absence of the control rule contained in the corresponding cell in the RB being the chromosome phenotype.

The GA proposed emploies an elitist selection scheme and the genetic operators used are of different nature. While the crossover operator is the standard two -point crossover, the mutation operator is specifically designed for the process. When it is applied over an alelle different from the blank symbol, changes it either up or down one level or to the blank code. When the previous gen value is the blank symbol, it selects a new value at random.

Finally, the fitness function is based too on an application specific measure. A Measure of Convergence is considered for this task. The fitness of an individual is determined by applying the FLC employing the RB coded in its genotype to the controlled system with several different starting points and computing the convergence of it to the desired equilibrium point.

The RB derivation method proposed by Karr

Karr proposes a method for deriving FLC RBs [Kar91a] where the complete RB in form of decision table is not coded in each chromosome. The approach of Karr is based in the fact that the number of rules is provided by an expert, together with many complete rules forming the RB and the antecedents for the reamining ones. He presents an GA derivation method for learning the rule consequents of this last type rules.

As it can be viewed, the method needs the existence of a deep knowledge about the controlled system. As Karr comments, his approach is useful when the controller human designer knows all the input states relevant for the problem (in this way, he knows the number of rules needed to control it) and the control action that have to be performed in order to obtain an optimal control from several of them (that is, he knows several complete antecedent-consequent control rules), but the actions that have to be associated to the remaining input states are not obvious for him. Usually, the known rules are those describing the extreme conditions, which are easy to write in many cases.

The coding outline will associate a binary number to each label belonging to the output variable term set in the way that each of them will be mapped to a binary number according to its order in the label set. Thus, considering seven linguistic terms describing the control variable, the possible control actions for one rule will be represented as a three-bit string (000 represents action 1 (NB), 001 represents action 2 (NM), and so on). A chromosome is obtained by joining the partial codings of the rules for what there is not known the control action to apply. Therefore a chromosome codes only a little part of the complete RB. In the commented example, calling n the number of not determinated rules, a string of length $3n$ represents every possible configuration for this partial FLC rule set.

The GA employed is the same used for the Karr's DB definition method that we have commented in the previous subsection, that is, a Simple GA with binary coding, proportional selection mechanism, simple crossover and random mutation. In the same way, the fitness function is defined by using an application specific measure.

Finally, it is important to note that the author proposes to combine their two FLC genetic design methods, the DB definition method and the RB derivation method, in order to design a complete KB. In this way, the RB derivation method is applied in a first step and, once an acceptable RB have been learned by the GA, the selection of high-performance membership functions (an accurated DB) is carried out using the above described tuning process.

The RB derivation method proposed by Bonarini

The method proposed by Bonarini [Bon93], called ELF (Evolutionary Learning of Fuzzy Rules), is quite different to the other approaches developed into this group. The author considers the high computationally cost needed to derive a RB by applying an GA with a population of individuals coding the whole decision table and presents a model based on learning only the control rules that the FLC will use.

In order to develop this task, the GA used by Bonarini maintains a population of rules This new coding will allow to learn the complete structure of the control rules, that is, their antecedent and consequent and the optimal number of rules forming the FLC RB. The author wants to learn with ELF what states occur in the controlled system and which are irrelevant for it, obtaining optimal RBs for the application.

The fitness function judges the state reached at each activation of the rules. Each individual of the population, that is, each rule, will have associated information about several questions: how good it has been judged (its strength), when it has been generated, when it has been triggered the last time and how much it contributed to past control actions performed by the controller. ELF will modify the strength associated to a rule according to the performance of the action it contributed to. This performance is evaluated by the fitness function.

Other important characteristic of this method is that it is designed in order to run in a real enviroment. The process first selects among the rules matching the current state of the controlled system, the rules matching better than a degree given by the FLC designer. The rules belonging to this set will compete between them in order to propose the best control action for the current state.

Several genetic operations are applied over this set of individuals. Some of these rules are considered tested enough since they contributed to past control actions more than a given reference. If some of them have low strength, it means that the control actions proposed by them do not perform well. These rules are substituted by others performing better and the consequents of such rules are modified with a probability proportional to their strength. This last step constitute the GA mutation operator and makes ELF to look for new rules in a neighboorhood of the good ones previously learned.

If much time have passed from the last rule modification, this means that the population of rules matching well the current state have stucked and all of them have almost the same strength. In this case, ELF selects the worst rule and mutates its consequents in order to continue the search looking for a better configuration.

If there exists few rules matching the current state (the parameter representing the enough number of rules is changed dinamically), ELF generates a new rule covering it and proposes a control action at random. This is the only mechanism introducing new antecedents and it is called *cover-detector*. It may introduce with a given probability also "dont care" symbols as values for some variables in the new rule generated. Rules whose antecedents contain these kind of symbols match different states and compete with different groups of rules, one for each of the matching states.

The dual situation is in which there are too many rules matching a state. In this case, the genetic operator applied is called *rule-killer* and it simply eliminate the worst rule matching this state from the population.

One time performed the selection and applied the genetic operators, ELF uses the FLC with the RB coded in the population taking the current state as input. The fitness function will then evaluate the new state obtained, giving it a reward. There exists a sharing process in the distibution of this reward, each individual gets part of the total reward according to the contribution of the rule that it codes gave to the control action applied. This process results in a modification of the strenght associated to the rules coded in the population.

Finally, several questions have to be noted. On one hand, Bonarini proposes application specific performance measures to define the fitness function. On the other hand, ELF gives different RBs as output of the learning process. It is due to each time the performance of the system is higher than a "satisfactory" value given by the controller designer, it saves the current RB and forces modifications in the current population in order to obtain a better solution.

3.3.3 Learning the Fuzzy Logic Controller Knowledge Base using Genetic Algorithms

This last group is the one to with more contributions have been made in the last years. There exists many approaches for genetic learning complete FLC KBs such as [CV93, HLV94, HLV95a, LT93, LP94, LM94, NL94, NHW92, SK94, VM95] and all of them present different characteristics. Belonging to these great group of works, we find approaches presenting variable cromosomal length [CV93, LP94], others making use of many expert knowledge in order to improve the learning process [HLV95a, LT93] and several working with chromosomes encoding single control rules instead of complete KBs [HLV95a, VM95]. Many of them define the fitness function by means of a single application specific measure, usually a measure of convergence, while others include more objective to optimize for obtaining more robust KBs [HLV95a, LT93]. In the following we are going to study five different approaches belonging to this family.

The KB learning method proposed by Lee and Takagi

In [LT93] Lee and Takagi introduce an GFS design method that allows to learn automatically a complete FLC KB. The method is based on an RB formed by TSK control rules. In this way, the derivation of the rule consequents will consist on learning the weights w_i used for combining linearly the input values in order to obtain the outputs.

For the linguistic terms in the rule antecedents, the authors consider triangular-shaped membership functions although they note that the method can work with any kind of parametrized membership functions, such as Gaussian, bell, trapezoidal or sigmoidal ones. They suppose a fuzzy partition of the input space. Each triangular membership function associated to each one of the primary fuzzy sets (linguistic terms) is represented by three parameters. The first one is the center point of the triangle, that is, its modal point. Only the center of the triangle associated to the first primary fuzzy set is given as an absolute position, while the parameters associated to the other ones represent the distance existing from the center point of the current triangle to the previous one. The other two parameters represent the left and right point of the triangle base respectively and they both present membership function value 0.

The GA used emploies a binary coding. First, the membership functions are encoded by taking the binary values of the three associated parameters and joining them into a binary substring. Eight bits are used to encode each parameter value. Then, the complete DB is encoded by joining the partial codings of the membership functions associated to each one of the input variables primary fuzzy sets one after other. The last part of the chromosome is built by coding the parameters w_i associated to each combination of the input values and joining them into a new binary substring. Eight bits are used again to encode the values of these parameters. In this way, each chromosome represents a complete KB. The number of rules forming the KB will depend on the number of primary fuzzy sets associated to each one of the input variables and it will be equal to the product of them.

The coding used allows to decide the optimal number of control rules forming the RB in the following way. Those linguistic terms in what the center point of their associated triangular membership functions are out of a concrete bound, obtained from the known knowledge about the system, are not considered to belong to any rule in the RB. Thus all rules whose antecedent should be a combination of the valid input values coded in the first part of the chromosome, will belong to the FLC RB. As an example, let us consider a chromosome coding the KB of a system with two input variables having associated m and n valid linguistic labels respectively. The whole number of rules of the RB encoded in it will be $m \times n$.

The fitness function is based on optimizing two different criteria. On one hand, an application specific measure is considered (concretely, in the example proposed it is used a measure of convergence and the FLC is applied from different initial states). On the other hand, chromosomes coding RBs with a large number of rules are penalized in order to obtain others with less rules.

Finally, it is very important to note that the authors propose two different ways of incorporating previous known knowledge about the system to the proposed GFSs design method in order to improve its behavior. They comment that this previous knowledge will make the GA gain significant speedup if the solutions designed by means of it are approximatelly correct. On one hand, it is possible to incorporate knowledge via the initial settings for the FLC KB parameters. This knowledge is used to generate the GA initial population. Thus, the individuals forming it are not all generated at random but several of them are obtained by equally partitioning the input spaces into varying number of linguistic terms. This knowledge can also be used to initially set appropriately the parameters w_i of the rule consequents. On the other hand, the previous knowledege can be incorporated via structural representation of the KBs. For example, in a problem of symmetric nature as the Inverted Pendulum, the FLC can be constrained in order to simetrically partition the input space about the origin. This will reduce the size of the search space because the number of membership functions is reduced by half.

The KB learning method proposed by Ng and Li

The method proposed by Ng and Li [NL94] is very specific since it is designed for applying in two-inputs-one-output systems all whose input and output spaces are partitioned into exactly seven primary fuzzy sets (the term set of the three variables is the same: $\{zero, \pm small, \pm medium, \pm large\}$). On the other hand, it is not able to learn the number of rules that will constitute the optimal RB. As we are going to see in the following, the whole decision table is represented in the part of chromosome corresponding to the RB. Therefore, the way for derivating the RB is the same used by the methods belonging to the second group commented in this work. The look-up table will allways present dimmension 7 × 7 and all the rules contained in it will belong to the final RB.

The DB considered is designed by using symmetrical exponential membership functions defined by the following expression:

$$\mu_{\pm i}(x) = exp\left(-\frac{|x \pm \alpha_i|^{\beta_i}}{\sigma_i}\right) \tag{3.8}$$

considering that $\mu_{+large} = 1$, if $x > \alpha_{+large}$ and $\mu_{-large} = 1$, if $x < \alpha_{-large}$.

Hence, it is possible a parametric representation of the membership functions by means of the set of parameters (α, β, σ), representing respectively the position, shape and scaling parameters.

The most important characteristic of this approach is that the FLC design space is coded in base-7 chromosomes. As in the great majority of the methods belonging to this group, each chromosome represents a complete KB. Each chromosome is built by joining five substrings coding the different parameters of the problem. The first one of them represents the FLC RB and, as we have commented yet, the process is similar to this developed in the methods deriving only the FLC RB. The fourty nine rule consequents contained in the decision table cells are encoded one after other in a substring presenting a base-7 value for each one of them.

The second substring is constituted by the scaling parameters (σ) associated to the seven linguistic labels of the two input variables. It presents fourteen base-7 values, two for each one of the following parameters: $\sigma_{\pm large}$, $\sigma_{\pm medium}$, $\sigma_{\pm small}$ and σ_{zero}. The next substring of eight base-7 values represent the positions (α) of the fuzzy sets associated to the terms "small" and "medium" (each one requiring two bits), whilst the positions of those giving meaning to "large" and "zero" are fixed ($\alpha_{-large} = -3$, $\alpha_{+large} = 3$ and $\alpha_{zero} = 0$). The fourth substring represents K_1 and K_2 as gains of the two inputs variables, error and change of error, with three base-7 bits associated to each and, finally, the last group of eigth integer characters encodes the shape coefficients β of the fuzzy sets associated to the term set of both variables, requiring a base-7 value for each parameter.

The fitness function employed by the GA is application specific. It is based directly on computing the value obtained by the FLC in a measure of convergence when controlling the system by using the concrete KB encoded in the individual.

The KB learning method proposed by Leitch and Probert

The more important characteristic presented by the method proposed by these authors [LP94] is the specific coding outline that they design in order to represent FLC KBs into chromosomes in an efficient way. The coding proposed will make the genetic variation to be increased, reducing in this way premature convergence and avoiding the need for complex crossover operators.

Leitch and Probert begin studying the previous approaches for designing GFSs and comment that the coding schemes employed by the different GAs developed for this task in the specialized literature represent the KB in a very fixed form. In order to avoid this shortcoming they develop a coding scheme more flexible than the position dependent ones commented. In their *context depending coding* the meaning of a piece of chromosome is not determined by the absolute position that it presents into the genotype, as usual, but it is determined by surrounding genes contained into it. In this way, the chromosomal length is variable.

They consider a DB with spherical fuzzy sets each one determined by its concrete center point presenting height 1, and all of them by a global radius value. All inputs and output are scaled to the interval $[0, 1]$ and the values of the commented parameters are encoded by using binary substrings of variable length. The parameter value is decoded in the following way. Let it be a binary substring of length n whose integer value is $m \leq n$. The real number encoded in it is $\frac{m}{2^n - 1}$.

The method is able to learn the optimal number of rules forming the RB, the input variables involved in each rule and the shape of the membership functions associated to the input and output variables. As it can be viewed, it does not use a previously defined fuzzy partition of the input and output spaces but learn directly the membership functions associated to each variable involved in each concrete rule. Therefore, the KB obtained from the learning process clearly present an approximative behavior.

Each chromosome encodes a complete KB by using the context depending coding commented. The alphabet of the chromosome consists of the integers 0, 1 and the letter E, used to indicate the end of a number (the value E is associated to a surrounding gen). The encoding of a rule is obtained by joining several substrings separated by E symbols. The first one presents a bit for each system input variable, taking value 1 if the concrete variable is involved in the rule and 0 otherwise. The surrounding genes contained in this substring are ignored. Next substring encodes the center of the membership functions associated to the input variables involved in the rule, each one separated from each other by a surrounding gen. An empty value is associated to the variables not taking part in the rule. The third substring presents only a binary number coding the membership functions radius. The value encoded represents a percentage from a maximum radius value given by the controller designer. The last substring encodes the central point of the membership function associated to the output variable.

A chromosome is obtained by joining the partial codings of the rules contained in the KB. The evolutionary process can make the end of a chromosome not coincide with the end of a rule (it rarely does). In this case, the incomplete rule is ignored and the genes coding it (called junk genes) are used to limit the effect of the disruption caused by the genetic operators.

Regarding to these ones, single point crossover is employed with a little modification in its usual form: the crossover point is randomly selected on each of the chromosomes involved. On other hand, classical mutation operator is used. Due to the coding scheme employed by the GA, both operators can not produce illegal chromosomes.

We finish noting several remarks about the method. The initial population formed by chromosomes of random initial length is generated at random. The fitness function emploies a measure of convergence since the authors enunciate that this is the more efficient way to design GFSs. Several noise is introduced in the simulations in order to obtain more robust KBs from the learning process.

The KB learning method proposed by Cooper and Vidal

The underlying idea in the approach proposed by Cooper and Vidal [CV93] is that the excesive length of the chromosomes encoding the KB can make the GA not able to find accurated solutions due to the high complexity of the search space. In this way, they propose a novel encoding scheme which maintains only those rules necessary to control the target system, allowing to obtain an RB with optimal number of rules.

In this case, the membership functions contained in the DB are triangles characterized by the location of its center and the half-length of its base. A single rule, therefore, consists of the concatenation of the one-byte unsigned characters (assuming values from 0 to 255) specifying the centers and half-lengths of the membership functions. The rule descriptions for a single KB are then concatenated into a single bit string where the number of rules is not restricted. Hence, the GA employs a integer coding and the chromosomes present variable-length.

This GA does not use the classical genetic mutation operator. In this approach, this operator include the inversion of the copied bit and the addition or deletion of an entire rule. These latter two mutations permit the size of a FLC KB to evolve. The cycle of evaluation and reproduction continues for a predetermined number of generations or until an acceptable performance level is achieved.

To be meaningful, the genetic paradigm requires that the rules in the two strings should be aligned so that similar rules are combined with each other. Simply by combining the strings in the order they appear it does not preserve much information about either KB encoded and produces nearly random results, rather as a child KB that performs in a manner similar to its parents. For example, this can make that the child chromosome will present repeated rules. Therefore, before reproduction, both strings must be aligned so that the centers of the input variables match as closely as possible. The most closely matching rules are combined first, followed by the next most closely matching rules from those remaining and so on. Any rules forming a longer string that is not matched are added at the end.

The fitness function is designed by using a measure of convergence. The FLC is used to control the system from twenty test cases and its accuracy is measured during sixty seconds each trial.

The KB learning method proposed by Herrera et al.

The KB learning process proposed in [HLV95a] presents several important differences with respect to the other methods belonging to the same group as we are going to see in the following. First, it can be considered the existence of previous control rules derivated directly from the human process operator and include them in the learning process, being able to combine them with other rules automatically learned and to detect incorrect rules into this previous set. Moreover, the approach is based on several steps and not in a single process such in the other methods. The chromosomes of the main GA represent single rules and not complete KBs. On other hand, the fitness function is based not only in an FLC performance measure but it include several criteria in order to obtain optimal KBs. Finally, the approach will obtain a KB with an approximative nature even in the case in which there exists several descriptive fuzzy control rules provided by the expert.

The approach is based on the use of GAs under the following hypotheses:

- There is some linguistic information from the experience of the human controller but linguistic rules alone are usually not enough for designing successfully a control system or could not be available.
- There is some numerical information from sampled input-output (state-control) pairs that are recorded experimentally.
- The combination of these two kinds of information may be sufficient for

a successful design of a FLC KB.
- The possibility of not having any linguistic information and having a complete numerical information is considered.

Taking into account the aforementioned hypothesis a learning process is designed according to the following goals:

- to develop a KB generating process from numerical data pairs; and
- to develop a general approach combining both kinds of information, linguistic information and fuzzy control rules obtained by the generating process, into a common framework using both simultaneously and cooperatively to solve the control design problem.

In order to reach these goals, it is proposed a methodology based on the design of the three following components:

1. a KB generating process of desirable fuzzy control rules able to include the complete knowledge of the set of examples,
2. a combining information and simplifying rules process, which finds the final KB able to approximate the input-output behaviour of a real system,
3. a tuning process of the final KB DB,

all of them developed by means of GAs.

As it is possible to have some linguistic $IF - THEN$ rules given by an expert, it is used a linguistic fuzzy rules structure to represent them. That is, a previously defined DB representing fuzzy partitions with real-world meaning is considered. On other hand, there are sampled input-output pairs and a free fuzzy rules structure to generate the fuzzy rules covering these examples is used (the meaning presented by a linguistic term is different when it belongs to different rules, that is, the example set are used to generate rules with an approximative behavior). The membership functions of the linguistic labels involved in the control rules are trapezoidal-shaped. Then both kind of rules are combined, by using a simplification method based on a GA, and finally a tuning method is applied over the simplified set of rules that will make all of them finally present an approximate behavior.

The generating fuzzy rules process consists of a *generating method* of desirable fuzzy rules from examples using GAs together with a *covering method* of the set of examples.

- The generating method of fuzzy rules is developed by means of a real coded GA (RCGA) where a chromosome represents a fuzzy rule and it is evaluated by means of a frequency method. The RCGA finds the best rule in every running over the set of examples according to the following features which will be included in the fitness function of the GA.
- The covering method is developed as an iterative process. It permits to obtain a set of fuzzy rules covering the set of examples. In each iteration, it runs the generating method choosing the best chromosome (rule), assigns to every example the relative covering value and removes the examples with a covering value greater than a value ϵ provided by the controller designer.

Because we can obtain two similar rules in the generating process or one rule similar to another given by an expert, it is necessary to combine and simplify the complete KB obtained from the previous process for deriving the final KB allowing to control the system. Finally, the tuning method presented in [HLV95b] and commented in a previous subsection is applied over the simplified KB for obtaining a more accurated one.

3.4 FINAL REMARKS

In this chapter we have reviewed some approaches of GFSs in which GAs have been used for designing FLCs KBs. Three different modes to cope this problem have been attached and different methods developing each one of them have been analyzed.

There are many others researchs that have contributed to these three areas as we have cited above. Moreover, there are other researchs that have contributed to various aspects of GFSs, not included in the three GFS designing modes aforementioned, which we have not a change to deal with but are worthy to mention. They are commented in the following.

In [SKG93] it is proposed an automatic design method combining self-organizing feature maps and GAs. In a first step, the fuzzy control rules and linguistic variables are extracted from a referential data set by using a self-organizing process. Then a GA is used to find optimal membership functions (that is, to tune the DB).

George et al. [GSR94] propose a method based on GAs and Neural Networks (NN) for defining FLC DBs by learning the membership functions associated to the primary fuzzy partitions provided by the expert. The combination of the two search process allows to combine the best characteristics of each one of them in order to solve the problem. Other papers combining both GA and NN are [IFSA95, SFH95].

Hoffman and Pfister present two different genetic based methods for designing hierarchical FLCs by learning the RB [HP94, HP95]. While the first of them uses a Simple GA with fixed-length binary coding and usual genetic operators [HP94], the second one employs a Messy GA with variable-length binary coding and the crossover operator replaced by two simple ones: splice and cut.

Several approaches based on fuzzy classifier systems (FCSs) have been developed [BK94, CF94, PB93, VR91b, VR91a]. The FCS is a genetic based machine learning system which integrates a fuzzy RB, the bucket-brigade learning algorithm (the learning algorithm commonly used by the classifier systems) and a GA [GS92, VR91b, VR91a]. Each classifier represents a fuzzy rule and the FCS employs a GA to evolve adequate rules running over the population of classifiers searching new improved ones.

Finally, to point out that although the application of GA for designing fuzzy systems is recent, it has an increasing concern over the last years that will allow to obtain fruitful researches in the building of fuzzy logic based intelligent systems.

REFERENCES

[BK94] Bäck T. and Kursawe F. (July 1994) Evolutionary algorithms for fuzzy logic: A brief overview. In *Proc. Fifth International Conference on Information Processing and Management of Uncertainty in Knowledge Based Systems (IPMU'94)*, pages 659–664. Paris.

[BMU95] Braunstingl R., Mujika J., and Uribe J. P. (March 1995) A wall following robot with a fuzzy logic controller optimized by a genetic algorithm. In *Proc. Fourth IEEE International Conference on Fuzzy Systems (FUZZ-IEEE'95)*, pages 77–82. Yokohama.

[BN95] Bolata F. and Nowé A. (March 1995) From fuzzy linguistic specifications to fuzzy controllers using evolution strategies. In *Proc. Fourth IEEE International Conference on Fuzzy Systems (FUZZ-IEEE'95)*, pages 1089–1094. Yokohama.

[Bon93] Bonarini A. (September 1993) Elf: Learning incomplete fuzzy rule sets for an autonomous robot. In *Proc. First European Congress on Fuzzy and Intelligent Technologies (EUFIT'93)*, pages 69–75. Aachen.

[Cas95] Castro J. L. (March 1995) Fuzzy logic controllers are universal approximators. To appear in IEEE Transactions on Systems, Man and Cybernetics.

[CCC+94] Cárdenas E., Castillo J. C., Cordón O., Herrera F., and Peregrín A. (January 1994) Influence of fuzzy implication functions and defuzzification methods in fuzzy control. *BUSEFAL* 57: 69–79.

[CCC+95] Cárdenas E., Castillo J. C., Cordón O., Herrera F., and Peregrín A. (January 1995) Applicability of t-norms in fuzzy control. *BUSEFAL* 61: 28–37.

[CF94] Carse B. and Fogarty T. C. (1994) A fuzzy classifier system using the pittsburgh approach. In Davidor Y., Schwefel H. P., and Mäanner R. (eds) *Parallel Problem Solving from Nature - PPSN III*, pages 260–269. Springer-Verlag, Berlin.

[CHP95a] Cordón O., Herrera F., and Peregrín A. (March 1995) Applicability of the fuzzy operators in the design of fuzzy logic controllers. Technical Report DECSAI-95111, University of Granada, Department of Computer Science and Artificial Intelligence, University of Granada, Granada, Spain.

[CHP95b] Cordón O., Herrera F., and Peregrín A. (July 1995) T-norms versus implication functions as implication operators in fuzzy control. To appear in Proc. Sixth International Fuzzy Systems Association World Congress (IFSA'95).

[CV93] Cooper M. G. and Vidal J. J. (1993) Genetic design of fuzzy logic controllers. In *Proc. Second International Conference on Fuzzy Theory and Technology (FTT'93)*. Durham.

[DHR93] Driankov D., Hellendoorn H., and Reinfrank M. (1993) *An Introduction to Fuzzy Control*. Springer-Verlag.

[FTH94] Fathi-Torbaghan M. and Hildebrand L. (July 1994) Evolutionary strategies for the optimization of fuzzy rules. In *Proc. Fifth International Conference on Information Processing and Management of Uncertainty in Knowledge Based Systems (IPMU'94)*, pages 671–674. Paris.

[GS92] Geyer-Schulz A. (1992) Fuzzy classifier systems. In Lowen R. (ed) *Fuzzy Logic: State of the Art*. Kluwer Academic Publishers, Dordretch.

[GSR94] George S. M., Saxena A., and Rambabu P. (September 1994) Genetic algorithm in the aid of fuzzy rule deduction. In *Proc. Second European Conference on Soft Computing and Intelligent Technologies (EUFIT'94)*, pages 1130–1133. Aachen.

[HLV94] Herrera F., Lozano M., and Verdegay J. L. (July 1994) Generating fuzzy rules from examples using genetic algorithms. In *Proc. Fifth International Conference on Information Processing and Management of Uncertainty in Knowledge Based Systems (IPMU'94)*, pages 675–679. Paris.

[HLV95a] Herrera F., Lozano M., and Verdegay J. L. (February 1995) A learning process for fuzzy control rules using genetic algorithms. Technical Report DECSAI-95108, University of Granada, Department of Computer Science and Artificial Intelligence, University of Granada, Granada, Spain.

[HLV95b] Herrera F., Lozano M., and Verdegay J. L. (1995) Tuning fuzzy logic controllers by genetic algorithms. *International Journal of Approximate Reasoning* 12: 293–315.

[HMB93] Harris C. J., Moore C. G., and Brown M. (1993) *Intelligent Control. Aspects of Fuzzy Logic and Neural Nets*. World Scientific Publishing.

[HP94] Hoffmann F. and Pfister G. (September 1994) Automatic design of hierarchical fuzzy controllers using genetic algorithms. In *Proc. Second European Conference on Soft Computing and Intelligent Technologies (EUFIT'94)*, pages 1516–1522. Aachen.

[HP95] Hoffmann F. and Pfister G. (July 1995) A new learning method for the design of hierarchical fuzzy controllers using messy genetic algorithms. To appear in Proc. Sixth International Fuzzy Systems Association World Congress (IFSA'95).

[HT94] Hwang W. R. and Thompson W. E. (June 1994) Design of fuzzy logic controllers using genetic algorithms. In *Proc. Third IEEE International Conference on Fuzzy Systems (FUZZ-IEEE'94)*, pages 1383–1388. Orlando.

[HTS93] Hu H.-T., Tai H.-M., and Shenoi S. (1993) Fuzzy controller design using cell mappings and genetic algorithms. In *Proc. Second International Conference on Fuzzy Theory and Technology (FTT'93)*. Durham.

[IFSA95] Ishigami H., Fukuda T., Shibata T., and Arai F. (May 1995) Structure optimization of fuzzy neural networks by genetic algorithm. *Fuzzy Sets and Systems* 71(3): 257–264.

[Kar91a] Karr C. (March 1991) Applying genetics. *AI Expert* pages 38–43.

[Kar91b] Karr C. (February 1991) Genetic algorithms for fuzzy controllers. *AI Expert* pages 26–33.

[KB93] Kropp K. and Baitinger U. G. (September 1993) Optimization of fuzzy logic controller inference rules using a genetic algorithm. In *Proc. First European Congress on Fuzzy and Intelligent Technologies (EUFIT'93)*, pages 1090–1096. Aachen.

[KKS85] Kiszka J. B., Kochanska M. E., and Sliwomska D. S. (January 1985) The influence of some fuzzy implication operators on the accuracy of a fuzzy model - parts i and ii. *Fuzzy Sets and Systems* 15: 111–128, 223–240.

[Lee90] Lee C. C. (March 1990) Fuzzy logic in control systems: Fuzzy logic controller - parts i and ii. *IEEE Transactions on Systems, Man and Cybernetics* 20(2): 404–435.

[LM94] Liska J. and Melsheimer S. S. (June 1994) Complete design of fuzzy logic systems using genetic algorithms. In *Proc. Third IEEE International Conference on Fuzzy Systems (FUZZ-IEEE'94)*, pages 1377–1382. Orlando.

[LP94] Leitch D. and Probert P. (1994) Context depending coding in genetic algorithms for the design of fuzzy systems. In *Proc. IEEE/Nagoya University WWW on Fuzzy Logic and Neural Networks/Genetic Algorithms*. Nagoya.

[LT93] Lee M. A. and Takagi H. (July 1993) Embedding apriori knowledge into an integrated fuzzy system design method based on genetic algorithms. In *Proc. Fifth International Fuzzy Systems Association World Congress (IFSA'93)*, pages 1293–1296. Seoul.

[LW91] Liaw C. M. and Wang J. B. (July 1991) Design and implementation of a fuzzy controller for a high performance induction motor drive. *IEEE Transactions on Systems, Man and Cybernetics* 21(4): 921–929.

[MA75] Mamdani E. H. and Assilian S. (1975) An experiment in linguistic systhesis with a fuzzy logic controller. *International Journal of Man-Machine Studies* 7: 1–13.

[NHW92] Nomura H., Hayashi I., and Wakami N. (1992) A learning method of simplified fuzzy reasoning by genetic algorithm. In *Proc. International Fuzzy Systems and Intelligent Control Conference (FSIC'92)*, pages 236–245. Louisville.

[NL94] Ng K. C. and Lee Y. (June 1994) Design of sophisticated fuzzy logic controllers using genetic algorithms. In *Proc. Third IEEE International Conference on Fuzzy Systems (FUZZ-IEEE'94)*, pages 1708–1712. Orlando.

[PB93] Parodi A. and Bonelli P. (July 1993) A new approach to fuzzy classifier system. In *Proc. Fifth International Conference on Genetic Algorithms (ICGA'93)*, pages 223–230.

[SFH95] Shijojima K., Fukuda T., and Hasegawa Y. (May 1995) Self-tuning fuzzy modeling with adaptive memebership function, rules, and hierachical structure based on genetic algorithm. *Fuzzy Sets and Systems* 71(3): 295–309.

[SK94] Satyadas A. and Krishnakumar K. (June 1994) Ga-optimized fuzzy controller for spacecraft attitude control. In *Proc. Third IEEE International Conference on Fuzzy Systems (FUZZ-IEEE'94)*, pages 1979–1984. Orlando.

[SKG93] Surmann H., Kanstein A., and Goser K. (September 1993) Self-organizing and genetic algorithms for an automatic design of fuzzy control and decision systems. In *Proc. First European Congress on Fuzzy and Intelligent Technologies (EUFIT'93)*, pages 1097–1104. Aachen.

[Thr91] Thrift P. (1991) Fuzzy logic synthesis with genetic algorithms. In *Proc. Fourth International Conference on Genetic Algorithms (ICGA'91)*, pages 509–513.

[TS85] Takagi T. and Sugeno M. (January 1985) Fuzzy identification of systems and its applications to modeling and control. *IEEE Transactions on Systems, Man and Cybernetics* 15(1): 116–132.

[VM95] Velasco J. R. and Magdalena L. (July 1995) Genetic learning applied to fuzzy rules and fuzzy knowledge bases. To appear in Proc. Sixth International Fuzzy Systems Association World Congress (IFSA'95).

[VR91a] Valenzuela-Rendón M. (1991) The fuzzy classifier system: A classifier system for continuously varing variables. In *Proc. Fourth International Conference on Genetic Algorithms (ICGA'91)*, pages 346–353.

[VR91b] Valenzuela-Rendón M. (1991) The fuzzy classifier system: Motivations and first results. In Männer R. and Manderick B. (eds) *Parallel Problem Solving from Nature - PPSN II*, pages 330–334. Springer-Verlag, Berlin.

[Zad65] Zadeh L. A. (1965) Fuzzy sets. *Information and Control* 8: 358–353.

[Zad73] Zadeh L. A. (1973) Outline of a new approach to the analysis of complex systems and decision processes. *IEEE Transactions on Systems, Man and Cybernetics* 3: 28–44.

4

The Science of Breeding and its Application to Genetic Algorithms

Heinz Mühlenbein [1]

4.1 INTRODUCTION

The aim of population genetics is to understand the evolution of genetic populations. Genetic algorithms use an artificial genetic population for optimization purposes. Therefore it should be of no surprise that population genetics can be used to analyze the behavior of genetic algorithms. Unfortunately the behavior of genetic population is very difficult to model mathematically. Therefore population genetics developed a set of models and a set of approaches which investigate specific aspects of genetic populations. Three different approaches have been tried:

- the phenotypic approach by the biometricians (Galton, Pearson)
- the genotypic approach by the Mendelians
- the statistical approach used by breeders

The biometricians introduced the concept of *correlation* and *regression* to quantify the relation between offspring and parent. Their analysis centers on quantitative traits. The Mendelians use Mendel's *genetic chance model* to compute the change of the gene frequencies in the population. Mendel's model is restricted to discrete genes. The scientific way of breeding starts with the equation for the *response to selection*. It tries to predict the outcome of selection experiments. Modern textbooks about population genetics describe the Mendelian approach mainly.

For the theory of genetic algorithms all three approaches are useful. There are at least five parameters necessary to describe the initial state of the population and the selection process of a genetic algorithm. They are:

- the population size N
- the initial frequency of the desirable allele p_0
- the number of loci n
- the mutation rate m

[1] GMD Schloss Birlinghoven D-53754 Sankt Augustin Germany
muehlenbein@gmd.de

- the intensity of selection I

It would be futile to investigate the genetic algorithm with all five parameters variable. Therefore we will investigate simpler models with one or more parameters fixed. A similar approach has been used in population genetics. For some problems with a large population size we will be able to compute the expected number of generations until convergence.

In this paper and also in previous papers I distinguish between empirical laws and theorems. Empirical laws are derived from carefully performed computer experiments. Theorems are obtained by purely mathematical reasoning. Empirical laws are by no means less true than theorems. They are laws carefully deduced from the results of numerical experiments. This procedure was and is successfully used in physics. A historical example are the laws describing the movements of the planets. Kepler derived his famous laws empirically. They explained all the available data, in addition they could be used for prediction. Newton was able to derive the same laws by postulating a gravitational force between the sun and the planets. Thus in Newton's theory Kepler's laws can be proven mathematically. In my terminology Newton converted an empirical law to a theorem by a theory. The main source of confusion is that the word "empirical" has one sense in which it refers to something based purely on observation without theoretical depth. But I use the word in its classical sense. My thesis is that the classical empirical approach is a viable alternative that should be pursued more consciously and more rigorously.

Most of the results reported in this paper have been obtained in cooperation with other researchers. Therefore I will use we instead of I to acknowledge the common work.

The outline of the paper is as follows. First we introduce the equation for the response to selection used by breeders. Then we analyze the genetic algorithm using selection and recombination only for two recombination methods and binary genes. We show that a new recombination method called gene pool recombination performs better than the usual recombination of two parents. Then we investigate evolution without selection, also called genetic drift. A genetic population will converge to just one genotype because of random sampling with replacement. In the final sections we investigate the Breeder Genetic Algorithm **BGA** for continuous genes.

4.2   RESPONSE TO SELECTION

In this section we summarize the theory presented in Mühlenbein and Schlierkamp-Voosen (1993,1994). Let $\bar{f}(t)$ be the average fitness of the population at generation t. The response to selection is defined as

$$R(t) = \bar{f}(t+1) - \bar{f}(t). \tag{4.1}$$

The amount of selection is measured by the selection differential

$$S(t) = \bar{f}_s(t) - \bar{f}(t), \tag{4.2}$$

where $\bar{f}_s(t)$ is the average fitness of the selected parents. The equation for the response to selection relates R and S:

$$R(t) = b(t) \cdot S(t). \tag{4.3}$$

The value $b(t)$ is called the *realized heritability*. For many fitness functions and selection schemes, the selection differential can be expressed as a function of the standard deviation σ_p of the fitness of the population. For *truncation selection* (selecting the $T \cdot N$ best individuals) and for normally distributed fitness, the selection differential is proportional to the standard deviation (Falconer, 1981):

$$\frac{S(t)}{\sigma_p(t)} = I.$$

The value I is called the *selection intensity*. For arbitrary distributions one can show the following estimate (Nagaraja, 1982):

$$\frac{S(t)}{\sigma_p(t)} \le \sqrt{\frac{1-T}{T}}.$$

For normally distributed fitness the famous equation for the response to selection is obtained (Falconer, 1981):

$$R(t) = I \cdot b(t) \cdot \sigma_p(t). \tag{4.4}$$

The above equation is valid for a large range of distributions, not just for a normal distribution. The response depends on the selection intensity, the realized heritability, and the standard deviation of the fitness distribution. In order to use the above equation for prediction, one has to estimate $b(t)$ and $\sigma_p(t)$. The equation also gives a design goal for genetic operators — to *maximize the product of heritability and standard deviation*. In other words, if two recombination operators have the same heritability, the operator creating an offspring population with larger standard deviation is to be preferred.

For proportionate selection as used by the simple genetic algorithm (Goldberg, 1989) it was shown by Mühlenbein and Schlierkamp-Voosen (1993) that

$$\frac{S(t)}{\sigma_p(t)} = \frac{\sigma_p(t)}{\bar{f}(t)}.$$

Before using these equations however, we will derive exact difference equations for two and three loci directly.

4.3   TWO LOCI

For simplicity we restrict the discussion to two loci and proportionate selection. In this case there are four possible genotypes: $(0,0), (0,1), (1,0)$, and $(1,1))$ which we index by $i = (0,1,2,3)$. We denote their fitness values m_0, m_1, m_2, and $m_3)$ respectively. Let $q_i(t)$ be the frequency of genotype i at generation t. We assume an infinite population and *uniform crossover*. With uniform crossover the allele of the offspring is obtained randomly from one of the parents. In the following we will call this method two parent recombination (TPR). For proportionate selection the exact equations describing the evolution of the frequencies q_i can be derived easily. These equations — known for diploid chromosomes in population genetics (Crow and Kimura, 1970) — assume an infinite population.

$$q_i(t+1) = \frac{m_i}{\bar{f}(t)} q_i(t) + \epsilon_i \frac{D(t)}{2\bar{f}(t)^2} \quad i = 0, 1, 2, 3 \tag{4.5}$$

with $\epsilon = (-1, 1, 1, -1)$. $\bar{f}(t) = \sum_{i=0}^{3} m_i q_i(t)$ is the average fitness of the population. $D(t)$ defines the deviation from linkage equilibrium

$$D(t) = m_0 m_3 q_0(t) q_3(t) - m_1 m_2 q_1(t) q_2(t). \tag{4.6}$$

Note that $D(t) = 0$ if $q_0(t) q_3(t) = q_1(t) q_2(t)$ and $m_0 m_3 = m_1 m_2$. The first condition is fulfilled if the genotypes are binomially distributed. This assumption is called the *Hardy-Weinberg equilibrium* in population genetics. The general nonlinear difference equations have not yet been solved analytically, (see the discussion by Naglyaki (1992)), but it is possible to derive an exact expression for the realized heritability. By summation we obtain

$$R(t) = \bar{f}(t+1) - \bar{f}(t) = \frac{V(t)}{\bar{f}(t)} - (m_0 + m_3 - m_1 - m_2) \frac{D(t)}{2\bar{f}(t)^2}, \tag{4.7}$$

where $V(t) = \sigma^2(t) = \sum q_i(t)(m_i - \bar{f}(t))^2$ denotes the variance of the population. Using $S(t) = V(t)/\bar{f}(t)$ we obtain the exact equation for the heritability,

$$b(t) = 1 - (m_0 - m_1 - m_2 + m_3) \frac{D(t)}{2\bar{f}(t)V(t)}. \tag{4.8}$$

In general, $b(t)$ depends on the genotype frequencies. Note that $b(t) = 1$ if $D(t) = 0$ or $m_0 + m_3 = m_1 + m_2$. The second assumption is fulfilled for the function ONEMAX(2) which has the fitness values $m_0 = 0, m_1 = m_2 = 1, m_3 = 2$. From (4.5) we obtain

$$\begin{aligned}
\bar{f}(t+1) &= q_1(t+1) + q_2(t+1) + 2q_3(t+1) \\
&= \frac{q_1(t) + q_2(t) + 4q_3(t)}{\bar{f}(t)} \\
&= 1 + \frac{2q_3(t)}{\bar{f}(t)}.
\end{aligned}$$

Let $p(t)$ denote the frequency of allele 1. Then by definition $\bar{f}(t) = 2p(t)$. Therefore we obtain

$$R(t) = 1 - p(t) + \frac{B_3(t)}{p(t)}, \tag{4.9}$$

where $B_3(t)$ denotes how far $q_3(t)$ deviates from the frequency given by the binomial distribution:

$$B_3(t) = q_3(t) - p^2(t). \tag{4.10}$$

The exact difference equation for $p(t)$ can be written as

$$p(t+1) = p(t) + \frac{1}{2}(1 - p(t)) + \frac{B_3(t)}{2p(t)}. \qquad (4.11)$$

This equation has two unknown variables, $p(t)$ and $q_3(t)$. Therefore $p(t)$ cannot be directly computed. Selection leads the population away from the binomial distribution, and TPR is not able to recreate a binomial distribution for the offspring population.

We now discuss a function where $D(t) = 0$ if the population starts in a Hardy-Weinberg equilibrium. An example is MULT(2) with fitness values $m_0 = 1, m_1 = m_2 = 2, m_3 = 4$. In this case the difference equation for $p(t)$ is given by

$$p(t+1) = p(t)\frac{2}{1 + p(t)}, \qquad (4.12)$$

which can be solved easily.

In summary, even linear fitness functions lead to difficult systems of difference equations. The genetic population moves away from Hardy-Weinberg equilibrium. A class of multiplicative fitness functions with $m_0 m_3 = m_1 m_2$ leads to simpler equations, because the population stays in Hardy-Weinberg equilibrium.

4.4 GENE POOL RECOMBINATION

The exact analysis of recombination together with selection leads to difficult nonlinear differential equations. Recombination of two genotypes creates a linkage between the genes at different loci. This linkage is very hard to describe mathematically. Therefore we decided to look for a recombination operator that leads to simpler equations, like those we used as an approximation. This operator must create a binomial distribution. Fortunately, there is a simple recombination scheme that fulfills this condition; we call it *gene pool recombination (GPR)*.

The idea of using more than two parents for recombination is not new. Already Mühlenbein (1989) used eight parents; the offspring allele was obtained by a majority vote. Multi-parent recombination has also been investigated recently by Eiben et al. (1994) though their results are somewhat inconclusive. For binary functions the bit-based simulated crossover (BSC) of Syswerda (1993) is similar to GPR. However, his implementation merged selection and recombination. An implementation of BSC which separates selection and recombination was empirically investigated by Eshelman and Schaffer (1993). GPR is an extension of BSC, it can be used for any representation — discrete or continuous.

Definition: *In gene pool recombination the two "parent" alleles of an offspring are randomly chosen with replacement from the gene pool given by the parent population selected before. Then the offspring allele is computed using any of the standard recombination schemes for TPR.*

For binary functions GPR is obviously a Bernoulli process. Let $p_i^s(t)$ be the frequency of allele 1 at locus i in the **selected** parent population. Then GPR creates offspring with allele frequency $p_i(t+1) = p_i^s(t)$ and variance $p_i(t+1)(1 - p_i(t+1))$ at locus i.

In order to analyze GPR we will derive difference equations for the gene frequencies, valid for arbitrary fitness functions and infinite populations. As before, we restrict the analysis to the case of two loci and proportionate selection.

Let $q_i(t)$ be the frequency of genotype i at generation t. For $n = 2$ loci, the marginal gene frequencies $p_1(t)$ and $p_2(t)$ can be obtained from

$$\begin{aligned} p_1(t) &= q_2(t) + q_3(t) \\ p_2(t) &= q_1(t) + q_3(t). \end{aligned}$$

We assume that the initial population has a binomial distribution. This means that

$$\begin{aligned} q_0(0) &= (1 - p_1(0))(1 - p_2(0)) \\ q_1(0) &= (1 - p_1(0))p_2(0) \\ q_2(0) &= p_1(0)(1 - p_2(0)) \\ q_3(0) &= p_1(0)p_2(0). \end{aligned}$$

Then the following theorem holds:

Theorem 1 *Let the initial population have a binomial distribution. For an infinite population with GPR and proportionate selection, the marginal frequencies $p_1(t)$ and $p_2(t)$ can be obtained from*

$$p_1(t + 1) = p_1(t)\frac{m_2(1 - p_2(t)) + m_3 p_2(t)}{\bar{f}(t)} \tag{4.13}$$

$$p_2(t + 1) = p_2(t)\frac{m_1(1 - p_1(t)) + m_3 p_1(t)}{\bar{f}(t)}. \tag{4.14}$$

The realized heritability, $b(t)$, is given by

$$b(t) = 1 - (m_0 m_3 - m_1 m_2)(m_0 - m_1 - m_2 + m_3)\frac{p_1 p_2(1 - p_1)(1 - p_2)}{V\bar{f}}, \tag{4.15}$$

where p_1, p_2, V, and $\bar{f}$ depend on t.

Proof: Proportionate selection selects the genotypes for the parents of population $t + 1$ according to

$$q_i^s(t) = \frac{m_i}{\bar{f}(t)}q_i(t).$$

From q_i^s the marginal frequencies p_1^s and p_2^s can be obtained from the two equations $p_1^s(t) = q_2^s(t) + q_3^s(t)$ and $p_2^s(t) = q_1^s(t) + q_3^s(t)$. For each locus, GPR is a Bernoulli process; therefore, the marginal gene frequencies of parents and offspring remain constant

$$p_1(t + 1) = p_1^s(t) \quad p_2(t + 1) = p_2^s(t).$$

Combining these equations gives equations 4.13 and 4.14. The expression for the realized heritability can be obtained after some manipulations.

Remark: Theorem 1 can be extended to arbitrary functions of size n, or genetically speaking to n loci. This means that the evolution of an infinite genetic population with GPR and proportionate selection is fully described by n equations for the marginal gene frequencies. In contrast, for TPR one needs 2^n equations for the genotypic frequencies. For GPR one can in principle solve the difference equations for the marginal gene frequencies instead of running a genetic algorithm.

Note that $b(t) = 1$ if $m_0 m_3 = m_1 m_2$ or if $m_1 + m_2 = m_0 + m_3$. Let us first consider ONEMAX(2). The average fitness is given by $\bar{f}(t) = p_1(t) + p_2(t)$. If $p_1(0) = p_2(0)$, we have $p_1(t) = p_2(t) = p(t)$ for all t. From $\bar{f}(t) = 2p(t)$ we obtain

$$R(t) = 1 - p(t) \tag{4.16}$$

and

$$p(t+1) = p(t) + \frac{1}{2}(1 - p(t)). \tag{4.17}$$

This equation is similar to the equation obtained for TPR. It can be solved easily. Both equations become equal if $B_3(t) = 0$. This shows that for linear fitness functions, GPR and TPR give similar results — with a slight advantage for GPR, which converges faster.

Let us now turn to the function MULT(2). Combining $\bar{f}(t) = (1 + p(t))^2$ with equation (4.13), we obtain equation (4.12). For Mult(2) TPR and GPR lead to the same difference equation. One can show that in general for multiplicative functions ($m_0 m_3 = m_1 m_2$) TPR and GPR are equal.

For many loci the above analysis can easily be extended to fitness functions which are called "unitation" functions. For these functions the fitness values depend only on the number of $1's$ in the genotype. Again for simplicity we consider three loci only. Let u_i denote the fitness of a genotype with i $1's$. Under the assumption that $p_1(0) = p_2(0) = p_3(0)$, all marginal frequencies have the same value, which we denote as $p(t)$. Then we obtain for the marginal frequency $p(t+1) = c \cdot p(t)$

$$c = \frac{u_1(1-p)^2 + 2u_2 p(1-p) + u_3 p^2}{u_0(1-p)^3 + 3u_1 p(1-p)^2 + 3u_2 p^2(1-p) + u_3 p^3}, \tag{4.18}$$

where $p = p(t)$. If $c > 1$ the marginal frequency $p(t)$ increases, if $c < 1$ it decreases. As a specific example we analyzed a "deceptive" function of 3 loci, (Goldberg, 1989). Let the fitness values of this function be $u_0 = 28$, $u_1 = 26$, $u_3 = 30$. The global optimum is at 111, the local optimum at 000. The fitness value for u_2 can be varied. Let us take $u_2 = 0$ as an example. If $p_0 > 0.639$ the population will converge to 111. If $p_0 < 0.639$ the population will converge to 000. At $p_0 = 0.639$ we have $c = 1$. This point is an *unstable equilibrium point.* In figure 4.1 c is shown for $u_2 = 25$ and $u_2 = 0$. For $u_2 = 25$ c is around 1, meaning a very slow change of $p(t)$.

Remark: The analysis of unitation functions of three or more loci shows that a genetic algorithm using selection and recombination only is **not a global optimization** method. Depending on the frequency distribution of the genotypes and the fitness values, a genetic algorithm with infinite population size will deterministically converge to one of the local optima. The equations derived in this

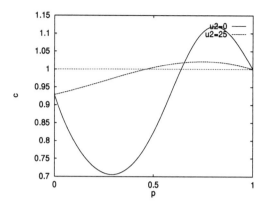

Figure 4.1   The line $c = 1$ divides the attractor regions for 000 and for 111.

chapter can be used to determine the optima to which the genetic algorithm will converge.

As a last example we will analyze a fitness function where the results of TPR and GPR are very different. For simplicity we take two loci. The fitness values are defined as $m_0 = 0.99, m_1 = 0, m_2 = 0$, and $m_3 = 1$. Table 4.1 shows that GPR very slowly changes the gene frequencies at the beginning. In fact, if $m_0 = 1$ the population would stay in equilibrium. After three generations GPR changes the gene frequencies very quickly. In contrast, TPR dramatically changes the frequencies in the first generation. The population immediately goes to the equilibrium points for the symmetric fitness function $m_0 = m_3 = 1, m_1 = m_2 = 0$. It takes TPR a long time to leave this equilibrium point and march to the optimum.

Table 4.1   Results for TPR and GPR for a bimodal function

t	REC	q_0	q_1	q_3	$\bar{f}$	Var
0	TPR	0.250	0.250	0.250	0.4975	0.2475
1	TPR	0.372	0.125	0.378	0.7463	0.1857
2	TPR	0.369	0.125	0.381	0.7463	0.1857
3	TPR	0.365	0.125	0.385	0.7464	0.1856
9	TPR	0.287	0.121	0.471	0.7556	0.1819
0	GPR	0.250	0.250	0.250	0.4975	0.2475
1	GPR	0.247	0.250	0.252	0.4975	0.2475
2	GPR	0.242	0.250	0.257	0.4977	0.2476
3	GPR	0.233	0.250	0.268	0.4983	0.2477
6	GPR	0.120	0.226	0.427	0.5457	0.2467
9	GPR	0.000	0.006	0.988	0.9883	0.0115

Proportionate selection should not be used for a genetic algorithm, its selection intensity is far too low, (Mühlenbein and Schlierkamp-Voosen, 1994). Unfortunately, for other selection methods the equations for the marginal gene frequencies are difficult to obtain. For truncation selection an approximate analysis can be done by using the equation for the response to selection. The following theorem was proven by Mühlenbein and Schleirkamp-Voosen (1993) in a different context.

Theorem 2 *Let $p(t)$ be the frequency of allele 1 in the population at generation t. Let the fitness have a binomial distribution, i.e.,*

$$\sigma_p(t) = \sqrt{n \cdot p(t) \cdot (1 - p(t))}.$$

If the population is large enough to converge to the optimum, and if the selection intensity I is greater than 0, then the number of generations needed to reach equilibrium is approximately

$$GEN_e = \left(\frac{\pi}{2} - \arcsin(2p_0 - 1)\right) \cdot \frac{\sqrt{n}}{I}, \tag{4.19}$$

where $p_0 = p(0)$ denotes the probability of allele 1 in the initial population. The value $p(t)$ can be approximated as

$$p(t) \approx 0.5 \left(1 + \sin(\frac{I}{\sqrt{n}}t + \arcsin(2p_0 - 1))\right). \tag{4.20}$$

Remark: The assumptions of the theorem are fulfilled for the ONEMAX(n) function. In this case the theorem is also approximately valid for TPR, (Mühlenbein and Schlierkamp-Voosen, 1994). For TPR GEN_e is about 25% larger, independent of n.

4.5   GENETIC DRIFT

The analysis of the previous sections is valid for very large populations. This leads to deterministic equations for averages. In finite populations the chance introduced by finite sampling has to be modelled. The mathematical analysis can be done in principle with Markov chains, but, unfortunately, the number of transitions scales exponentially with the number of loci and the size of the population. The analysis gets simpler if we assume no selection. This case is called *genetic drift*. The finiteness of the population causes convergence to a single genotype, even without selection. It is a result of sampling with replacement.

Asoh and Mühlenbein (1994) have analyzed genetic drift for TPR. They showed that a genetic algorithm with TPR but without selection converges surprisingly quickly. This means that just by sampling with replacement, the variance of the population is continuously reduced. At the equilibrium the population consists of a single genotype only. For TPR the following proposition was obtained by fitting numerically obtained values:

Proposition 1 *Let the number of loci be n. Let each gene have two alleles. Using TPR the mean convergence time $\tau_n(N)$ of a population of size N is approximately*

$$\tau_n(N) \approx 1.4N \left(0.5 \ln(n) + 1.0\right) \text{ for } p(0) = 1/2. \tag{4.21}$$

Table 4.2   Genetic drift for GPR (n loci, N popsize).

n/N	5	11	51	101	401
1	5.7	13.9	68.9	137.9	553.1
2	8.0	19.2	94.8	189.5	759.0
4	10.6	25.4	124.4	248.4	993.5
8	13.4	32.0	156.3	311.9	1247.0
16	16.4	38.9	189.5	378.0	1510.0

Genetic drift for GPR is similar to a standard statistical process. Given n independent sampling processes with replacement, each process having a population N, what is the expected time for *all* n processes to converge, i.e to consist of copies of one member only? This is a classical statistical question which can be answered for certain distributions. We state the following theorem without proof.

Theorem 3 *Let the expected mean convergence time of the process be exponentially distributed with mean $\tau = 1/\lambda$. Then the expected mean time for all n processes to converge is given by*

$$\tau_n = \frac{1}{\lambda} \sum_{\nu=1}^{n} \frac{1}{\nu}. \tag{4.22}$$

The sum can be approximated by $\ln(n) + \gamma$ with $\gamma = 0.577\ldots$. By setting $1/\lambda = 1.4N$ equations 4.21 and 4.22 have the same asymptotic order. We have not been able to verify that the expected mean convergence time of GPR is exponentially distributed. Therefore we present numerical simulations obtained by Markov chains in table 4.2.

For $n = 1$ GPR and TPR are equal. For large N the mean time to convergence is approximately $\tau_1 = 1.39N$, as predicted by the proposition. The increase of τ with n is smaller than given by equation 4.22. A comparison of the numerical values for TPR presented by Asoh and Mühlenbein and table 4.2 shows that τ_n is only slightly larger for GPR than for TPR. This demonstrates that TPR and GPR give on the average very similar results, despite the fact that the underlying statistical processes are different.

Genetic drift is an important factor for genetic algorithms, especially in small populations. Whereas large populations converge deterministically to an equilibrium, may small populations converge to any of the many optima.

We next apply the theory to continuous genes.

4.6   CONTINUOUS PARAMETER OPTIMIZATION

Let an unconstrained optimization problem be given on a domain $D \subset R^n$

$$min(F(\mathbf{x})) \qquad a_i \leq x_i \leq b_i \qquad i = 1, ..., n \tag{4.23}$$

The breeder genetic algorithm **BGA** was designed to solve the above problem (Mühlenbein and Schlierkamp-Voosen, 1993). The BGA for continuous parameter optimization consists of several components which we summarize shortly. The selection is done by *truncation selection* described before. In the first version the following genetic operator have been implemented.

Discrete recombination
Let $\mathbf{x} = (x_1, \ldots, x_n)$ and $\mathbf{y} = (y_1, \ldots, y_n)$ be the parent strings. Then the offspring $\mathbf{z} = (z_1, \ldots, z_n)$ is computed by

$$z_i = \{x_i\} \ or \ \{y_i\}$$

x_i or y_i are chosen with probability 0.5.

BGA mutation
A variable x_i is selected with probability p_m for mutation. The BGA normally uses $p_m = 1/n$. At least one variable will be mutated. A value out of an interval $[-range_i, range_i]$ is added to the selected variable. $range_i$ defines the *mutation range*. It is normally set to 0.5 times the domain of definition of variable x_i.
   Given x_i a new value z_i is computed according to

$$z_i = x_i \pm range_i \cdot \delta \qquad (4.24)$$

The $+$ or $-$ sign is chosen with probability 0.5. δ is computed from a distribution which prefers small values. This is realized as follows

$$\delta = 2^{-k\alpha} \quad \alpha \in [0, 1]$$

k is called the precision constant. α is obtained from an uniform distribution.

BGA line recombination
The BGA line recombination uses components from both, mutation and recombination. It creates new points in a direction given by the two parent points. The placement of the point is done by the BGA mutation scheme. It works as follows: Let $\mathbf{x} = (x_1, \ldots, x_n)$ and $\mathbf{y} = (y_1, \ldots, y_n)$ be the parent strings with $\mathbf{x}$ being the one with better fitness. Then the offspring $\mathbf{z} = (z_1, \ldots, z_n)$ is computed by

$$z_i = x_i \pm range_i \cdot \delta \cdot \frac{y_i - x_i}{\|\mathbf{x} - \mathbf{y}\|} \qquad (4.25)$$

The $-$ sign is chosen with probability 0.9. The offspring is placed more often in the descending direction.
   The rationale behind the three operators is as follows. The BGA mutation operator is able to generate *any* point in the hypercube with center $\mathbf{x}$ defined by $x_i \pm range_i$. But it tests much more often in the neighborhood of $\mathbf{x}$. In the above standard setting, the mutation operator is able to locate the optimal x_i up to a precision of $range_i \cdot 2^{-(k-1)}$. Discrete recombination is a breadth search. New points are created at the edges of the hypercube defined by the two parents. The BGA line recombination tries new points in a direction defined by the parent points.
   In the next three sections we will analyze mutation and recombination. In order to do this we have to introduce some mathematical definitions.

4.7   ORDER OF CONVERGENCE AND COMPUTATIONAL COMPLEXITY

The definition of acceptable norms by which to evaluate and compare the efficiency of random search techniques remains a major research question. The best method seems to be the study of the distribution of the number of steps required to reach the essential infimum. The algorithms can then be evaluated by comparing the expected number of steps and/or higher moments of this distribution. To do this, we must rely on idealized benchmark situations. Clearly, not all possible functions can serve as test functions for such an investigation.

As a first step in this direction we proposed in (Mühlenbein and Schlierkamp-Voosen, 1994) to investigate the scaling of a given algorithm for a suite of test functions. Scaling defines the *computational complexity* of the algorithm. If a test function is defined for an arbitrary number of variables n, then the expected number of steps required to reach the essential infimum as a function of n has to be computed. Computational complexity has been a very useful concept in computer science.

In numerical analysis the *order of convergence* is used as a first criterion for evaluation. The order of convergence measures how fast the approximations converge to the infimum. The function is held fixed. There are two convergence measures, one defined for the function values $f(x)$, one defined for the x values. For simplicity we restrict our definition to the case that the infimum is unique.

Definition 1: *Let x^* denote the infimum and $f^* = f(x^*)$. Let $\|x\|$ be a norm in R^n. Then the order of convergence is* **linear** *in f (or in $\|x\|$), if there exists a constant $c_f < 1$ (or $c_x < 1$) such that*

$$|f(x_{t+1}) - f^*| \leq c_f \cdot |f(x_t) - f^*| \tag{4.26}$$

or

$$\|x_{t+1} - x^*\| \leq c_x \cdot \|x_t - x^*\| \tag{4.27}$$

The difficult relation between convergence in f and in $\|x\|$ will not be discussed here. Algorithms that converge as a higher power, i.e.

$$|f(x_{t+1}) - f^*| \leq c_f \cdot |f(x_t) - f^*|^m \quad m > 1$$

are said to converge super-linearly. A famous example is the Newton-Raphson algorithm which converges quadratically. But linear converge is not bad at all. Linear convergence means that successive significant digits are won linearly with computational effort. In other contexts the above linear convergence would be termed "exponential" or "geometrical". We believe that linear convergence is the best one can achieve for random search methods which do not use the derivative of the given function.

Unfortunately linear convergence is differently defined if the function has to be approximated by some series of known functions. Here it means

$$\|f - s_t\| \leq \frac{c}{t} \|f - s_{t-1}\|$$

where s_t is the approximation obtained with e.g t data points.

The following lemma can be easily proven. It gives the average number of steps required to reduce the error by a factor of ϵ.

Lemma 1: *If the order of convergence is linear in f or in x, then the avarage number of steps s_f or s_x required to reduce $|f(x_t) - f^*|$ or $\|x_t - x^*\|$ by a factor of ϵ is bounded by*

$$s_f \le \frac{\ln(\epsilon)}{\ln(c_f)}, \quad and \ s_x \le \frac{\ln(\epsilon)}{\ln(c_x)}. \tag{4.28}$$

We will show in the next section that for specific recombination operators the BGA converges linearly in f for a class of unimodal functions. Furthermore we will estimate the computational complexity.

4.8   ANALYSIS OF THE BGA MUTATION

In (Mühlenbein and Schlierkamp-Voosen, 1993) we have proven that a BGA with popsize $N = 1$ (1 parent, 1 offspring, the better of the two survives) using only mutation has approximate *linear order of convergence* for unimodal functions. The proof of this theorem is technically difficult. Therefore we just state it as empirical law and show its validity by simulations.

Empirical law 1: *Given a point x_l with distance $range_i \cdot 2^{-k/2} \le r \le range_i$ to the optimum. For simplicity let the optimum be at $x^* = 0$. Then the expected progress $E(\|x_l\|)$ of the simple BGA with mutation only is given by*

$$E(\|x_l\|) \approx \frac{c(k)}{n} \cdot \|x_l\| \tag{4.29}$$

The number of iterations IT needed to approximate the optimum up to a precision of ϵ is given for $n \gg 0$ by

$$IT \approx \frac{n}{c(k)} ln \frac{\|x_0\|}{\epsilon} \tag{4.30}$$

For $k \ge 8$ we have
$$c \propto 1/k.$$

The empirical law shows that a BGA with mutation converges linearly for the quadratic sphere function. The progress depends on k. The larger k, the smaller the progress. But in order to locate the optimum with a given precision ϵ, the value $range_i \cdot 2^{-(k/2)}$ has to be less than ϵ. Therefore a large k may be necessary.

We show the dependence of the BGA on the precision constant k in table 4.3. The task is to minimize the hypersphere of dimension $n = 1$ and $n = 10$,

$$F_0(x) = \sum_{i}^{n} x_i^2.$$

The values are averages over 100 runs. The initial value is given by $x_0 = (1, ...1)$, the range is set to 2. The precision value means, that the function value is for the first time below the given precision.

Table 4.3 Number of trials needed to achieve the above precision

$n = 1$	0.1	0.01	0.001	0.0001	0.00001
$k = 16$	14	23	32	43	54
$k = 24$	21	34	45	56	69
$k = 32$	29	47	63	81	99
$n = 10$	0.1	0.01	0.001	0.0001	0.00001
$k = 16$	371	536	696	845	1023
$k = 24$	552	762	1011	1231	1462
$k = 32$	758	1035	1297	1571	1851

Note that for every fixed k the BGA needs about the same number of function evaluations to improve the precision by a factor of 10. This clearly shows the linear order of convergence. The numerical values have a high variance, therefore any conclusions have to be made with care.

Remark: *For $n = 1$ and $k = 16$ the BGA needs about 10 evaluations to improve the solution by a factor of 10. For $k = 24$ and $k = 32$ there are 15 resp. 20 evaluations needed. For $n = 10$ the numbers are about 160,230 and 280. For $n = 100$ the numbers are about 2000,3000 and 4000. The computational complexity of the BGA mutation scheme is therefore about $O(n\ln(n))$.*

The above result is qualitatively also true for multimodal functions. We take as example the popular function of Rastrigin

$$F_0(x) = 20n + \sum_i^n (x_i^2 - 20cos(2\pi x_i)).$$

In table 4.4 numerical data from simulations is given. For these simulations the initial point was at $x_0 = (10, ..10)$. The range was set to 20. In this area Rastrigin's function has quite a number of local minima. This reduces of course the probability of a mutation success. The BGA needs quite a number of function evaluations to locate the attractor region of the global minimum. The attractor region is reached if the function values are less than 0.1. Now the BGA proceeds as fast as for the hypersphere.

Table 4.4 Number of trials needed to achieve the above precision

$n = 1$	1	0.1	0.01	0.001	0.0001
$k = 16$	56	126	137	148	166
$k = 24$	79	197	211	225	237
$k = 32$	93	242	261	280	299
$n = 10$	1	0.1	0.01	0.001	0.0001
$k = 16$	2986	4360	4486	4616	5529
$k = 24$	4115	6093	6265	6456	6684
$k = 32$	4945	7163	7412	7681	7905

The increase of the number of trials with k is at least for $n = 1$ the same as for the hypersphere.

4.9   ANALYSIS OF RECOMBINATION OPERATORS

A number of different recombination operators have been proposed for continuous genes. Some of the most popular are discrete recombination (Schwefel, 1981; Mühlenbein and Schlierkamp-Voosen, 1993), intermediate recombination (Schwefel, 1981), extended intermediate recombination (Mühlenbein and Schlierkamp-Voosen, 1993), extended line recombination (Mühlenbein and Schlierkamp-Voosen, 1993), and BLX-O.a crossover (Eshelman and Schaffer, 1992). A thorough evaluation of these recombination operators has not yet been done. In this section we will analyze discrete, intermediate and extended intermediate recombination and a new soft recombination scheme gleaned from fuzzy set theory.

Let $(x_1, \ldots, x_n)$ and $(y_1, \ldots, y_n)$ be the selected parent chromosomes. With discrete recombination (DR) the offspring variable z_i is chosen randomly from x_i and y_i. With intermediate recombination (IR) the offspring variable is given by $(x_i \leq y_i)$

$$z_i = x_i + \alpha_i \cdot (y_i - x_i)$$

where α_i is either fixed to 0.5, chosen randomly in the interval $[0, 1]$ (IR in the narrow sense) or chosen randomly in the interval $[-d, 1 + d]$ (extended intermediate recombination (EIR). The rationale behind EIR is to introduce more variance. Fuzzy recombination (FR) is inspired from fuzzy set theory. The probability that the offspring has the value z_i is given by a bimodal distribution,

$$p(z_i) \in \{\phi(x_i), \phi(y_i)\}, \tag{4.31}$$

with triangular probability distributions $\phi(r)$ having the modal values x_i and y_i with

$$
\begin{aligned}
x_i - d \cdot |y_i - x_i| \leq r \leq x_i + d \cdot |y_i - x_i| \\
y_i - d \cdot |y_i - x_i| \leq r \leq y_i + d \cdot |y_i - x_i|
\end{aligned}
\tag{4.32}
$$

for $x_i \leq y_i$ and $d \geq 0.5$. We mainly used $d = 0.5$ for the simulations. All recombination operators are volume oriented. They create offspring randomly within a hyper-rectangle defined by the parent points.

The equation for the response to selection leads to a design criterion for genetic operators. In order to maximize the cumulative response, a recombination operator should *maximize the product of the realized heritability* and *the standard deviation* of the offspring generation. This design goal we will subsequently use to analyze the recombination operators defined above.

For the evaluation the following unimodal test functions will be used. They consist of a standard function (*sphere*), a function where the variables have different importance for the fitness function (*ellipse*), a function where the minimum is at the boundary (*sum*) and a function which is not differentiable at the infimum (*pyramid*).

$$
\begin{array}{ll}
F_{sphere}(x) = \sum_{i=1}^{n} x_i^2 & |x_i| \leq 1 \\
F_{ellipse}(x) = \sum_{i=1}^{n-1} x_i^2 + 10^4 x_n^2 & |x_i| \leq 1 \\
F_{sum}(x) = \sum_{i=1}^{n} x_i & 0 \leq x_i \leq 1 \\
F_{pyramid}(x) = (\sum_{i=1}^{n} (1 - |x_i|))/n & |x_i| \leq 1
\end{array}
$$

The performance of the recombination operators is shown in Figure 4.2. If a good approximation is required, then FR or EIR with $d = 0.5$ should be used. They linearly converge to the solution up to a precision of 10^{-12}. EIR with $d = 1$ does not converge at all. Selection reduces the search space, but EIR places the offspring in the whole area. So selection is counterbalanced by this recombination operator. In contrast, DR converges very early. DR would need a huge population size to achieve a good approximation. For IR the mean fitness decreases the fastest, but it also converges prematurely. IR reduces the variance too fast. Before we analyze the results in more detail, we formulate the most important result of our simulations as a law.

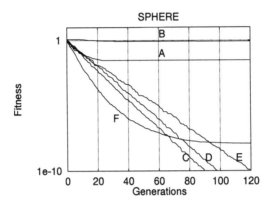

Figure 4.2 *Sphere*, n=32, popsize N=512, I=1.4; A: DR, B: EIR with $d = 1.0$, C: FR with $d = 0.5$, D: EIR with $d = 0.5$, E: FR with $d = 1.0$, F: IR

Empirical law 2: *The early generations of the BGA simulation always converge linearly in f, for all recombination operators, excluding EIR with $d = 1$. The slope of this linear region is independent of the size of the population N. A larger N increases the length of the linear region (but not the slope) and locates the optimum with higher accuracy.*

Our subsequent analysis will be restricted to the region giving linear convergence. The challenge for the GA designer is to find a recombination operator which gives a large linear region and a steep slope. This means that this operator has to introduce the right portion of variance into the offspring population. If the variance is too large then the algorithm does not converge at all, if it is too small then it converges prematurely.

This problem is investigated for EIR in more detail in figure 4.3. Here we vary d from 0 to 1. As already known, IR (EIR with $d = 0$) converges the fastest, but the convergence stops early. EIR with $d = 0.25$ gives the best results, if a solution accuracy of at least 10^{-6} is required. EIR with $d = 1$ does not converge at all. The value of the best d gets larger if the infinum is not contained in the interior. Here it would be obviously better to have a larger d. Simulations confirm this statement. We just report the results for the function *sum*. For $d = 0.25$ we have premature convergence, $d = 0.5$ gives the best results. Therefore it seems that $d = 0.5$ is a good choice for a large class of functions. The same is true for fuzzy recombination FR.

Next we make a more detailed analysis of the recombination operators by using the design criterion for recombination operators defined earlier. First we compute the realized heritability $b_t = R(t)/S(t)$ for DR, IR, EIR and FR with $d = 0.5$. In figure 4.4 the results for the linear function *sum* are displayed. The heritability of DR and IR

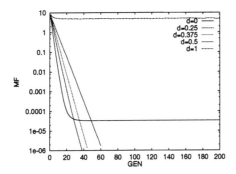

Figure 4.3   *Sphere, n = 32,* popsize $N = 1024$, $I = 1.2$: Mean fitness MF

is about 1.0. FR has a heritability of about 0.83 and EIR of about 0.73. Heritability favours DR and IR, but we have already seen that DR and IR converge prematurely. This confirms our statement made in the introduction that a large heritability is not enough. The operator has also to create enough variance. A closer analysis of the reduction of variance will be made for fuzzy recombination in the next section. Figure 4.2 indicates that FR performs about 10% better than EIR. The reason for this fact shows figure 4.3(b). The heritability of FR is more than 10% larger than for EIR.

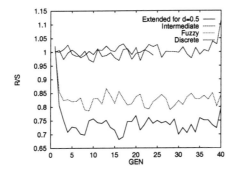

Figure 4.4   *Sum, n = 32, $N = 1024$, $I = 1.6$*: realized heritability R/S

Our simulations have shown that the behavior of the fuzzy recombination algorithm is almost deterministic. In particular the average function value decreases similarly for runs with the same set of parameters but with different initial population. As stated in empirical law 2, the linear convergence part is independent of the population size N, if N is greater than a critical population size N^*. N^* is the size needed to approximate the infinum with the required precision. It depends om I, n and ϵ. The determination of N^* is extremely difficult and cannot be discussed in this paper. Instead we will empirically determine the computational complexity of the algorithm.

Definition 2: *Let N be the size of the population used by the breeder genetic algorithm. Then $gen^*(N, I, n, \epsilon, f)$ is defined as the number of generations needed to obtain an approximation $\tilde{f}$ such that $|f^* - \tilde{f}| \leq \epsilon$.*

We will first investigate the dependence of $gen^*(N, I, n, \epsilon, f)$ on I, afterwards on

n and on f. We assume that we have a sufficiently large population, so that the required approximation accuracy can be obtained. Then gen^* is independent from N by empirical law 2. Therefore we will subsequently write $gen^*(I, n, \epsilon, f)$.

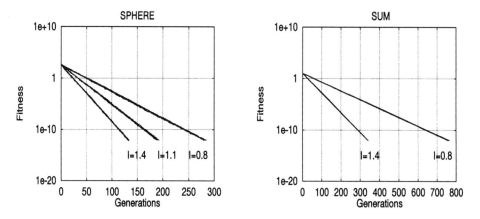

Figure 4.5   Average fitness of (a) *Sphere* (b) *Sum*, $n = 32$, $N = 512$, 5 runs.

Some of our simulation results are displayed in Figure 4.5. Note that the behavior of the algorithm is almost deterministic. There is no visible difference between the five runs. The figure suggests an inverse proportionate dependence of gen^* on I. Using additional simulation runs, we applied $Mathematica^{TM}$ to fit the data and got the relation:

$$gen^*(I, n, \epsilon, f) \approx \frac{c_1(n, f, \epsilon)}{I^{1.4}}. \qquad (4.33)$$

In the same way, we also determined that

$$gen^*(I, n, \epsilon, f) = c_2(f, I, \epsilon) \cdot n^{0.7}. \qquad (4.34)$$

These two empirical results can be combined, giving an estimate for the computational complexity

$$gen^*(I, n, \epsilon, f) = c(f, \epsilon) \frac{n^{0.7}}{I^{1.4}} \qquad (4.35)$$

This estimate is surprisingly similar to the one obtained for the discrete ONEMAX function (Mühlenbein and Schlierkamp-Voosen, 1994)[9]

$$gen^* = c \frac{\sqrt{n}}{I}.$$

The difference of the two formulas can be partially explained. A more precise look at the heritability of the recombination operators shows, that the heritability decreases with the number of dimensions. This decrease is very small ($O(n^{-0.2})$). Therefore we first did not notice this second order phenomenon.

The same analysis can be made also for convergence in $\|x\|$. We have used the usual Euclidian norm in R^n. This measure is independent from the function values. Therefore it is more general. Figure 4.6 shows that the BGA with FR also converges linearly in $\|\bar{x}\|$, where $\bar{x}$ is the average of the x values.

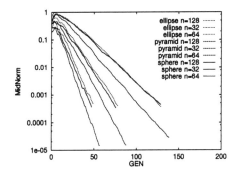

Figure 4.6 $N = 1024$, $I = 1.2$: Average Euclidian distance $Midnorm$

All curves obey the same scaling law formulated in equation (4.34) for f. The convergence behavior is only different at the very first generations. Here $\|x\|$ increases slightly. But the scaling laws are not effected by this behavior. Note that the results of *sphere* and *ellipse* are almost identical, despite that these functions are very different. Only the function *pyramid* converges faster. This has to be expected, because its level set is a hyper-rectangle which fits better to our recombination operator. Our recombination operator creates offspring within a hyper-rectangle defined by the selected parents. If the level set is spherical, selection will select parent points lying on spherical discs. The recombination operator will create offspring in a hyper-rectangle which contains the spherical disc.

Fuzzy recombination has also been used together with gene pool recombination. It converges for the sphere function about 25% faster than two parent mating.

4.10   GPR FOR CONTINUOUS FUNCTIONS

For discrete genes it was shown that a genetic algorithm with gene pool recombination converges faster than two parent recombination. In this section we will compare *uniform fuzzy two parent recombination* (UFTPR) with *uniform fuzzy gene pool recombination* (UFGPR). First they will be compared for the sphere function with a random initial population seed within a cube containing the optimum at the center. The results are shown in Figure 4.7.

Note that the realized heritability for UFGPR is about 10% higher than for UFTPR. This gives a higher convergence speed. Corresponding to the operator design criterion we have to analyze the realized heritability $b(t)$ and the phenotypic standard deviation $\sigma(t)$. From Figure 4.7 it can be seen that $b(t)$ is on average a constant after a short adaptation phase. $b(t)$ depends more on n than on t. In Figure 4.8 values for $b_n(t)$ from simulations are given for $I = 1.4$ for up to $n = 300$ variables (dotted graph). The realized heritability decreases with n. A fit reveals the relation

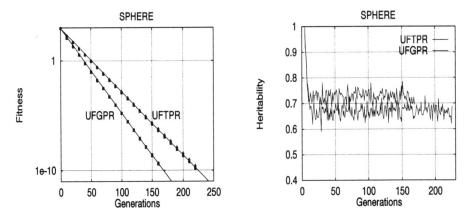

Figure 4.7   Comparison of two parents (UFTPR) and gene pool (UFGPR) recombination for the sphere function, $\overline{f}(t)$ (left) and $b(t) = R(t)/S(t)$ (right), $n = 100$, $N = 1000$, $I = 1.4$, • simulation, 5 runs overlaid, – model prediction

$b_n(t) \approx 1 - erf(0.03578 \cdot \sqrt{n/2})$ (solid graph) with erf the error integral. This fit is rather accurate for $n > 40$.

In the left part of the figure, the ratio $\overline{f}(t)/\sigma_p(t)$ is shown for two selection intensities. The inverse of this ratio is called the coefficient of variation (CV) in statistics. The coefficient of variation CV for a selection intensity $I = 1.4$ is larger than for $I = 0.8$. For $n > 40$ the relations $CV(I = 0.8) = 1.3059 \cdot \sqrt{2/n}$ and $CV(I = 1.4) = 1.4757 \cdot \sqrt{2/n}$ are fitting the experimental results very well. At present we are not able to show how the constants of these equations are related to the selection intensity.

In the next section we will derive a scaling law from a more theoretical view.

4.11   APPROXIMATE THEORETICAL ANALYSIS

It has been observed empirically for quite a time that the equation for the response to selection is valid for a large class of problems in quantitative genetics (Falconer, 1981; Verrier et al., 1991; Turelli and Barton, 1994)). Therefore our analysis starts with this equation

$$R(t) = b(t)I\sigma_p(t) \tag{4.36}$$

$b(t)$ was already numerically computed in the previous section. In order to solve the above equation, $\sigma_p(t)$ has to be estimated. For simple binary functions, we successfully approximated $\sigma_p(t)$ by a binomial distribution. In general, an estimate of $\sigma_p(t)$ for a genetic population under selection is very difficult.

In population genetics the following approach has been tried. The variance $V(t)$ and not the standard deviation σ is investigated. The variance is decomposed into a number of terms, usually into two terms. These are the variance of the selected

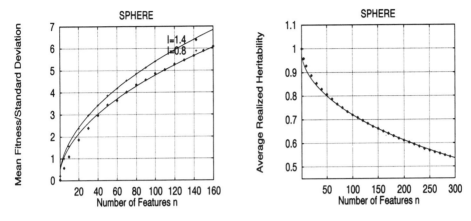

Figure 4.8 Inverse coefficient of variation $1/CV = \overline{f(t)}/\sigma(t)$ (left) for $I = 0.8$ and $I = 1.4$, and the average realized heritability $b_n(t) = R(t)/S(t)$ for $I = 1.4$ (right) for the sphere function with UFGPR

parents and the additional variance introduced by random mating. Selection reduces the variance, and mating with recombination increases the variance. These two forces have to be balanced.

We just outline the behavior for a population with normal distributed fitness values. The detailed theoretical analysis will be described elsewhere. For normal distributed fitness values the variance V_s of the selected parents can be computed analytically (Falconer, 1981). It is given by

$$V_s(t) = (1 - I(I - X))V(t). \qquad (4.37)$$

Here, X denotes the abscissa of the truncation point. The variance of the offspring is given by

$$V(t) = 0.5V_s(t - 1) + V_r(t - 1),$$

where V_r denotes the variance introduced by mating with recombination. The factor 0.5 has to be introduced because two parents give just one offspring, the fitness value of which depends on the midparent value and some "noise" introduced by mating. If we do not select, then we have $V_s(t - 1) = V(t - 1)$ and $V(t) = V(t - 1)$ because the population is in equilibrium. This gives $V_r(t - 1) = 0.5V(t - 1)$. We assume that this equation is valid also if selection is done. Combining the equations we obtain

$$V(t) = (1 - 0.5I(I - X))V(t - 1). \qquad (4.38)$$

Numerical simulations have confirmed this equation. A BGA with fuzzy recombination decreases the variance by a constant. This constant depends on I, but also on n and the function to be optimized. For the following theorem we will assume

$$\sigma(t) = c\sigma(t - 1),$$

where $\sigma = V^{1/2}$. This equation has the solution

$$\sigma(t) = \sigma(0)c^t. \tag{4.39}$$

It is difficult to estimate c, especially its dependence on the selection intensity I and on the size of the problem n. We have made intensive simulations for $n = 32, 128, 512$ and $I = 2.6, 2.0, 1.6, 1.2$. Due to space limitations we are not able to show the numerical data. The results are summarized by the following law.

Empirical law 3: *For the fitness function sum the constant c_{sum} is given by*

$$c_{sum} \approx \sqrt{1 - \frac{I^{1.4}}{n^{0.7}}} \tag{4.40}$$

We are now able to prove the following theorem.

Theorem 4 *Under the assumption of empirical law 3 the BGA converges linearly in f. The constants are given by*

$$c_{sum} = \sqrt{1 - \frac{I^{1.4}}{n^{0.7}}} \qquad s_{sum} \approx -2\ln(\epsilon) \cdot \frac{n^{0.7}}{I^{1.4}}. \tag{4.41}$$

Proof: From $R(t) = -\bar{f}(t+1) + \bar{f}(t)$ we obtain from the equation for the response

$$|\bar{f}(t+1)| \leq (1 - bI\frac{\sigma(t)}{\bar{f}(t)})|\bar{f}(t)|. \tag{4.42}$$

For notational convenience we assume $f^* = 0$. The average fitness is obtained from the sum

$$-\bar{f}(t) + \bar{f}(0) = \sum_{s=1}^{t-1} R(s).$$

This equation can be solved easily

$$\bar{f}(t) = \bar{f}(0) - bI\sigma(0)\frac{1 - c^t}{1 - c}.$$

We assume that the population converges to the infimum, i.e $\bar{f}(t) \rightarrow 0$. Then

$$\bar{f}(0) = bI\sigma(0)\frac{1}{1 - c}.$$

Inserting this equation, we obtain

$$\bar{f}(t) = bI\sigma(0)\frac{c^t}{1 - c} = bI\frac{\sigma(t)}{1 - c}.$$

Therefore we get the equation

$$\frac{\bar{f}(t)}{\sigma(t)} = \frac{bI}{1 - c} \tag{4.43}$$

Inserting this expression into the first equation we obtain $\bar{f}(t+1) \leq c \cdot \bar{f}(t)$. This proves the first conjecture.

The constant s_{sum} is obtained from lemma 1 by replacing $\ln \sqrt{1-x}$ in 4.28 with $-0.5x$.   □.

The estimate given in equation (4.41) is quite accurate. If we set $\epsilon = 10^{-10}$, $I = 1.4$, $n = 32$, we obtain $s_f \approx 325$. From figure 4.4 we get $s_f \approx 290$. For different functions the constant c_f is different, but the constant s_f shows always the same asymptotic behavior concerning I and n.

Equation 4.43 is an interesting result by itself. It shows how the variance has to be balanced. In order to achieve linear convergence the mean distance to the optimum divided by the variance has to be a constant. Unfortunately the mean distance to the optimum is not known for a real BGA run, therefore this quotient cannot be directly used.

4.12 CONCLUSION

In this paper we have investigated genetic algorithms for discrete and continuous genes. In both cases a new recombination operator, which we call gene pool recombination has been shown to be superior to the usual two parent recombination. Furthermore a BGA with fuzzy recombination converges linearly for a set of benchmark functions. An exact mathematical definition of the class of functions where the BGA will converge linearly is difficult. But some comments can be given describing the class of functions where the BGA with FR alone will not converge. Fuzzy recombination is obviously a volume oriented search. The volume to be searched is a hyperrectangle defined by the parents. The hyperrectangle is parallel to the axes. If the minimum of the function is located at the end of a very steep and curved valley, this recombination method is obviously not able to locate the minimum. The steep valley is a very small part of the hyperrectangle. This problem can be solved by introducing additional operators, like line recombination (Mühlenbein and Schlierkamp-Voosen, 1993). Another solution is to transform the axes, so that the direction of the valley is one of the axis.

In the paper the classical science of breeding has been used as the theoretical foundation for genetic algorithms. It lead to new design goals of genetic operators. We iterate our hope (Mühlenbein and Schlierkamp-Voosen, 1993) that researchers proposing new GA implementations investigate the order of convergence and the computational complexity of their algorithms. This is the only way to change the research from mystic and individual belief to a science with a solid foundation.

Acknowledgement: The breeder genetic algorithm has been developed together with D. Schlierkamp-Voosen. The research concerning fuzzy recombination and gene pool recombination has been done together with H.-M. Voigt.

REFERENCES

Asoh H. and Mühlenbein H. (1994) On the mean convergence time of evolutionary algorithms without selection and mutation. In Davidor Y., Schwefel H.-P., and Männer R. (eds) *Parallel Problem Solving from Nature*, Lecture Notes in Computer Science 866, pages 88–97. Springer-Verlag.

Crow J. F. and Kimura M. (1970) *An Introduction to Population Genetics Theory*. Harper and Row, New York.

Eiben A., Raue P.-E., and Ruttkay Z. (1994) Genetic algorithms with multi-parent recombination. In Davidor Y., Schwefel H.-P., and Männer R. (eds) *Parallel Problem Solving from Nature*, Lecture Notes in Computer Science 866, pages 78–87. Springer-Verlag.

Eshelman L. and Schaffer J. (1992) Real-coded Genetic Algorithms and Interval-schemata. In Whitley L. (ed) *Foundations of Genetic Algorithms*, pages 187–202. Morgan-Kaufman, San Mateo.

Eshelman L. J. and Schaffer J. D. (1993) Crossover's niche. In Forrest S. (ed) *Fifth Int. Conf. on Genetic Algorithms*, pages 9–14. Morgan Kaufmann, San Mateo.

Falconer D. S. (1981) *Introduction to Quantitative Genetics*. Longman, London.

Goldberg D. (1989) *Genetic Algorithms in Search, Optimization and Machine Learning*. Addison-Wesley, Reading.

Mühlenbein H. and Schlierkamp-Voosen D. (1993) Predictive Models for the Breeder Genetic Algorithm I. Continuous Parameter Optimization. *Evolutionary Computation* 1: 25–49.

Mühlenbein H. and Schlierkamp-Voosen D. (1994) The science of breeding and its application to the breeder genetic algorithm. *Evolutionary Computation* 1: 335–360.

Mühlenbein H. (1989) Parallel genetic algorithm, population dynamics and combinatorial optimization. In Schaffer H. (ed) *3rd Int. Conf. on Genetic Algorithms*, pages 416–421. Morgan Kaufmann, San Mateo.

Nagaraja H. (1982) Selection differentials. In Kotz E., Johnson N., and Read C. (eds) *Encyclopedia of Statistical Science*, pages 334–336. Wiley, New York.

Naglyaki T. (1992) *Introduction to Theoretical Population Genetics*. Springer, Berlin.

Schwefel H.-P. (1981) *Numerical Optimization of Computer Models*. Wiley, Chichester.

Syswerda G. (1993) Simulated crossover in genetic algorithms. In Whitley L. D. (ed) *Foundations of Genetic Algorithms 2*, pages 239–255. Morgan Kaufmann, San Mateo.

Turelli M. and Barton N. (1994) Genetic and statistical analyses of strong selection on polygenic traits: What, me normal. *Genetics* 138: 913–941.

Verrier E., Colleau J., and Foulley J. (1991) Methods for predicting response to selection in small populations under additive genetic models: a review. *Livestock Production Science* 29: 93–114.

5

NATURAL AND ARTIFICIAL NEURAL NETS

R. MORENO-DIAZ [1]

Objectives

Historically, the concepts of artificial neural nets arose from the efforts to formalize detected properties of the nervous system according the mathematical and logical tools available at that time. Soon after the pioneer work of McCulloch and Pits in 1943, the modular or granular aspect of the theoretical frame advanced by them was taken over by computational theorists like Kleene who gave rise to the algebraic theory of automata. From the pure theoretical side, the "granular machine" concepts flourished quickly in the 50's and 60's, as well as the interest in using methods and concepts from logic, computation, information processing and transmission with the aim to gain a better understanding of the behaviour of the nervous system.

At the same time, the concepts associated to granular machines have developed to be of intrinsic interest, independently of wether they help to describe neural functions. Because the recent technical possibility of constructing non trivial, relatively large nets of computing units, the whole subject regained momentum to form the large body of what is known today as artificial neural nets.

The objectives of this paper are to explore the interplay between that formal description of natural neural nets and the artificial ones. Part I will deals with more global properties of the nervous system, in line with the original proposals by McCulloch and Pitts. Part II treats the interaction between brain theory and what is

---

[1] R. Moreno-Diaz
   Universidad de Las Palmas de Gran Canaria, 35008 and
   Instituto Tecnologico de Canarias, 35003
   Canary Islands - Spain

today common in high level Artificial Intelligence and Behaviour. Part III treats that interplay in relation to more specific real neural nets at different levels of the nervous pathway, with emphasis, as it will be argued, in those peripheral nets where the neural code is better understood.

PART I. Formal Neural Nets. Biological Counterparts of Logical Machines

5.1 The Aims of the Classical Theory

Even since the original paper by McCulloch and Pitts *(McCulloch, Pitts, 1943)*, it is very difficult to mistake the basic modular unit as a model for a neuron. In the authors own words, formal but not factual equivalences were sought, so that many formal neurons might be needed to embody a single property of a real living neuron. The relevant points are the synthesis theorems, where the problem is to find a modular net of their formal neurons that will behave in a prescribed manner, if it exists. The formal language they used is absolutely opaque, and to comprehend their logical contributions it is much recommended to refer to later reformulations, at the light of the developments in automata theory. Anyway, the authors manage to reach the conclusions that a net of their formal neural nets, with loops - feedback -, provided with scanners and a tape can do just what a Touring Machine can, though they know that brains can do much better.

The original theorems were, in no way, obvious and minimal, that is, even since there always is a "waste" of formal neurons when synthesizing even simple operations, that waste was too large. For example, to perform a simple contrast detection of the type of the exclusive or, they were required two layers of neurons already. That is simply because the primitive formal unit was a linear machine followed by a threshold.

It is very illustrative to consider how the interplay between neurophisiology and logic came into scene for the so called presynaptic inhibition, which permitted McCulloch and Blum *(Blum, 1961)* to cope with the possibility of the "universal logical" formal neuron, capable of computing any logical function of its inputs.

5.1.1   Interaction of afferents

Interaction of afferents is the simplest logical formulation for the presynaptic inhibition found in Rana pipiens by Shypperheyn. In essence, it consists that fibres reaching a neuron bifurcate in a way that they may drastically inhibit other fibres to the cell, prior to the synapsis. In what follows we shall use the clearer descriptions by Monroy *(Monroy, 1981)* and others. A simple illustration is in fig. 5.1 (a), where each fibre, when stimulated, completely blockes propagation in the other, before it reach the synapsis. For $t = 1$, it is easily verified that the cell computes the "exclusive or" of x_1 and x_2.

The trick is then to synthesize any nonlinear logical function by allowing the interaction, which in fact corresponds to a new layer of logic. The systematic approach is illustrated in fig. 5.2 (b), where α_i, α_{ij}, $\alpha ijk \ldots$ are the synaptic weights, and θ is the threshold.

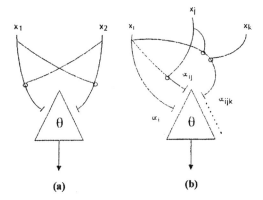

Figure 5.1   (a) Illustration of simple presynaptic inhibition. (b) Systematic
formulation of presynaptic inhibition for fiber x_i

The firing condition is then:

$$\sum \alpha_i x_i + \sum \alpha_{ij} x_i x_j' + \sum \alpha_{ijk} x_i x_j' x_k' \geq \theta \tag{5.1}$$

The number of degrees of freedom for a unit with a fixed threshold is, for M inputs
fibres

$$M + M(M-1) + M \binom{M-1}{2} + \ldots + M \binom{M-1}{M-2} + M = M2^{M-1} \tag{5.2}$$

which is larger than 2^M, and therefore shows the redundant effect of the interaction
of afferents.

The synthesis problem is then, what is the minimal neurophysiological paradigm
for a neuron computing an arbitrary function of their inputs ?.

Now, the typical Blum-McCulloch procedure is to assign a threshold to each
minterm of the inputs, with the condition that minterm $x_1', x_2', \ldots, x_m'$ must have
a zero threshold. There are systematic ways to proceed in the synthesis, but here we
are interested rather in the implications of this apparent dendritic redundancy (or
dendritic profusion).

From the logical point of view, the redundancy shows up from the possibility that a
neuron can compute a whole set of different logical functions as its threshold changes.
That is, the limiting case of no-redundancy is when the neuron computes $2M + 1$
different functions, including tautology and contradiction, which is the maximum
number of logical functions that a neuron having a fixed anatomy can compute. The
number of different thresholds is precisely $2M$. But this is the more unreliable neuron,
as we shall see.

In the fifties, McCulloch and Von Newman were engaged in the problems of
reliability. For neural nets, McCulloch and later Winograd and Cowan, worried about

the stability of the function of a neuron when its threshold changes. They try to solve the problem by a multilayered network with interaction of afferents (multilayered redundancy) or by the classical multichannel redundancy of Von Newman.

As we can see, dendritic profusion provides and additional mechanism for logical stability, since for all (or part) of possible threshold values the same logical function can be assigned. The two limiting cases (maximal functionality, minimal logical stability and viceversa) are illustrated in fig. 5.2 for the simple case of the "exclusive or". It is also interesting how this type of arguments helps dilucidating the role of retinal lateral interaction cells.

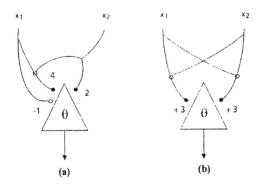

(a)   (b)

Figure 5.2 Illustration of maximal funcionality versus minimal logical stability for threshold changes. Neuron (a) computes a diferent logical function for each threshold ("exlcusive or" for threshold 2) while neuron (b) computes "exclusive or" for threshold 1,2,3. (From Monroy, 1981)

5.1.2 Oscilations in Nets

In 1965, the neurophysiology group of MIT was after the problems of how more economically and efficiently can a network (granular if one wishes) store dynamic patterns of activity for later retrieval. The basic unit, of course, was the formal neuron with interacting afferents, capable of computing any logical function of its inputs.

First, there is a problem of counting the maximum number of oscillating modes that a net of N neurons can engage in. A mode of oscillation is a circular sequence of states for the net. Schnabel *(Schnabel, 1966)* counted them to be, for a net of N neurons:

$$N_0 = \sum_{K=2}^{2N} (K-1) \binom{2^N}{K} \tag{5.3}$$

a number with increases fantastically with the number of neurons. In fact, it applies to any finite automaton of 2^N states.

A second question was whether this astronomical number (for N a little large) could be stored and retrieved in-from a fixed anatomy. Da Fonseca *(da Fonseca, 1967)* saw it non practical and started to explore the potentiality of non-linear feedback shift registers as alternatives for dynamic storage. His conclusions were practical in the sense that the corresponding networks had a much larger possibility to be embodied somewhere in the nervous systems. In fact, formal nets as they were at the mid'sixties showed such a logical potentiality that all deterministic and most of probabilistic automata where particular cases, as it happened to da Fonsecas's dynamic storage models. But, as it is the case for general abstract theories when referring to real word, they may be describing all in a zero depth.

Anyway, the interplay between natural and artificial proceeded. It was already obvious that logical and computational tools available falled too short to provide a frame where to progress from the nervous system description beyond perhaps sensorial systems and primary cortex.

There it came the search for "intension" as it was a contraposition to the required "extensional" descriptions, impractical for complex behaviour, in which any one automata theory minded will necessarily fall in. Efforts by da Fonseca, Mira and others *(da Fonseca, Mira, 1969)*, went in that direction, but it remains a strictly open mater.

Let us state it in the following way: By the early seventies it was already quite well stablished that any behaviour that can be expressed as a result of the interaction of one or various probabilistic or deterministic automata had a very cute counterpart in a logically minimal network having a nice neurophysiological look of thresholds and dendrites interaction. From the automata theory points of view, there was only one point missing, due to an error in a theorem on the synthesis of arbitrary probabilistic automata by McCulloch *(Moreno-Diaz, McCulloch, 1969; McCulloch et al, 1969)* and Moreno-Diaz. For a net of N probabilistic neurons it is not possible to synthesize an arbitrary probabilistic automaton, so it looked that other neurophysiological paradigm was behind the logical constructs.

5.2   Present Status of the Theory

As it was pointed above, formal neuronal net theory as it is contemplated as potential biological counterparts of logical machines, suffered from the sixties, the influence of automata theory, influence which has been both limiting and enriching. It has limited the scope and practical applicability because the very nature in "extenso" of anything derived from automata theory to refer the real world. And enriching, because the new points of view permitted to state new questions to the theory.

In our opinion, the theory has stopped its vertical development, though lateral branches are still allow to grow to provide for concrete applications and perhaps for new theorems carrying some intellectual satisfaction.

In a rather general sense, there is only one type of questions that matter to modular or granular brain theory, from top to bottom and even from bottom to top. This questions have to do with the "level of description" that some of us and others have systematically defended. It is something like stabilising the "ceiling" from where we are to go down in the interpretation - explanation. Or in the way up being aware of the power of the rocket that will allow us to reach a certain height in the description. In both cases, the ways might be full of delusions.

For the case of the already classical theory of formal neural nets, the ceiling was apparent to be that of the formalisms of probabilistic arbitrary automata. Without stopping in fuzzy floors for the moment, we shall consider, with quick notation, the way from cell bodies, thresholds, afferent and efferent interactions to probabilistic automata and viceversa.

5.2.1   *The formalism from naive neurons to probabilistic automata and back. Some side-effects*

Without paying any attention to the physiological significance of subjects such as temporal and spatial integration (linear or non linear), space-time equivalence, multiplicative effects, all and some more expressable in terms of what follows we preceded to the next fast-food definitions.

1. AUTOMATA. Except for an output encoder (a net without loops):

   (a) A deterministic automaton of 2^N states and M inputs lines is a set of 2^M transitions matrices, $M_{ij}(x)$, dimensions 2^Nx2^N, such that for each input configuration X X, there is one an only one 1 per row. A functional matrix is a state transition matrix in which each term contains the logical sum of the input configuration that makes that term equal to 1. Obviously $M_{ij}(x) = 1$ (i is the row)

   (b) A probabilistic automaton is the same that above, except that each term (ij) is preceded by a number P_{ij} between 0-1, with the condition that, for each $X = X_0$

$$P_{ij}M_{ij}(x_0) = 1$$

2. NEURONS

   (a) A McCulloch Pits formal neuron (F.N.) is a linear boolean function of its inputs followed by a threshold and a delay.

   (b) A F.N. with interaction of afferents (F.N.A.) is a F.N. capable of computing any boolean function of the inputs.

   (c) A probabilistic F.N. is a F.N.A. where each input configuration has a probability $(0 \leq p \leq 1)$ of firing it.

3. THEOREMS. As mentioned before, there are two types of general theorems for the theory which, in general, are more difficult to state than to prove.

   (a) Analysis theorems (constructive)

      i. An arbitrary neural net, of N deterministic neurones and M external inputs with any type of feedback, is equivalent to a functional matrix of 2^Nx2^N, and therefore, to a deterministic automaton. There is a straight-forward procedure to go from the functions performed by each neuron to the expression of the functional matrix.

      ii. The same for probabilistic neural nets with feedback, with respect to probabilistic automata.

(b) Synthesis Theorems (constructive)

    i. Any arbitrary deterministic automaton of $2N$ states and M logical inputs lines is equivalent to a neural net of N neurons with feedback, and M external inputs. The neurons must have, in general, interaction of afferents. There is a straight forward procedure to obtain the net, given the functional matrix of the automaton. [Ref.9]

    ii. There is not, in general, a probabilistic neural net which is equivalent to an arbitrary probabilistic automaton.

There are three side effects of this explicit connection between formal neurons and automata, which were made apparent by the end of the sixties. First, there is a quick way to show that all possible modes of oscillation for a net of N neurons can be embodied and retrieved by/from a net of N neurons with feedback, though for N a little large, the connectivity of the net is out of hand. Obviously, a number of inputs lines, M, is required such that 2^M equals or superates Schnabels number, and the proof goes as follows:

- First, for each mode of oscillation, construct a transition matrix. States not involved in the oscillation are made to enter a transient ending in one state of the mode. Second, assign mutually exclusive input configurations to each transition matrix. Built the functional matrix and proceed to the neural net according theorem 3.b.i.
- The second side effect is that, since linear or non-linear shift registers for memory are particular cases of automata having relatively simple functional matrices, new effective and rapid ways for their synthesis were available.
- And third, it started some people to think why result of negative theorem 3.b.ii.

The fact of an axon-axonal interaction in real neurons at least at the level of neurons in the retina had became recognized by the early seventies, the interaction being much faster than that which involves the cell body.

As it happened with dendritic interaction in Blum's formulations, axon-axonal interaction could account for the "missing" layer which a minimal synthesis theorem required for arbitrary probabilistic automaton. First, it was recognized that the limitations came for considering that firing probabilities for each neuron were independent. They are not, in general. But, what type of dependence will produce the appropriate results more simply ?. It was found that some type of hierarchical dependence at the axonal level will solve the problem *(Moreno-diaz, Hernandez Guarch, 1983)*, so that neurons higher at the hierarchy, influence the rest, not viceversa. It seemed that McCulloch's program no. 1 was over. The equivalences are diagrammed in fig. 5.3. For each case the minimal theorems are constructive, that is, they provide for effective ways for performing the analysis and the synthesis. In fact, CAST systems for formal neural nets synthesis *(Suarez Araujo, Moreno-Diaz, 1989)* are under development.

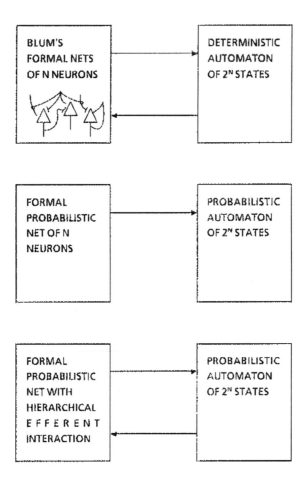

Figure 5.3 Diagram of Minimal Analysis and Synthesis Theorems for the clasical theory of Formal Neural Nets. Arrows to the right, signify analysis theorems. To the left, synthesis theorems

PART II. Towards Higher Level Representations

5.3   Theoretical frames for neuronal nets at higher levels

Since the times of McCulloch, it had became apparent that theorization on neural nets and the nervous system is "a priori" limited by the nature of the formal and conceptual tools used to describe processes. In fact, in his own group, two additional research efforts were carried out at the end of the sixties, which coexisted with the more formal and perhaps traditional described in the previous sections.

One such an effort tried to model the command and control capabilities of neurons at the reticular formation of vertebrates, *(Sutro, 1966)* which required that much more complex concepts of basic units or neurons should be used, such that it make no sense to try to reduce them to formal nets with afferent and efferents interactions. Also, the efforts to find an appropriate model for the behaviour of the group 2 Ganglion cells int the frog's retina, such as the finding of Maturana, Lettvin et al had shown *(Lettvin et al, 1959)*.

The situation had became clearer when trying to model higher parts in the nervous system, other than sensory levels *(Mira, Delgado, 1983)*, where a high level meaning has to be associated to the neuronal code. Several proposals have been made to treat ad hoc areas such a primary and secondary cortex and cerebellum.

It is obvious that the neuronal function is a really complex process and it is a priori required to define the level a which the formals tools being used apply and not to move too far from that level to reach conclusions. This questions have been clearly stated by Mira *(Mira, 1983)* within the general frame of an appropriate cybernetic methodology which fully applies when referring to neural nets. They treat to the problem of neural communication at the cortical level, which treatment requires a general model which will select the data pertinent to pragmatic aspects of neural coding. It seems that a neuron or a pool of them must at least send messages which have the nature of descriptions orders or questions. Each message must be broken down into components, which are assigned significances according semantic tables, related to the features of the descriptions and to their relationships. Were the neuron to be characterized by some type of operator, this operator should have the nature of composition rules for generating messages and to move from messages to interpretations. As it follows, the difficulties for "measuring" meaning in neural codes are enormous, but some wise guessing can be done on it. Even at the sensory level, Lettvin already found multiple meanings in the coded output of retinal ganglion cells. This would be some type of minimal model for a neuron in the primary or secondary sensory cortex (see fig. 5.4), which strongly contrasts with the naive interpretation, say, of visual cortex by some of the current accepted neurophysiology.

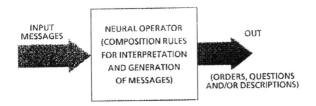

Figure 5.4   Minimal conceptual model for a neuron in primary cortex.

Certain types of frames, appropriate for "operating neurons" which mostly generate descriptives messages are the one proposed by Mira et al *(Mira, Moreno-Diaz, 1984)*, and which are illustrated in fig. 5.5.

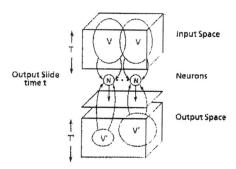

Figure 5.5   Generalized neuron.

It consists essentially in the generalization of the layered computation concepts, widely used when theorizing on visual functions and processing. As it is shown in the figure, there are previous definitions of input and output spaces as well as the nature of the operator for each neuron (N) to perform.

Within the concepts of layered computation, input and output spaces consist of stacks of "images", having in the illustration, two pure space variables plus time (vertical axis). For the input space, the stack consists of the inputs to every neuron back to a time T, which is the input memory. For the output space, the stack keeps all outputs from all neurons, up to a time T'. The output space is a type of "cooperative" memory since all the stack is available to all neurons to generate the present output ("slide" at time t). As illustrated in the figure, each neuron takes a decision based on the information contained in volumes V and V', to generate a "point" value in the output slide, which the next time will go down to the output stack.

Notice that there are properties which do not depend upon the nature of the decision rules or operators of the neurons of the layer. That is, if the decision rule is the same for all neurons, and volumes V and V' move to occupy the same relative position with respect to the neuron, there is spatial invariance and the operation corresponds to a "generalized convolution" *(Bolivar et al, 1989)*. Generalized convolutions have proved of usefulness in visual processing and, in general, in operations on data fields.

Time invariance requires constancy in time for volumes V and V' and constancy for the decision rule for the neuron.

A first selection for the decision rule level and tools is made by said authors, which permit a transparent illustration of the power of the frame, and will be commented later. We refer here to the representation frames where input and output spaces do not coincide with the physical extension (time- space), and to the general requirements for such type of models.

5.4   Neural Nets at representation spaces

To characterize social functional properties of neurons, such as cooperativity and collective decision making (such as in models for the reticular formation), a higher level language is necessary that those of the previous sections. In general, there seems to be a compromise between the level of symbolism at the input and output spaces and the level of symbolism of the operator or decision rule for the neuron, such that higher the one, the lower is the other. Some properties are common:

a) In this type of formulation, the input and output spaces do not have to have any physical interpretation, unless one goes down to the microscopic level. It is something similar to what happens in band theory in solid state physics, where it has no meaning to physically localize the band.

b) Concepts such as memory, cooperative interaction, decision and others have the character of theoretical constructs, not strictly localized nor can they be assigned to concrete physical structures. Said constructs are of use inasmuch as they provide for representation spaces, which in turn are valid as long as they "provide explanations" of the phenomena and serve to predict results in new "experimental" situations.

c) Physiological basal constructs are not so important as those which have evolutive significance.

d) Input and output spaces contain symbols, in a situation similar to that of taking decisions in artificial image identification processes.

A crucial point *(Mira, Delgado, 1986)* is how the representation input space is generated. As it has been proposed, sources of external or internal information, from where properties (which are independent) plus a fuzzy synthesis are used to generate an "input hypersurface" X[P1.P2... PN,t] where independent variables are those properties plus proper time, (see fig. 5.6). To keep in mind what is here intended for the neurons to operate on, we can resort to the typical situation of making a decision by a structured artificial visual system.

We finally refer to a recent high level neural model proposed also by Mira and Delgado *(Mira, Delgado, 1992)* aimed to clarify cooperative processes, within what can be considered McCulloch Program II. Here, higher concepts from the most sophisticated formal tools are used to cope with some formal properties of the neurvous tissue. In summary, each neuron is described in terms of a "local computation frame", with a series of slots (in the A.I. sense), which include input and output spaces volumes, meanings in the proper domain, linear programs (convolution like and delays), algorithms for dialogue and local consensus, acting and learning and control slots. The neural model seems to be a powerful tool to investigate neural reliability, from constructs such as "lesion", "redundancy", residual functions, reconfigurability transforms, as well as to attack the old, and still not satisfactorily solved, problem of the command and control system (reticular formation). The net is configured as an alternation of analytic and algorithmic layers.

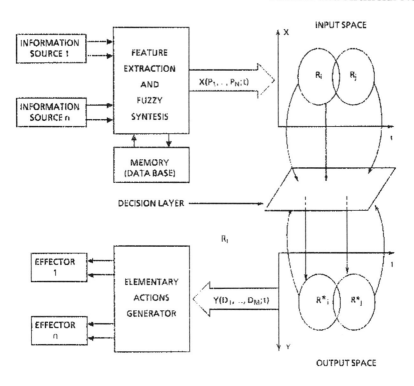

Figure 5.6 High Level Neuronal Operation (From Mira et al., 1983).

5.5   Global versus microscopic: Synthesis of what neural nets ?

McCulloch Program no. 2 was a much more "down to earth" and somehow a reversal
of Program 1. Program 2 is stated much more in line with the sciences of the Artificial
in that it goes from global to microscopic. Four main group of questions : What is
there in the functions of different parts of the nervous system? What is a "granular"
system that will perform the functions ? What are the functions of the grains ? Is it
there something like "the universal grain" ? As it can be seen, this is much closer to
an engineer approach than that of a standard neurophysiologist or neuroanatomist. In
fact, the first clear paper produced by his Neurophysiology Laboratory of MIT, with
Lettvin, Maturana and Pitts went in that direction *(Lettvin et al, 1959)*.

Soon after this type of questions were posed, it become clear that much powerful
tools to go from global to grains than that available were necessary, in line of the
new tools described in previous sections. Like a designer of a complex robotic system,
it became necessary to distinguish among subsystems much bigger that neurons or
pools of neurons, something that he called epistemological neurophysiology. From the

anatomical point of view, and from global neurophysiology the parts were there, but with very little knowledge about them, most of which were -and are- mere hypotheses.

McCullogh provided block diagrams of a kind of generalized nervous system, under the pressure of L. Sutro of the Instrumentation Laboratory of MIT, who at that time (1965), leadered a project for a Robotic Visual System. His block diagrams need no additional comment, but a number of brain theoretical questions that arise about structures, communications protocols, nature of messages and mostly how we can imagine machines that performs those vaguely stated functions. L. Sutro provided "engineer" versions where the same questions were still present.

The central issue is how sensing, perception, planning decision and action are , first, performed and thereafter integrated. The solution is not unique and could be broken down for specific tasks. But, obviously, we will like the system to do more, something like a generalized artificial robot to compare with the natural ones. Figure 5.7 shows ours very much simplified version of the complex architecture of McCulloch's and Sutro's proposal, but which allows for some descriptions in terms of recent tools from artificial perception, artificial intelligence and robotics.

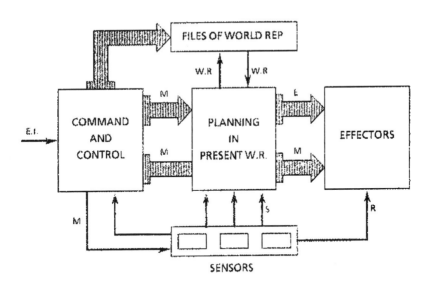

Figure 5.7 Simplified version of a block diagram employing engineering nomenclature (From Sutro et al., 1967).

First, the overall system presents a set of "modes of behaviour" that mimic the accepted model of behaviour of the vertebrates. The selection of a particular mode is performed by the command and control system, based mostly in present sensorial information (S.D.) and the status of the system. An external input (E.I.) is allowed from the external world (in practice, it cames from an operator's console) to modulate the selection of a mode.

Information concerning the selected mode (M) is sent to the sensors which are to be tuned to optimize the sensors data acquisitions pertinent to the mode of action. It is also sent to the component labelled Planning in Present World Representation. Finally, the selected mode commands and controls the process of stablishing goals according to the mode, the process of planning and the execution of the plan, by taking into account continuous highly processed sensory data (S). Actualized world representations are sent back through W.R. lines when the mode changes.

There are direct lines from sensors to effectors (line R) which are equivalent to reflexes. Line E provides for high level instructions to the effectors according to the plan of action, which are to be decoded into concrete motor-effector actions.

The basic function of a command and control system is to commit the whole system to one overall mode of behaviour belonging to a not very large set. This is what enables it to behave as a well integrated unit instead of a loose connection of separate sensors, effectors and processors. In this sense, a command and control computer is as it was pointed out before a close paradigm to the operation of the reticular formation in vertebrates. All modes of behaviour are mutually exclusive for such a computer. First, it receives relatively unprocessed information from all of the sensors situated in sensory and effector subsystems. Second, it gives signals which control, tune and set the filters of all external inputs. In McCulloch words, "this is the structure that decide what to look and having lokeed, what to heed". It also controls all the information flow from and to higher level computers.

From a structural point of view, the command computer must have a modular architecture, or, at least, it must simulate it. The basic idea is that a set of computing units (C.U.) it such that each computing module receives information only from a reduced part of the overall, little processed sensory inputs.

Each computing unit is capable of both general diagnostics about overall input situations and of specialized diagnostics according to the values of a concrete subset of the input lines.

A crucial point is that a consensus of the diagnostics -which corresponds to the selection of a single mode of behaviour- must be reached by the computing units in a relatively short time.This requires a very strong crosstalk among the computing units, which is a peculiar feature of the so called cooperative processors. There are two basic properties of the computing modules, that can be stated easily by means of the terminology common in expert systems.

In fact, we can look at the computing units as if they were simplified expert systems working on their own data bases and with their own inference engines on their specialized domain of sensory inputs. But they are capable also of giving up before the evidence in diagnostics by other units which show to have more relevant information for the case. This "giving up" must be understood in the sense of a recruiting of the rest of the modules by those having more confidence about their diagnostics. As it was stated by McCulloch, modules having the information pertinent to the case "cry louder", and doing so, they recruit the rest. The result of this strong crosstalk is that the system converges in one mode, in the sense that practically all the units decide the same mode of behaviour, though with perhaps different degree of confidence.

Modularity and division of expertise, with overlapping, among the computers units are the two basic features of a rules are the necessary addendum to achieve convergence. This architecture is supossed to provide for two main goals: first, to speed up the decision process by which a mode of behaviour is selected; second, the system is supposed to present high reliability, in such a way that it will arrive into an appropriate consensed mode, even when some of the expert units are destroyed.

This second aspect, that is, the reliability intrinsic to distributed expertise, precludes any decision based upon a single majority organ, because its malfunction will imply total inoperativity. That is, the conclusion that a consensus has been reached cannot be the output of any special testing unit receiving its inputs from the expert units. Instead, the decided modes must be appropiatly labelled according to their procedence to prevent mixing, and be present in a non-computing structure, that is, a set of wires -or axons- or, in other words, in a kind of decision bus.

From this, it becomes clear that reaching a rapid consensus in the mode of behaviour at the command and control computer is absolutely necessary for the rest of the system to operate coherently, because otherwise, the various higher and lowersubsystems to be controlled, will have a high probability to picking up operating instructions from the decision bus, which belong to different exclusive modes of behaviour, such that a kind of neurotic situation will be created.

For the world representation processes, both in files and the one presently used it is needed a multisensorial environment mapping prior or after transforms. There are two basic ways to structure multisensorial information which, in turn, admit different levels of representation, from geometric to highly symbolic, at least for artificial systems. These two ways have a correlate with representations, which tend to be optimal for discriminating among environmental patterns or representations to optimize the acquisition of clues for actions. These correspond to:

a) Integrated representation, both at low levels of acquisition and a high
   levels of processing sensory data.
b) Step by step representation, in which integration only occurs at high
   levels -that is in symbolic or language- structures.

In other words, and putting aside for the moment all natural systems, we may either represent the sensed environment by means of a multidimensional space where all sensory modalities are present with their own resolution at low level, while all high level processing is performed directly in this space. Or we can construct a high level intersensorial representation space by previously extracting properties, classifying and labelling each sensory modality separately. We finally make some comments on planning.

Planning takes place in a word representation scenery or stage, which correspond to the selected mode of behaviour. It is useful to consider that this may be performed at various levels of sensory data elaboration. The question of generating plans for complex problem solving as a matter of artificial intelligence techniques, suggests two practical-artificial problem domains that are of immediate use. First, there is the problem of moving solid shapes through a restricted world, and second, there is planning in blocks world.

The domain of the first problem is very close to that of navigating a robot by avoiding obstacles, and it is based on a scenery whose representation may be geometric

-like, having low symbolic content. The second problem domain may contain symbolic scenaries and the planning be performed at higher levels.

This suggests a first dichotomy in planning, that can be cascaded under command and control of the selected mode of behaviour.

Planning on the low level scenery can be formulated in quasi- geometric terms. The scenery itself is, in most of cases, determined by a set of n independent variables, defining a "space", plus a number of restricted zones. These restrictions are volumes V_i, with decision rules defining the membership to V_i for any point of the space. Objects are sets of point $P_1, P_2, \ldots, \ldots P_n$ not belonging to V_i which are subjected to K relations $R_k(P_i)$. Generating a plan consists essentially in finding trajectories for P_i, in agreement which restrictions and relations, under a "goal field".

At higher levels, planning can be stated in a similar way, by appropiatly changing the level of signification of the concepts, and subsequently defining the rules for handling sceneries, objects, goal fields and trajectories.

So much as a brief summary of so many questions and subquestions. Anyway, if shows the way to proceed according Program 2. Take a subsystems, define the most precisely its functions and try to find a cooperative reliable granular structure to perform said functions. Then one might be in the right to name the "grains" artificial neurons.

PART III. From Facts to Artificial Neurons

5.6   Neural Codes and Levels of Description

A crucial point in deciphering functions associated to neural structures is to determine what messages are sent by the neurons and how are they coded. Practically, the only way to find out microscopically what is going on in a cell is to record its electrical activity, which have been a central aim of neurophysiology since the 40's. Unfortunately, this gives good results only for the front and back ends of the system, that is, for sensor and motor subsystems, because the need to well define and specify the inputs to well interpret the outputs or viceversa. As we proceed along the nervous system, deeper layers of neurons receive information which code is unknown, and which carries a heavy load of semantics. Also, the degree of cooperativity and the subsequent degree of reliability increases. An analogue representation of the semantic and cooperative complexity of various nervous structures, as one proceeds from receptors to central areas to effectors is illustrated in figure 5.8. That is, as we move from the receptors, the degree of the semantic complexity of the input and output spaces for neural layers increases, and cooperativity is performed at higher symbolic levels.

Thus, conventional electroneurophysiology will provide for pertinent results when exploring zones like the ones in between E and C, or B and F. It will result without much sense for the rest of the zones, at least in what respects to brain theory (we do not talk about potential clinical usefulness). Intrusive stimulation shall provoke the most spectacular results in zones like D (cortex) or A, which should correspond to structure related to the reticular formation. Stimulation in C shall provoke illusions. In B (motor cortex) shall produce motor actions more or less uncoordinated, all

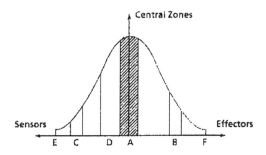

Figure 5.8   Analog representation of Semantic and Cooperative complexity of various nervous structures (From Mira et al., 1984).

this because the interpretation of the input space that each neural stage performs. Apart from microscopic questions, which are to be treated with physico-chemical tools all through the nervous system, global functional characterization will require more and more elaborated tools and models as one goes central. Conventional information coding, transmission and processing seems to be of applicability just for the sensorial and effector ends. Thus, for the retina it is accepted that there is some type of coding at the ganglion neurons axons of the type of pulse frequency, though a multiple coding (multiple meaning) and decoding along the axonal fibres have been found by Lettvin some twenty years ago.

5.6.1   Something on Artificial Neurons

In the early sixties there was a move to built hardware that simulated the membrane properties of a neuron, as well as its spatio-temporal integrative capabilities. Harmon *(Harmon, 1961)* was one of the first to build an Artificial Neuron and an ArtificialNeural Net that he used to duplicate some of the inner ear filtering functions. There was the time of the so called Perceptron Superstition. Also, in Spain, there were developed some artificial neuron models and nets *(Santesmases, Moreno-Diaz, 1965)* of a similar structure. A completely different approach came, the analytical-software approach, from the group of the MIT, to benefit from the experimental results on frog's eye in the classical papers of Lettvin et all, already mentioned.

   It is not intended here a revision of the many types of computing units that have appeared since and that have been named artificial neurons or artificial neural nets, but just a few illustration, so that it becames reasonable, within the idea of proceeding from facts about functions to systems, to admit a wide class of neural models as artificial, in the same sense that one talks about artificial vision, artificial intelligence or artificial behaviour.

5.7   Artificial Neurons for Vision and lateral implications

We are here interested in the interplay of retinal cells and artificial neurons. As it is well known, the retina has, as cortex, the structure of a layered computer, in the sense that it was expressed in Section 5.3 Thus, the arguments about the general frame in figure 5.5 are pertinent to retinal processing. In fact, said frame was historically developed from a generalization of the processes known to take place at the retina.

Layers at the retina are the photoreceptors, bipolar cells and ganglion cells, laterally interconnected by horizontal cells and amacrines. Recording are made mainly from ganglion cell axons. Ganglion cell total processing ranges from linear spatio-temporal contrast detection to highly specialized property detectors, which imply relatively local spatio-temporal non linearities followed for more global spatio-temporal operations. It has been reasoned in various theoretical papers *(Moreno-Diaz, Martin, 1984)* that specialized operations take place mainly at the inner plexiform layer which is the site of ganglionar dendrites and inner amacrines. Specialization goes up as we go down in the vertebrate phylum. That is, it is apparent that in the higher vertebrates, specialization, if any, takes place at higher levels, i.e., it is pushed back to cortex, with much more elaborate coding and operations.

Going back to the frame of figure 5.5, the nature of visual processing allows for a clear selection for the dimensions of the spaces, and the nature of input and outputs as well as operations by the neurons. Thus, we admit a certain quantity (a measure), $Y(x, y, t)$, as the output of neuron N at point xy at time t. Also, the input space, each point of the three dimensional "cube" as an input variable X. Now:

$$Y(x, y, t) = R_t [X, Y] \tag{5.4}$$

where X represents the values of the inputs quantities within volume V of present and past inputs slides; Y are the output quantities in volume V' of past output slides and R_t is a decision algorithm which assigns a value Y to the sets X, Y, which may vary as x, y, t do.

The main advantage of this representation is that it permits to visualize immediately, in terms of neuron-like elements, a large set of systems tools which range from probabilistic automata, digital filters to classical non-linear systems, as well as retinal cells behaviour and image processing. A list of illustrations can be found in *(Mira, Moreno-Diaz, 1984)*. We shall consider here two illustrations. First, let us consider one dimensional input spaceand not output space (pure parallel machine) as illustrated in figure 5.12.

The example can be extended easily to more dimensions and time. Now:

$$F_j = R_j [f_i] \tag{5.5}$$

where rule R_j applies to all $x_i \in R$ (the receptive field). Subscripts j and i go over the ordered natural numbers.

Then, we can visualize a generalized convolution as a transformation rule such that:

$$F_{j+k} = R_j\{f_{i+k}\} \, for \, all \, k, \, for \, all \, i \in R$$

Rule R_j is the same for all j so, it can be that of $j = 0$, that is:

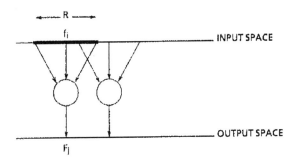

Figure 5.9   Illustration of an unidimensional input-output Spaces with no memories.

$$F_k = R_0\{f_{i+k}\} \, for \, all \, i \in R$$

For example, assume that the R_2 rule $(j = 2)$ is

$$R_2 \quad = \quad \{compute \, a = f_8 * f_{16}, \, if \, a > f_{24}, then \, F_2 = f_{24}$$
$$else F_2 = 0\}$$

Then

$$R_0 \quad = \quad \{compute \, a = f_6 * f_{14}, \, if \, a > f_{22}, then \, F_0 = f_{22}$$
$$else F_0 = 0\}$$

and

$$F_k = R_0\{f_{6+k}, \, f_{14+k}, \, f_{22+k}\} \tag{5.6}$$

It is immediate that if R_0 is a linear algebraic operation one has the conventional convolution. If R_0 is an analytical function of fi (actually a functional), a power expansion of R_0 gives

$$F_k = R_0(0,0,0) + \sum_i \left(\frac{\delta R_0}{\delta f_i}\right)_0 + \sum_i \sum_j \left(\frac{\delta^2 R_0}{\delta f_i f_j}\right)_0 f_i f_j + \cdots \tag{5.7}$$

where

$$W_i = \left(\frac{\delta R_0}{\delta f_i}\right)_0 f_i \, and \, W_{ij} = \left(\frac{\delta^2 R_0}{\delta f_i f_j}\right)_0$$

are the first, second, etc, order Kernels in a Wiener-Volterra expansion.

A very simple interesting case is for differential operators, that can be regarded as local convolutions. For example, the Laplacian operator (contrast detector, with centre periphery organization) is, in terms of convolution,

$$F_k = f_k - \frac{1}{2}[f_{k-1} + f_{k+1}] \tag{5.8}$$

As the order of the derivative increases, the "receptive field" also increases, so that the nth derivative will cover a receptive field of size $n + 1$. There is a simple rule to generate the coefficients that correspond to higher derivatives. For large n, the operation is not longer local.

A powerful algorithmic rule that have been used in visual processing corresponds to algorithmic lateral inhibition, in which the receptive field is divided in centre-periphery zones. Center zone is C and periphery is P. Process A takes place in C and process B in P, as follows

```
Select Max (C) = Cm
Select Max [Abs(P)] = Pm

Decision         if Cm > Pm
                 Then F(x,y) = Cm(x,y)
                 Else F8x,y)= -Pm
```

As it can be seen, the decision rule has the form, in general, of a program. More elaborated versions of the algorithmic lateral inhibition have been used to provide for convergence in cooperative decision systems.

As a second illustration, we refer to the Center-Periphery structured receptive fields found for a large family of retinal ganglion cells. The natural generalization of center-periphery leads to concentric quasi spherical volumes in the input spaces. The input space is now of four dimensions for each retina (two spatial plus time plus chromatic components). Processes in these volumes are as follows:

```
Compute a weighted average in C, Ac
Compute a weighted average in P, Ap

Compute |Ac-Ap|
```

Output $F(x, y, t)$ is a colour-coded (colour contrast sensitive) signal $S(x, y, t)$, obtained through a local non linearity from $F(xyt)$ (of the saturation type). Depending upon the relative position of centre and periphery, all types of ON, OFF, ON-OFF, symmetric and asymmetric spatial contrast, colour filtering or colour contrast coding are obtained.

For the cases of highly specialized retinal cells, to take into account the effect of the inner plexiform layer, centre and peripheral volumes are again divided into smaller volumes, which overlap, in each of which a process similar to the above is carried out. Then, the globalizing process cited before gives the final output of ganglion layer.

5.7.1   *Invariances by non linear transforms*

We consider now another example of the interplay between artificial processes and retinal nets. A typical problem in image preprocessing is to obtain cuasi invariances of the results when there are generalized changes of the illumination. One way to obtain such invariances is to high-pass filter the image, which gets ready of additive changes. Homotetic changes, however, are not superated, and some type of non-linear filter is required. A theorem by Munoz *(Munoz, 1987)* states that if T_1 and T_2 are two linear transforms, such that kernels K_1 and K_2 are

$$\int_R K_1 dx dy = 0 \; ; \int_R K_2 dx dy = 0$$

on the retina, the non linear transform T

$$T = \frac{T_1}{T_2}$$

provides for homotetic invariances.

The key non linearity is a divisive process. In terms of neural nets, divisive processes correspond to presynaptic inhibition. This suggests that such inhibitory mechanism might be used in a parallel computing structure to obtain said invariances. Such an structure is shown in figure 5.10. Input data are taken from the receptive fields by two different types of neurons, which perform something like transformations T1 and T2. The layer of presynaptic inhibition is a plexiform layer and the invariances are obtained after computational layer II, which may perform a simple linear transform.

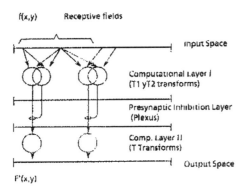

Figure 5.10   Parallel Computing Structure for intensity homotetic invariances.

This mechanism provides for simple explanation of neuronal adaptation in the retina. In this case, the input is the intensity of the image incident on the retina, or the outputs of the photoreceptors. Computational layer I corresponds to bipolars, that should be, at least, of two types and of which, there must be at least double than photoreceptors, as it happens in the fovea. Computational layer II, to the neuronal adaptation effects, is formed by the ganglionar cells. An interesting consequence is that, since the fovea is the only region in the retina where the number the bipolars is twice the photoreceptors it follows that neuronal light adaptation can only happens, in higher vertebrates, for the foveal regions.

5.8   More about automata and neural nets from auditive invariances

By developing alternative nets, according the anatomy,to detect auditory invariants (such as chords) where the cuasi- beginning and the cuasi-end of the chord are logical prerequisites for homotetical spatial invariance, a new formulation of automata and generalized neural nets in terms of alternating layers of analog and program-like decision layers was obtained *(Moreno-Diaz, 1989)*.

The idea is that there must be to paths for invariances. One computes the invariant parameters and the other applies it. That is, compute the "size" of the pattern and normalize it, in parallel.

We shall not go into details of how this is achieved in neural nets, but the important point is that a neuronal structure having layers of linear computation (LC) and local conditional branching (if then else type) LCB, in a double sequence, that is

$$LC \rightarrow LCB \rightarrow LC \rightarrow LCB$$

was the logical consequence of the prerequisites.

The important point is that any arbitrary automaton permits a decomposition of this type. These are the so called "Pinoso Theorems" because there were worked out at Pinoso, Alicante. They have been subsequently used in auditory and visual neuronal processing by Suarez-Araujo *(Suarez Araujo, 1990)*.

5.8.1   The "Pinoso Theorems"

Since neural formal nets (Generalized McCulloch) and automata are the same, we state the theorems for neural nets.

T1. For a net of N generalized neurons, with afferent interaction, without feedback there is a $LC \rightarrow LCB \rightarrow LC \rightarrow LCB$ structure that duplicates it. For one dimension, in the continuum, the function of neuron at site y, is, for a receptive field inputting $I(x)$, given through these four computational layers:

$$
\begin{aligned}
ANALOG.\,LAYER\,M(Z) &= \int_x W(x,z)I(x)dx \\
ALGOR.\,LAYER\,N(Z) &= \mu\,[M(Z) - \theta(Z)] \\
ANALOG.\,LAYER\,P(y) &= \int W(y,z)N(z)dz \\
ALGOR.\,LAYER\,Y(y) &= \mu\,[P(y) - 1]
\end{aligned}
$$

where μ is the step function.

T2. For a generalized neural net with feedback there is also a

$$LC \rightarrow LCB \rightarrow LC \rightarrow LCB$$

structure that duplicates it, but taking into account time effects. The formulation ends as follows

$$
\begin{aligned}
ANALOG.\ M(z,t) &= \int_x I(x,t)W(x,z)dx + \int_y O(y,t)W(y,z)dy \\
ALGOR.\ N(z,t) &= \mu\left[M(y,z) - \theta(z)\right] \\
ANALOG.\ P(z,t) &= \int N(z,t)W(y,z)dz \\
ALGOR.\ Y(z,t) &= \mu\left[P(z,t) - 1\right] \\
where\ O(y,t) &= \int_0^t K(y,t)Y(y,t)dt
\end{aligned}
$$

and K is the Kernel for the temporal axonic effect of the "neurons" (perhaps a simple delay).

T3. Since any arbitrary automata is equivalent to a neural net with feedback plus a neural net without feedback, any automata is equivalent to a cascade of two layered machines of structure a given by T2 and T1. These theorems are constructive, i.e. provide for effectives ways to go from one to another.

Figure 5.11 illustrates the layered equivalence of T1 and figure 5.12 indicates the equivalences developed by T1, T2 and T3.

For probabilistic or fuzzy machines, everything applies "mutatis mutandis".

5.9   On receptive fields and functions

Persons in picture processing worry about sizes of receptive fields and functions that should be performed on them by - if you wish - neuron-like elements. As well as neurophysiologists worry about what natural neurons receptive fields are and what they are doing.

In non structured artificial vision systems, what scientists and engineers seek is for kinds of "complete" transforms which provide for alternative descriptions of images that can be "truncated" for the purposes, so that a much low number of degrees of freedom are to be handle. (Candela, 1987) and Bolivar, 1989) after applying a whole set of standard transformations to pictures to obtain descriptors were aware that the search for orthogonality in the descriptors functions was really a handicap inherited from the times that global transforms (like Fourier, Hadamard, Haar) had to be made almost by hand. But with modern computer techniques it was not clear that there was any advantage except for the insurance of independence, that is, insuring that the corresponding transform results were non-redundant. They looked then for "complete" transforms, not necessarily orthonormal, but independent, so that the inverse transform will exist.

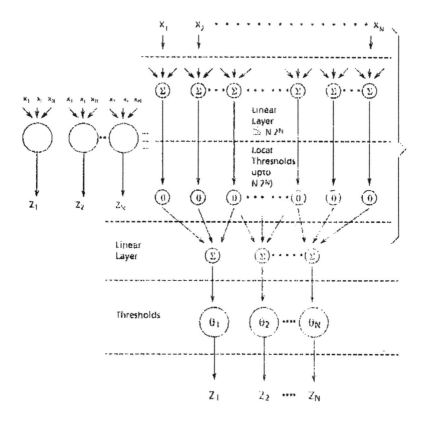

Figure 5.11   Layered computing structure to illustrate T1.

The first interesting result is that the position of the receptive field matter as much as the function you perform on them. Partitions means that one is computing on the sensory data in parallel, though with overlapping.

The concept of Progressive Resolution Transforms was introduced, in which it was realized that no matter what function it is performed, a vast class of partitions could be complete. That is, given an image (a "data field" in fact, because the image fills a data field in memory), then one can, and it is convenient, to separate receptive fields from functions, so that a data field partition can be, by itself, complete.

A particular class of partitions are the so called "foveal transforms". It is a progressive resolution transform (PRT) which provides for "maximum resolution" at a central or foveal area. This might be misleading, since the foveal PRT is a "complete" transform, that is, given time, all and each of the original pixels intensities could be recovered. But since we are after saving by truncation, that "given time" might be too

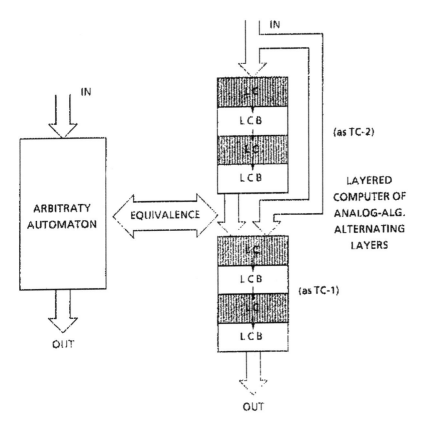

Figure 5.12   Illustration of "Pinoso Theorems".

long or to expensive for memory. The PRT fovea idea starts from the fact that for a "complete" partition of a data field, one can play with the initial number of degrees of freedom, to select, at will, receptive fields which are independent, not orthogonal. The dimension of a receptive field is the number of pixels it contains. Now, select a dimension d for a receptive field. The following theorem holds:

TA. For a data field of N degrees of freedom (pixels), given a dimension $d < N$ for receptive fields of computing units, there exist a complete PRT, which is formed by all independent receptive fields of dimension d plus $d - 1$ receptive fields which must be unitary (one pixel each) different, but otherwise arbitrary.

Now, those d-1, since they are arbitrary, can be chosen to be at the center of fovea of the retina (data field), which seems to be the case. But again, to that part, the theorem applies if needed for coding.

The important point is that, next, one asks what function is to be performed on each receptive field. It does not matter. Whatever function or functions are performed do not affect, so the simplest is to compute an average or perhaps a weighted average. Function and receptive fields are separated and independent questions. Figure 5.13 is an illustration of PRT foveal transform.

One nice thing is that, given its modular character, PRT's admit fast transforms. The equivalent neural net like equivalences, for parallel computation, are also almost immediate to find.

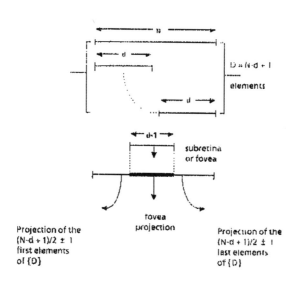

Figure 5.13   Illustration of a PRT foveal transform.

The next logical problem was one of trade off, which is probably what happens in the retina, since the resolution of fibers is not necessarily preserved, and there is a "multiple meaning" coding, which in this terminology will mean that each fibre of the optic nerve sends information pertaining to more that one "operation" on the receptive field.

There is a second theorem for that. It goes as follows TB. For a data field of N degrees of freedom, and an arbitrary partition giving L independent receptive fields such that $M = \frac{N}{L}$ is an integer, then, the computation of M functional coefficients which are linearly independent in each partition provides for a complete description of the data field.

Nature does not follow theorems exactly, but theorems provide for clear ways to approach nature. On any case, the parallel computing structures which result, are both illuminating to understand natural systems and to build artificial machines.

REFERENCES

Blum M. (1961) Properties of a neuron with many inputs. In *Von Foester H. , Gopf G. (eds.) Principles of Self Organization. NY. Pergamon*

Bolivar O. (1989) hacia la teoria de las transformaciones de campos receptivos y campos de datos. *Tesis Doctoral. Univ. Las Palmas de G. Canaria.*

Bolivar O., Candela S., Munoz. J.A. (1989) Non-linear data transforms in perceptual Systems. In *Computer Aided Systems Theory EUROCAST'89 Lecture Notes in Comp. Sciences. Pichler, F., Moreno-Diaz, R. eds. Springer*, pp. 301–309.

Candela S. (1987) Transformaciones de Campo Receptivo Variable en Proceso de Imagenes y Vision Artificial. *Tesis Doctoral. Univ. Las Palmas de G. Canaria.*

da Fonseca J.S. (1967) Synthesis and linearization of feedback shift register: basis for a model of memory. *QPR n86, Research Laboratory of Electronics, MIT, Cambridge Mass*, pp. 355–366.

da Fonseca J.S., Mira, J. (1969) Signification and Intension. In *Univ. of Lisbon Pub. Fac. of Medicine. Lisbon.*

Harmon L.D. (1961) Artificial Neurons. *Bell Teleph. System Monog. 4134, I.*

Lettvin J.Y., Maturana H.R., McCulloch W.S., Pitts W.H. (1959) What the frog's eye tells the frog's brain. *Proc. IRE*, 47: pp. 400–415.

McCulloch W.S., Pitts W.H. (1943) A logical calculus of the ideas inmanent in nervous activity. In *Bull. Math. Biophys*, 9: 127–247.

McCulloch W.S., Papert S.A., Blum M., da Fonseca J.S., Moreno-Diaz R. (1969) The fun of failures. *New York Academy of Sciences*, 156, 2.

Mira J., Delgado A. (1983) On cooperative languages for survival and neural decision making. *Proc. Int. Symp. on Man in High Tech. Env. Nammer*, pp. 569–581.

Mira J., Moreno-Diaz R. (1984) Un marco teorico para interpretar la funcion neuronal a altos niveles. In *Biocibernetica. Moreno- Diaz R., Mira. J. eds. Siglo XXI. Madrid*, pp. 149–171.

Mira J., Delgado A. (1986) A logical model od cooperative processes in cerebral dynamics. *Cybernetics and Systems'86. Reidel*, pp. 311–318.

Mira J. (1989) System Behaviour and Computing Structure. In *Computer Aided System Theory, EUROCAST 89. Lecture Notes in Comp. Sciences. Pichler, F., Moreno-Diaz, R. eds. Springer*, pp. 267–283.

Mira J., Delgado A (1991) A linear and algorithmic formulation of cooperative computation in neuronal nets. In *EUROCAST'91. Springer in press.*

Monroy J. (1981) Redes Neuronales: una reformulacion. *Tesis Lic. Facultad de Ciencias. UNAM. Mexico DF, Mexico.*

Moreno-Diaz R., McCulloch, W.S. (1969) Circularities in nets and the concept of functional matrices. In *Byocibernetics of the Nervous System. pp. 145–150. Proctor, L. ed. Little and Brown, MA.*

Moreno-Diaz R. (1971) Deterministic and probabilistic neural nets with loops. *Math. Biosciences*, 11: 129–136.

Moreno-Diaz R., Hernandez Guarch F. (1983) On the biological formal counterparts of logical machines. *Kybernetes*, 12: 183–185.

Moreno-Diaz R., Martin F. (1984) Concepts for a quantitative theory of retinal processing. In *Cyb. and Systems AFCET 84. Paris*, pp. 215–222, 1984.

GENETIC ALGORITHMS in ENGINEERING and COMPUTER SCIENCE
Editor J. Periaux, G. Winter   ©1995 John Wiley & Sons Ltd.

Moreno-Diaz R., Mira J., Suarez-Araujo C., Delgado A. (1989) Neuronal Net to Compute homothetic auditive invariances. *Proc. Medit. Conf. on Medical and Biol. Eng.*

Munoz J. A. (1987) Jerarquizacion de estructuras de nivel bajo y medio para reconocimiento visual. *Tesis Doctoral. Univ. Las Palmas de Gran Canaria.*

Santesmases J.G., Moreno-Diaz R. (1965) Neuronas artificiales para la simulacion del reflejo condicionado.*Ann. Real Soc. Esp. de Fis y Quim.*, LXI(A) 3-12.

Schnabel C.P.J. (1966) Number of modes of oscillation of a net of N neurons.*Q.P.R. n 80. Research Laboratory of Electronics, MIT, Cambridge Mass*, pp. 280–285.

Suarez Araujo CP., Moreno-Diaz Jr. R. (1989) Bases of a CAST System for formal neural nets. In *Computer Aided System Theory EUROCAST'98. Lecture Notes in Comp. Science. Pichler, F; Moreno- Diaz, R. ed. Springer*, pp. 232–242.

Suarez-Araujo C.P. (1990) Contribuciones a la integracion Multisensorial y Computacion Neuronal Paralela.*Tesis Doctoral. Univ. de Las Palmas de Gran Canaria.*

Sutro L. (1966) Sensory Decision and Control Systems. *Instrumentation Laboratory R-548. MIT, Candbridge MA.*

6

Evolution Strategies I: Variants and their computational implementation

Thomas Bäck[1] and Hans-Paul Schwefel[2]

6.1 INTRODUCTION

The first evolution strategy (ES) was developed in 1964 at the Technical University of Berlin (TUB) by Rechenberg and Schwefel as an experimental optimization technique. First applications of the strategy dealt with problems like driving a flexible pipe bending or changeable nozzle contour into a shape with minimal loss of energy [KS70]. This early variant of an evolution strategy, the so-called (1+1)-ES, works on the basis of two individuals only, i.e., one parent and one descendant per generation. The descendant is created by applying binomially distributed variations (with expectation zero and variance σ^2) to the parent, and either the descendant becomes parent of the next generation, if it is better than its parent, or the parent "survives." The (1+1)-ES was soon used also for continuous decision variables (with normally distributed variations), the latter presently being the major application domain. Rechenberg [Rec73] developed a theory of convergence velocity for the (1+1)-ES, and derived a rule for deterministically adapting the standard deviation of mutations according to the measured success frequency of mutations. According to an average of the optimal success probabilities for two simple model functions (corridor model and sphere model), the rule was called 1/5-success rule.

[1] Informatik Centrum Dortmund, Joseph-von-Fraunhofer-Str. 20, 44227 Dortmund, Germany. E-mail: baeck@ls11.informatik.uni-dortmund.de
[2] University of Dortmund, Department of Computer Science, 44221 Dortmund, Germany. E-mail: schwefel@ls11.informatik.uni-dortmund.de

Soon, the (1+1)-ES was substituted, especially in computer applications, by variants with $\mu > 1$ parents and $\lambda > 1$ descendants per generation. Nevertheless, for historical and completeness reasons, the (1+1)-ES is discussed in section 6.2 of this overview.

The first multimembered evolution strategy, the $(\mu+1)$-ES with $\mu > 1$, already introduced the concept of recombination to form one offspring individual from a mix of parental attributes. After mutation and fitness evaluation, the offspring replaces the worst parent individual if it performs better than that. Although it was never widely used, this strategy provided the basic idea to facilitate the transition to the modern $(\mu \overset{+}{,} \lambda)$-strategies, that represent the state-of-the-art [Rec94, Sch95]. In case of a $(\mu+\lambda)$-strategy, μ parents create λ descendants by recombination and mutation, and the μ best individuals out of parents plus descendants are deterministically selected to become parents of the new generation. In case of a (μ,λ)-strategy, the μ best individuals out of the descendants only form the new parent set; in this case, $\lambda \geq \mu$ is required, otherwise $\lambda \geq 1$ is sufficient.

Besides a population of individuals and recombination of parental information, the (μ,λ)-ES substitutes the deterministic step-size control by a self-adaptation process that learns step sizes (and, optionally, covariances) on-line during the evolutionary search process [HB92, Sch95]. The $(\mu \overset{+}{,} \lambda)$-strategy is presented in detail in section 6.4 of this paper.

Ongoing research efforts led to the development of a variety of special variants of evolution strategies, e.g. for exploiting parallel hardware architectures or for solving multiple criteria decision making problems. A brief overview of these variants is given in section 6.5.

Finally, we conclude by outlining a proposed generalized evolution strategy which allows for a gradual transition between $(\mu+\lambda)$-selection and (μ,λ)-selection by introducing a maximal life span κ of the individuals. The contemporary $(\mu, \kappa, \lambda, \rho)$-strategy also incorporates some other features which either theoretically or experimentally proved to be useful extensions of the algorithm [SR95].

6.1.1 Notation

Throughout this paper, we assume an n-dimensional, continuous parameter optimization problem of the form

$$f^* := f(\vec{x}^*) = \min\{f(\vec{x}) \mid \vec{x} \in M \subseteq R^n\} \quad , \tag{6.1}$$

where $M = \{\vec{x} \in R^n \mid g_j \geq 0 \ \forall j \in \{1,\ldots,q\}\}$ is the set of feasible points and $g_j : R^n \to R$ are inequality constraints. Individuals $\vec{a} \in I$ are elements of an individual space $I = R^n \times S$, where the set S of strategy parameters depends on the kind of evolution strategy. $P^{(t)} = \{\vec{a}_1,\ldots,\vec{a}_k\} \in I^k$ denotes a population[1] of $k \in \{\mu,\lambda\}$ individuals at generation t, where a population is interpreted as a multiset of elements from I (duplicates might occur in the population). $\mu, \lambda \in N$ denote the number of parents and of descendants, respectively. The genetic operators are denoted by

---

[1] Because the cardinality of all sets is known in advance and elements are often considered in a specific order (by objective function values) and need to be indexed, we take the freedom to interchange vector and set notation by writing $\{\vec{a}_1,\ldots,\vec{a}_k\} \in I^k$ (rather than $\subset 2^I$).

mappings

$$\begin{array}{rcll}
\textbf{rec} & : & I^\mu \to I & \text{recombination} \\
\textbf{mut} & : & I \to I & \text{mutation} \\
\textbf{sel}_\mu^k & : & I^k \to I^\mu & \text{selection}, k \in \{\lambda, \mu + \lambda\}
\end{array} \tag{6.2}$$

which also depend on additional, operator-specific parameters such as step length variabilities $\tau, \tau_0, \beta \in R_+$ for mutation and recombination type $\omega \in \{0, 1, 2, 3\}$.

A single iteration of the evolution strategy, i.e., a step from a population $P^{(t)}$ towards the next parent population $P^{(t+1)}$ is modeled by the mapping

$$opt_{ES} : I^\mu \to I^\mu \tag{6.3}$$

according to

$$opt_{ES}(P^{(t)}) = \textbf{sel}_\mu^k(\sqcup_{i=1}^\lambda \{\textbf{mut}(\textbf{rec}(P^{(t)}))\} \sqcup Q) \tag{6.4}$$

where $Q \in \{P^{(t)}, \emptyset\}$ depends on the selection operator $((\mu, \lambda)$-selection: $Q = \emptyset$, $k = \lambda$; $(\mu + \lambda)$-selection: $Q = P^{(t)}$, $k = \mu + \lambda)$, $\sqcup$ denotes the union operator on multisets, and $\mu = 1$ is supposed to imply no recombination, i.e., $\textbf{rec} = \textbf{id}$.

The notation $z \sim N(\zeta, \sigma^2)$ denotes a realization of a normally distributed random variable with expectation ζ and variance σ^2, and $u \sim U(\cdot)$ is a realization of a uniformly distributed random variable (the argument is either a continuous interval or a finite set of values). In contrast to uniformly distributed random numbers, most programming language libraries do not provide a function to generate normally distributed random numbers. From a pair $u_1, u_2 \sim U((0, 1])$ of uniformly distributed random numbers, however, one can easily generate two independent, normally distributed random numbers with expectation zero and variance one according to (see [Sch95], pp. 115-116):

$$z_1 = \sqrt{-2 \ln u_1} \sin(2\pi u_2) \quad , \quad z_2 = \sqrt{-2 \ln u_1} \cos(2\pi u_2) \quad .$$

$z_1, z_2 \sim N(0, 1)$ can be transformed by $z_i' = \sigma_i z_i$ to obtain random numbers distributed according to $N(0, \sigma_i^2)$.

6.2   THE $(1 + 1)$-EVOLUTION STRATEGY

As outlined in the introduction, the (1+1)-ES is characterized by $\mu = \lambda = 1$, the absence of recombination, and a deterministic rule for the modification of the (one and only) step size σ for all mutations. An individual $\vec{a} = (\vec{x}, \sigma) \in R^n \times R_+$ consists of the object variable vector $\vec{x}$ and one standard deviation[2] σ. The mutation operator is a composition of the (deterministic) modification[3] $\textbf{mu}_\sigma$ of σ and the mutation $\textbf{mu}_x$ of $\vec{x}$:

$$\textbf{mut} = \textbf{mu}_x \circ \textbf{mu}_\sigma \quad , \tag{6.5}$$

---

[2] We would like to add the remark that different step sizes $\sigma_i = \sigma \cdot s_i^{(0)}$ have already been used in the (1+1)-ES also, but the relations of all σ_i were fixed by the initial settings $s_i^{(0)}$ and only the common factor σ was adapted.

[3] As a notational convention, we drop the use of all projections of vectors to yield single components and simply define the notation

$$f_\sigma(\vec{x}, \sigma) = (\vec{x}, f_\sigma(\sigma)) \quad , \quad f_x(\vec{x}, \sigma) = (f_x(\vec{x}), \sigma) \quad ,$$

where

$$\tilde{\sigma} := \mathbf{mu}_\sigma(\sigma) = \begin{cases} \sigma/\sqrt[n]{c} & , \quad \text{if } p > 1/5 \\ \sigma \cdot \sqrt[n]{c} & , \quad \text{if } p < 1/5 \\ \sigma & , \quad \text{if } p = 1/5 \end{cases} \tag{6.6}$$

This definition reflects the 1/5-success rule after Rechenberg by updating the standard deviation σ at each generation[4], based on the measured relative frequency p of successful mutations. A choice of $c = 0.817$ was theoretically derived by Schwefel for the sphere model (see [Sch95], pp. 110–112).

The mutation of object variables proceeds by adding normally distributed variations with standard deviation $\tilde{\sigma}$ to the components of $\vec{x}$:

$$\tilde{x} := \mathbf{mu}_x(\vec{x}) = (x_1 + z_1, \ldots, x_n + z_n) \tag{6.7}$$

where[5] $z_i \sim \mathbf{N}_i(0, \tilde{\sigma}^2)$. Consequently, following (6.5) one obtains $\mathbf{mut}((\vec{x}, \sigma)) = \mathbf{mu}_x(\mathbf{mu}_\sigma(\vec{x}, \sigma)) = \mathbf{mu}_x(\vec{x}, \mathbf{mu}_\sigma(\sigma)) = \mathbf{mu}_x(\vec{x}, \tilde{\sigma}) = (\mathbf{mu}_x(\vec{x}), \tilde{\sigma}) = (\tilde{\vec{x}}, \tilde{\sigma})$.

The offspring individual $\tilde{a} = (\tilde{\vec{x}}, \tilde{\sigma})$ and the parent $\vec{a} = (\vec{x}, \sigma)$ are both involved in the selection operator $\mathbf{sel}_1^2 : I^2 \to I$, which yields a surviving individual according to an objective function value comparison of $\vec{a}$ and $\tilde{a}$:

$$\mathbf{sel}_1^2(\{\vec{a}, \tilde{a}\}) = \begin{cases} \{\tilde{a}\} & , \quad \text{if } f(\tilde{x}) \le f(\vec{x}) \\ \{a\} & , \quad \text{otherwise} \end{cases} \tag{6.8}$$

Using these mappings, the main loop of a (1+1)-ES reduces to

$$opt_{(1+1)-ES}(\{\vec{a}\}) = \mathbf{sel}_1^2(\{\mathbf{mut}(\vec{a})\} \sqcup \{\vec{a}\}) \quad , \tag{6.9}$$

and a pseudocode description of the algorithm is given below:

ALGORITHM 1   ((1 + 1)-ES)

> $t := 0;$
> *initialize* $P^{(t)} = \{(\vec{x}, \sigma)\};$
> *evaluate* $f(\vec{x});$
> **while** $(T(P^{(t)}) = 0)$ **do**   { T denotes a termination criterion }
> $(\tilde{\vec{x}}, \tilde{\sigma}) := \mathbf{mut}((\vec{x}, \sigma));$
>   *evaluate* $f(\tilde{\vec{x}});$ { determine objective function value }
>   **if** $(f(\tilde{\vec{x}}) \le f(\vec{x}))$ { select }
>   **then** $P^{(t+1)} := \{(\tilde{\vec{x}}, \tilde{\sigma})\};$
>   **else** $P^{(t+1)} := P^{(t)};$
> $t := t + 1;$
> **od**

and

$$(f_x \times g_\sigma)(\vec{x}, \sigma) = (f_x(\vec{x}), g_\sigma(\sigma)) \quad .$$

---

[4] An alternative consists in updating σ each n-th generation, using the constant c rather than $\sqrt[n]{c}$ as the update factor [Sch95].

[5] Assuming that different initial step sizes $s_i^{(0)}$ were defined, this has to be changed to $z_i \sim \mathbf{N}_i(0, (\tilde{\sigma} \cdot s_I^{(0)})^2)$.

Notice that the relative frequency p of successful mutations can easily be calculated by increasing a counter variable whenever selection chooses the offspring individual rather than the parent (i.e., a successful mutation has occurred). Schwefel formulates the corresponding algorithmic rule as follows (see [Sch95], p. 112):

> After every n mutations, check how many successes have occurred over the preceding $10 \cdot n$ mutations. If this number is less than $2 \cdot n$, multiply the step length by the factor $c = 0.85$; divide it by 0.85 if more than $2 \cdot n$ successes occurred.

The constant c was corrected by Schwefel in order to reflect the fact that the sphere model, for which the results were derived, is likely to require the fastest step size adaptation.

The (1+1)-ES is a kind of stochastic gradient technique, and the similarities to simulated annealing have been clarified by Rudolph [Rud93]. Though the algorithm is often useful, it is basically a local search strategy, and the 1/5 success rule may cause premature stagnation of the search due to the deterministic decrease of the step size whenever the topological situation does not lead to a sufficiently large success rate. One attempt to circumvent this problem, the $(\mu+1)$-strategy, introduced the population concept on the parent level and therefore facilitated the development of the $(\mu \overset{+}{,} \lambda)$-strategies.

6.3   THE $(\mu + 1)$-EVOLUTION STRATEGY

Besides the (1+1)-ES, Rechenberg also proposed the first multimembered evolution strategy, a $(\mu+1)$-ES (see [Rec73], chapter 9) with $\mu > 1$. The strategy was only outlined by Rechenberg, and he did not clarify how the modification of σ might work in this algorithm: Self-adaptation as used in the (μ,λ)-strategy definitely does not work in a $(\mu+1)$-strategy, and it is also not clear how the 1/5-success rule might be applied in the $(\mu+1)$-case, because the theoretical derivation only holds for the single parent strategy. Nevertheless, it is worthwhile to describe the algorithm here because it allows to introduce recombination in a straightforward way.

The recombination operator $\mathbf{rec} : I^\mu \to I$ is applied before mutation to create a single individual from the parent population. The recombined individual undergoes mutation, and the resulting offspring individual substitutes the worst individual of the parent population if it performs at least as well as the worst parent (elimination of the worst). In other words, the μ best individuals out of μ parents and one offspring are deterministically selected as the next parent population. This kind of selection strategy which substitutes (at most) one individual per iteration of the evolutionary loop (elimination of the worst instead of survival of the fittest) is termed *steady-state selection* in genetic algorithm research [Whi89].

In contrast to genetic algorithms, the recombination operator $\mathbf{rec} : I^\mu \to I$ creates just one individual per application. It works by first choosing ϱ $(1 \leq \varrho \leq \mu)$ parent vectors from $P^{(t)} \in I^\mu$ with uniform probability, and then mixing characters from the ϱ parents to create one offspring vector:

$$\mathbf{rec} = \mathbf{re} \circ \mathbf{co} \quad , \tag{6.10}$$

where $\mathbf{co} : I^\mu \rightarrow I^\varrho$ chooses ϱ parent vectors and $\mathbf{re} : I^\varrho \rightarrow I$ creates one offspring vector. The cases $\varrho = 2$ of bisexual recombination as well as $\varrho = \mu$ of global recombination are commonly applied, but other settings of ϱ are equally possible[6].

Depending on the recombination type ω, a variety of ways exist to recombine the parental vectors in order to create an offspring. The recombination type may differ for the various parts of a complete individual $\vec{a} = (\vec{x}, \vec{\sigma}, \vec{\alpha})$ of the $(\mu\overset{+}{,}\lambda)$-ES (see section 6.4 for the meaning of the components $\vec{\sigma}$ and $\vec{\alpha}$), such that we define the recombination types by referring to arbitrary vectors $\vec{b}$ and $\vec{b}'$, where $\vec{b}'$ denotes the part of an offspring vector to be generated and $b_{k,i}$ denotes the i-th component of a preselected individual number $k \in \{1, \ldots, \varrho\}$ out of the set of individuals chosen by co. The original $(\mu+1)$-ES applies recombination to the object variable vector $\vec{x}$ only.

The most commonly used recombination types are:

- $\omega = 0$: No recombination (holds always when $\mu = 1$ or $\rho = 1$). In this case, only a random choice $\mathbf{co} : I^\mu \rightarrow I$ of a single individual is performed, and $\mathbf{re} = \mathbf{id} : I \rightarrow I$ is just the identity mapping.
- $\omega = 1$: Global intermediary recombination, where the i-th vector component is averaged over all parents to obtain the corresponding offspring component value:

$$b'_i = \frac{1}{\varrho} \sum_{k=1}^{\varrho} b_{k,i} \tag{6.11}$$

- $\omega = 2$: Local intermediary recombination, which works by selecting two out of the ϱ parents for each vector component and calculating a weighted sum of the corresponding components of the two parents:

$$b'_i = u_i b_{k_1,i} + (1 - u_i) b_{k_2,i} \quad , \tag{6.12}$$

where $u_i \sim \mathbf{U}([0,1])$ or $u_i = 1/2$, and $k_1, k_2 \sim \mathbf{U}(\{1, \ldots, \varrho\})$ for each offspring.

- $\omega = 3$: Discrete recombination, where each vector component is copied from the corresponding component of an individual randomly chosen from the ϱ parents:

$$b'_i = b_{k_i,i} \tag{6.13}$$

where $k_i \sim \mathbf{U}(\{1, \ldots, \varrho\})$ at random for each i.

Of course, other recombination operators such as multiple point crossover as commonly used in genetic algorithms are also possible and could be added to the set of available operators.

Following equation (6.4), the main loop of the $(\mu+1)$-ES is formulated as follows:

$$opt_{(\mu+1)-ES}(P^{(t)}) = \mathbf{sel}_\mu^{\mu+1}(\{\mathbf{mut}(\mathbf{rec}(P^{(t)}))\} \sqcup P^{(t)}) \quad . \tag{6.14}$$

The selection operator simply yields the set of the μ best individuals of its argument, i.e., $\mathbf{sel}_\mu^k(P) = \tilde{P}$, where $|\tilde{P}| = \mu$, $|P| = k \geq \mu$, and

$$\forall \tilde{\vec{a}} \in \tilde{P} : \not\exists \vec{a} \in P - \tilde{P} : \quad f(\vec{x}) \leq f(\tilde{\vec{x}}) \quad . \tag{6.15}$$

---

[6] Recently, this concept has also been tested successfully in the context of genetic algorithm research [ERR95].

Algorithm 2 presents a pseudocode description of the $(\mu+1)$-ES as defined here formally.

ALGORITHM 2 $((\mu+1)$-ES)

> $t := 0$;
> $initialize\ P^{(0)} = \{\vec{x}_1, \ldots, \vec{x}_\mu\} \in I^\mu$;
> $evaluate\ f(\vec{x}_1), \ldots, f(\vec{x}_\mu)$;
> **while** $(T(P^{(t)}) = 0)$ **do**
>> $\tilde{\vec{x}} := \mathbf{mut}(\mathbf{rec}(P^{(t)}))$;
>> $evaluate\ f(\tilde{\vec{x}})$;
>> $P^{(t+1)} := \mathbf{sel}_\mu^{\mu+1}(\{\tilde{\vec{x}}\} \sqcup P^{(t)})$;
>> $t := t + 1$;
>
> **od**

The implementation of recombination is usually based on switching according to the recombination type ω. The parental individuals are chosen by means of a uniform random number generator.

As indicated before, the $(\mu+1)$-strategy does not offer an appropriate method to control the standard deviation σ, such that we did not further specify the structure of individuals and the working mechanism of the mutation operator. Even if the 1/5 success rule could be adapted to work, the strategy would still suffer from the disadvantages of this rule. The modern (μ,λ)-strategy, which is discussed in the next section, solves this problem by self-adapting the standard deviation(s) on-line during the search.

6.4   THE (μ, λ)- AND $(\mu + \lambda)$-EVOLUTION STRATEGY

The severe disadvantages of the 1/5 success rule for controlling the "step size" σ of the simple evolution strategy caused Schwefel to look for a more robust and general method to adjust the mutation parameters of the algorithm. Again, a solution to this problem was found by taking a closer look at the natural model, where the genotype itself incorporates mechanisms to control its own mutability (by means of genotype segments that encode repair enzymes, or by so-called mutator genes). Transferring this to the evolution strategy means that the standard deviation for mutation becomes part of the individual and evolves by means of mutation and recombination just as the object variables do — a process called self-adaptation of strategy parameters[7] [Sch77].

More precisely, individuals in a (μ,λ)-strategy are equipped with a set of strategy parameters which represent an n-dimensional normal distribution for mutating the individual:

$$I = R^n \times R_+^{n_\sigma} \times [-\pi, \pi]^{n_\alpha} , \tag{6.16}$$

i.e., $S = R^{n_\sigma} \times [-\pi, \pi]^{n_\alpha}$.

An individual $\vec{a} = (\vec{x}, \vec{\sigma}, \vec{\alpha}) \in I$ consists of the components

---

[7] Sometimes, the terms auto-adaptation or second-level learning are also used to denote this principle of evolving strategy parameters.

- $\vec{x} \in R^n$: The vector of object variables. Notice that this is the only part of $\vec{a}$ entering the objective function.
- $\vec{\sigma} \in R_+^{n_\sigma}$: A vector of standard deviations $(1 \leq n_\sigma \leq n)$ of the normal distribution.
- $\vec{\alpha} \in [-\pi, \pi]^{n_\alpha}$: A vector of inclination angles $(n_\alpha = (n - n_\sigma/2) \cdot (n_\sigma - 1))$, defining linearly correlated mutations of the object variables $\vec{x}$.

The strategy parameters $\vec{\sigma}$ and $\vec{\alpha}$ determine the variances and covariances of the n-dimensional normal distribution, which is used for exploring the search space (see part II for a discussion of the n-dimensional normal distribution and its properties).

The amount of strategy parameters attached to an individual can be varied by the user of an evolution strategy, depending on her feelings about the degree of freedom required by the objective function topology. As a rule of thumb, the global search reliability and robustness of the algorithm increases at the cost of computing time when the number of strategy parameters is increased. The settings most commonly used are:

- $n_\sigma = 1$, $n_\alpha = 0$: Standard mutations with one single standard deviation controlling mutation of all components of $\vec{x}$.
- $n_\sigma = n$, $n_\alpha = 0$: Standard mutations with individual step sizes $\sigma_1, \ldots, \sigma_n$ controlling mutation of the corresponding object variables x_i individually.
- $n_\sigma = n$, $n_\alpha = n \cdot (n - 1)/2$: Correlated mutations with a complete covariance matrix for each individual. Note that in this case the correspondence $\Delta x_i \propto \sigma_i$ is no longer valid.
- $n_\sigma = 2$, $n_\alpha = n - 1$: In one arbitrary direction of the search space the search is performed with variance σ_1^2 whereas σ_2^2 is the variance in all other directions perpendicular to the first one.

The basic idea of correlated mutations is illustrated for the case $n = 2$, $n_\sigma = 2$, $n_\alpha = 1$ in figure 6.1, where the lines of equal mutation probability density of the two-dimensional normal distribution are plotted. Notice that the standard deviations σ_1 and σ_2 determine the relation of the lengths of the main axes of the hyperellipsoid, and α_{12} represents the rotation angle of the hyperellipsoid. In the general case of correlated mutations, the mutation hyperellipsoid may align itself arbitrarily in the n-dimensional search space.

According to the generalized structure of individuals, the mutation operator $\textbf{mut} : I \rightarrow I$ is defined as follows[8]:

$$\textbf{mut} = \textbf{mu}_x \circ (\textbf{mu}_\sigma \times \textbf{mu}_\alpha) \quad . \tag{6.17}$$

The operator is applied after recombination to an individual

$$\hat{a} = (\hat{x}_1, \ldots, \hat{x}_n, \hat{\sigma}_1, \ldots, \hat{\sigma}_{n_\sigma}, \hat{\alpha}_1, \ldots, \hat{\alpha}_{n_\alpha})$$

and proceeds by first mutating the strategy parameters $\hat{\sigma}$ and $\hat{\alpha}$ and then modifying $\vec{x}$ according to the new set of strategy parameters obtained from mutating $\hat{\sigma}$ and $\hat{\alpha}$:

---

[8] Again, as a notational convention, we define

$$(\textbf{mu}_\sigma \times \textbf{mu}_\alpha)(\vec{x}, \vec{\sigma}, \vec{\alpha}) = (\vec{x}, \textbf{mu}_\sigma(\vec{\sigma}), \textbf{mu}_\alpha(\vec{\alpha})) \quad .$$

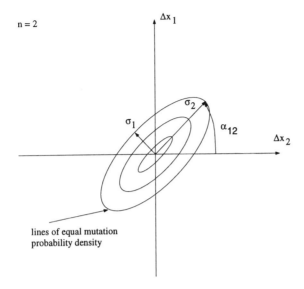

Figure 6.1: Illustration of the mutation ellipsoid for the case $n = 2$ (number of object variables), $n_\sigma = 2$ (number of variances), $n_\alpha = 1$ (number of covariances).

- **mu$_\sigma$** : $R_+^{n_\sigma} \to R_+^{n_\sigma}$ mutates the recombined $\hat{\vec{\sigma}}$:

$$\mathbf{mu}_\sigma(\hat{\vec{\sigma}}) := (\hat{\sigma}_1 \exp(z_1 + z_0), \ldots, \hat{\sigma}_{n_\sigma} \exp(z_{n_\sigma} + z_0)) =: \tilde{\vec{\sigma}} \qquad (6.18)$$

where $z_0 \sim N(0, \tau_0^2)$, $z_i \sim N(0, \tau^2)$ $\forall i \in \{1, \ldots, n_\sigma\}$.
To prevent the standard deviations from becoming practically zero, a minimal value of ε_σ is algorithmically enforced for all σ_i.

- **mu$_\alpha$** : $R^{n_\alpha} \to R^{n_\alpha}$ mutates the recombined $\hat{\vec{\alpha}}$:

$$\mathbf{mu}_\alpha(\hat{\vec{\alpha}}) := (\hat{\alpha}_1 + z_1, \ldots, \hat{\alpha}_{n_\alpha} + z_{n_\alpha}) =: \tilde{\vec{\alpha}} \qquad (6.19)$$

where $z_i \sim N(0, \beta^2)$ $\forall i \in \{1, \ldots, n_\alpha\}$. Empirically, $\beta \approx 0.0873$ ($\approx 5°$) has shown to yield good results.
Rotation angles are kept feasible (i.e., in the interval $[-\pi, \pi]$) by circularly mapping them into the feasible range whenever it is left by mutation.

- **mu$_x$** : $R^n \to R^n$ mutates the recombined object variable vector $\vec{x}$, using the already mutated $\tilde{\vec{\sigma}}$ and $\tilde{\vec{\alpha}}$:

$$\mathbf{mu}_x(\vec{x}) := (\hat{x}_1 + cor_1(\tilde{\vec{\sigma}}, \tilde{\vec{\alpha}}), \ldots, \hat{x}_n + cor_n(\tilde{\vec{\sigma}}, \tilde{\vec{\alpha}})) =: \tilde{\vec{x}} \qquad (6.20)$$

where $\vec{cor} := (cor_1(\tilde{\vec{\sigma}}, \tilde{\vec{\alpha}}), \ldots, cor_n(\tilde{\vec{\sigma}}, \tilde{\vec{\alpha}}))$ is a random vector with normally distributed, eventually correlated components. The vector $\vec{cor}$ can be calculated according to $\vec{cor} = \mathbf{T}\vec{z}$ where $\vec{z} = (z_1, \ldots, z_n)$ with

$z_i \sim \mathbf{N}(0, \tilde{\sigma}_i^2) \; \forall i \in \{1, \ldots, n_\sigma\}$ and

$$\mathbf{T} = \prod_{p=1}^{n_\sigma - 1} \prod_{q=p+1}^{n_\sigma} \mathbf{T}_{pq}(\tilde{\alpha}_j) \qquad (6.21)$$

with $j = \frac{1}{2}(2n_\sigma - p)(p+1) - 2n_\sigma + q$ [Rud92a]. The rotation matrices $\mathbf{T}_{pq}(\tilde{\alpha}_j)$ are unit matrices except that $t_{pp} = t_{qq} = \cos(\alpha_j)$ and $t_{pq} = -t_{qp} = -\sin(\alpha_j)$, i.e., the trigonometric terms are located in columns p and q, each. The multiplication from right to left provides an efficient way of calculating (6.21).

The use of a logarithmic normal distribution for the variation of standard deviations σ_i is motivated by the arguments that a multiplicative process guarantees standard deviations to remain positive, that the median equals one (i.e., the process is neutral when selection is disabled), and that smaller modifications are more likely than larger ones (see [Sch77], p. 168).

For recombination, all possibilities as described in section 6.3 for $\omega \in \{0, 1, 2, 3\}$ can be used in principle. The recombination operator modifies not only the object variables, but also the strategy parameters, and the operator might be different for the components $\vec{x}$, $\vec{\sigma}$, and $\vec{\alpha}$ of an individual. Consequently, the complete recombination operator is specified by vectors $\vec{\omega} = (\omega_x, \omega_\sigma, \omega_\alpha) \in \{0, 1, 2, 3\}^3$ and $\vec{\varrho} = (\varrho_x, \varrho_\sigma, \varrho_\alpha) \in \{1, \ldots, \mu\}^3$, where ω_x, ω_σ, and ω_α specify the recombination operator for object variables, standard deviations, and rotation angles, and ϱ_x, ϱ_σ, and ϱ_α are the number of potential parents involved in the recombination of object variables, standard deviations, and rotation angles. Consequently, recombination splits into three separate operator combinations for $\vec{x}$, $\vec{\sigma}$, and $\vec{\alpha}$, following equation (6.10):

$$\mathbf{rec} = (\mathbf{re}_x \circ \mathbf{co}_x) \times (\mathbf{re}_\sigma \circ \mathbf{co}_\sigma) \times (\mathbf{re}_\alpha \circ \mathbf{co}_\alpha) \quad , \qquad (6.22)$$

where $\mathbf{co}_\delta : I^\mu \to I^{\varrho_\delta}$ and $\mathbf{re}_\delta : I^{\varrho_\delta} \to I$ ($\delta \in \{x, \sigma, \alpha\}$).

Notice that, according to the separate application of potentially different recombination operators on the components of individuals, an object variable x_i and a standard deviation σ_i are usually not transferred to the offspring together, as a unit of information. Sometimes, the choice of a useful recombination operator for a particular optimization problem is relatively difficult and requires to perform some experiments, but in many cases the setting[9] $\vec{\omega} = (3, 2, 0)$ with $\vec{\varrho} = (\mu, \mu, 1)$ provides an appropriate parameterization.

Combining the operators mutation, recombination, and selection as defined here, the main loop of a (μ, λ)-ES is formulated according to equation (6.4) as follows:

$$opt_{(\mu, \lambda)-ES}(P^{(t)}) = \mathbf{sel}_\mu^\lambda(\sqcup_{i=1}^\lambda \{\mathbf{mut}(\mathbf{rec}(P^{(t)}))\}) \quad . \qquad (6.23)$$

According to definition (6.15), selection simply returns the set consisting of the μ best individuals of its argument set of size λ.

---

[9] I.e., discrete recombination on two randomly selected parents on $\vec{x}$ and local intermediary recombination using the whole parent population as a gene pool for the variances and no recombination for the inclination angles. Typically, the setting $u_i = 1/2$ is used for local intermediary recombination.

In case of a $(\mu+\lambda)$-ES, the main loop changes only slightly by taking into account the parent population, too, i.e., the argument set of **sel** has size $\lambda + \mu$:

$$opt_{(\mu+\lambda)-ES}(P^{(t)}) = \mathbf{sel}_\mu^{\mu+\lambda}(\sqcup_{i=1}^\lambda \{\mathbf{mut}(\mathbf{rec}(P^{(t)}))\} \sqcup P^{(t)}) \quad . \qquad (6.24)$$

Though it offers some theoretical advantage (see part II), this minor modification has the serious disadvantage that the self-adaptation of strategy parameters is hindered in working effectively, because misadapted strategy parameters may survive for a relatively large number of generations. Furthermore, the $(\mu+\lambda)$-selection mechanism fails in case of dynamically changing environments, and it tends to emphasize on local rather than global search properties. For these reasons, modern evolution strategies use (μ,λ)-selection, normally. Algorithm 3 presents the (μ,λ)-ES in pseudocode notation:

ALGORITHM 3 $((\mu, \lambda)$-ES)

> $t := 0;$
> *initialize* $P^{(0)} = \{\vec{a}_1, \ldots, \vec{a}_\mu\} \in I^\mu;$
> *evaluate* $f(\vec{x}_1), \ldots, f(\vec{x}_\mu);$
> **while** $(T(P^{(t)}) = 0)$ **do**
> > $\tilde{P} := \emptyset;$
> > **for** $i := 1$ **to** λ **do**
> > > $(\tilde{\vec{x}}, \tilde{\vec{\sigma}}, \tilde{\vec{\alpha}}) := \mathbf{mut}(\mathbf{rec}(P^{(t)}));$
> > > *evaluate* $f(\tilde{\vec{x}});$
> > > $\tilde{P} := \tilde{P} \sqcup \{(\tilde{\vec{x}}, \tilde{\vec{\sigma}}, \tilde{\vec{\alpha}})\};$
> > **od**
> > $P^{(t+1)} := \mathbf{sel}_\mu^\lambda(\tilde{P});$
> > $t := t + 1;$
> **od**

The self-adaptation process of strategy parameters is based on the existence of an indirect link between strategy parameters (the internal model of the individual) and the fitness of an individual as well as a sufficiently large diversity of internal models in the parent population: μ has to be chosen clearly larger than one, e.g. $\mu = 15$ [Sch92], and a ratio of $\lambda/\mu \approx 7$ is recommended as a good setting for the relation between parent and offspring population size[10]. So far, no theoretical results about the self-adaptation of strategy parameters have been obtained, but empirically three main conditions for a successful self-adaptation were identified [Sch92]:

- (μ,λ)-selection,
- a not too strong selective pressure (i.e., μ has to be clearly larger than one),
- recombination on strategy parameters.

To conclude this description of the (μ,λ)-ES, we have to mention the topics of initialization, constraint handling and termination criteria. So far, the basic method to handle constraints consists in repeating the processes of recombination and mutation as often as necessary to create λ feasible offspring individuals, i.e., with $g_j(\tilde{\vec{x}}) \geq$

---

[10] Typically, a $(15,100)$-ES is used.

$0 \ \forall j \in \{1, \ldots, q\}$. It is often necessary, however, to use a more intelligent method such as penalty terms, for instance, and constraint handling is a general topic of strong importance for all instances of evolutionary algorithms.

The simplest termination criterion $T : I^\mu \to \{0, 1\}$ stops evolution after a previously specified number of generations has passed, independently of the population diversity or the number of generations required for the last improvement that occurred. Alternative methods as proposed by Schwefel terminate the search if the objective function value improvement over a number of generations or the fitness difference between worst and best individuals falls below a certain threshold (see [Sch95], pp. 113-114).

Concerning initialization of the start population $P^{(0)}$, two possibilities are useful for comparing evolution strategies to classical optimization methods as well as genetic algorithms. First, with given lower and upper bounds $\underline{x}_i, \overline{x}_i \in R$, i.e., $\underline{x}_i \leq x_i \leq \overline{x}_i \ \forall i \in \{1, \ldots, n\}$, all individuals of $P^{(0)}$ are arbitrarily (most often uniformly) distributed within the bounded region. Second, one might also know a given start position $\vec{x}$, which is then assigned to one individual, and the remaining $\mu - 1$ individuals are generated by mutation of this individual with some enlarged step size $c \cdot \sigma^{(0)}$, $c > 1$.

The description presented in this section is certainly sufficient for the reader to implement a (μ, λ)-evolution strategy. More details on the realization of specific operators, however, can be found in [Sch95].

6.5   FURTHER VARIANTS OF THE EVOLUTION STRATEGY

The concept of self-adaptation of strategy parameters by means of extending the evolutionary operators mutation and recombination to the step sizes $\vec{\sigma}$ (and, later on, the covariances represented by rotation angles $\vec{\alpha}$) was introduced by Schwefel, who demonstrated the robustness and learning capabilities of the mechanism [Sch87, Sch92]. For the restricted case $n_\sigma = 1$, Rechenberg presented a simple alternative method called mutative step-size control, which proceeds by multiplying σ by a factor c' or $1/c'$ with equal probability ([Rec94], pp. 45-50). For an individual $\vec{a} = (\vec{x}, \sigma)$, the mutation operator $\mathbf{mu}_\sigma$ works as follows:

$$\tilde{\sigma} := \mathbf{mu}_\sigma(\sigma) = \begin{cases} \sigma \cdot c' & , \quad \text{if } u < 1/2 \\ \sigma/c' & , \quad \text{otherwise} \end{cases} \tag{6.25}$$

where $u \sim \mathbf{U}([0, 1])$. For the constant c', Rechenberg suggests a value of 1.3. Mutation of the object variables is performed as in equation (6.7). Rechenberg confirms Schwefel's observation that a $(\mu + \lambda)$-selection hampers the self-adaptation of strategy parameters, which is also true for the mutative step-size control. Although easy to implement, the mutative step-size control is insufficient for complicated optimization problems, because the internal models of individuals are too restricted by $n_\sigma = 1$ to explore arbitrary n-dimensional search spaces in an effective way.

Adapting a relatively large number of strategy parameters, however, requires to work with sufficiently large populations, which is hard to realize if the evaluation of objective function values is a computationally expensive task. Concerning correlated mutations, Rudolph already pointed out that it is worthwhile to make use of the history of the evolution process, e.g. by storing the phylogenetic tree of individuals up to a depth of n [Rud92a]. An accumulation of step-size information over the sequence of generations is also used by Ostermeier et al. to introduce a derandomized mutative step-size control of individual step sizes $\sigma_1, \ldots, \sigma_n$ in a $(1,\lambda)$-ES [OGH94]. These are interesting developments towards improving and generalizing the self-adaptation process, but more experience as well as comparisons of the different implementations and theoretical investigations are necessary to assess their pros and cons. And, moreover, the question of a proper test suite arises and whether one emphasizes robustness or efficiency.

Besides modeling evolution on the level of a single population, it is also possible to model the evolution of a species by introducing a higher-level competition between populations. Herdy uses the notation $[\mu'/\varrho', \lambda'(\mu/\varrho, \lambda)^\gamma]$ to describe an evolution scheme where μ' parent populations generate λ' offspring populations (involving ϱ' parent populations for creating one offspring population), and each population follows a (μ,λ)-evolution strategy[11] [Her92]. A recombination-mutation-selection step on the level of populations happens after each γ generations that have passed by on the level of individuals. Rechenberg points out that complete populations might be evaluated according to their average objective function value, but other measures of quality are also possible (see [Rec94], pp. 88-100). Both authors also indicate that even higher levels of hierarchical evolution strategies are possible.

The major advantage of evolving several populations in parallel and exchanging information between these populations only occasionally is clearly given by the higher convergence reliability in case of highly multimodal optimization problems: Such parallel algorithms with k (μ,λ)-evolution strategies usually succeed in finding better solutions than a single $(k \cdot \mu, k \cdot \lambda)$-strategy, because the subpopulations explore different regions of the search space and maintain genetic diversity by communicating information not too often to other subpopulations. In general, four additional parameters control a collection of parallel interacting subpopulations (also called demes): Exchange frequency, number of individuals to exchange, selection strategy for the emigrants, and replacement strategy to incorporate the immigrants. Due to the exchange of individuals, this is also often called a migration model. Such a migration-based evolution strategy (or evolutionary algorithm in general) is suitable for implementation on a coarse-grained parallel computer with a low communication bandwidth relative to the computational performance of each processor. An evolution strategy with asynchronously communicating parallel populations was presented by Rudolph for an application to a 100-city TSP [Rud91].

Besides parallel populations, the inherent parallelism of evolutionary algorithms can also be exploited on the level of individuals. The corresponding fine-grained parallel algorithm uses a neighbourhood structure (often a toroidal grid), where individuals interact only with other individuals in their neighborhood, i.e., recombination and selection are restricted to the neighborhood. In this model, favourable genetic information might spread over the population due to overlapping neighborhoods, and for this reason the model is sometimes called diffusion model. The corresponding implementation for an evolution strategy was also performed by Rudolph on a Connection Machine 2 [Rud92b].

Fine-grained parallel computers typically allow for a large number of processes of

---

[11] Additionally, μ and λ may differ for each of the λ' subpopulations.

low complexity, each, and have to provide a high communication bandwidth. Such machines are not suitable for implementing the migration model, where each process consists of a complete evolutionary algorithm. On the contrary, coarse-grained parallel machines with a small number of powerful processors allow to emulate a fine-grained neighborhood structure and to combine the migration and diffusion model in an elegant way, as shown by Sprave [Spr90, Spr93].

Finally, it is important to mention that the most time-consuming component of an evolution strategy may consist in the evaluation of objective function values or of the constraints — which is often the case for practical applications, especially when evaluating f and g_j requires to run a complete simulation model [BHS93]. Under such circumstances, it is an elegant solution to evaluate each function value on a single workstation in a local area network, because the communication requirements are small (only $\vec{x}$ and $f(\vec{x})$ have to be exchanged) and the requirement concerning computing power (and potentially other hardware resources such as main memory or disk storage) is high. Presently, the PVM software provides an elegant, hardware-independent method to create a virtual parallel computer by using a (heterogeneous) network of machines [GBD+94]. For a more detailed overview of the parallelization possibilities of evolutionary algorithms, the reader is referred to the work of Hoffmeister [Hof91].

Modifications of the original (μ,λ)-ES have also been developed for applying the algorithm to multiple criteria decision making (MCDM) problems (Kursawe presented a variant which is capable of generating arbitrarily many members of the Pareto-set by varying the selection criterion and using diploid individuals [Kur91]) and mixed-integer optimization (Bäck and Schütz discuss the extension of self-adaptation to discrete variables and the application of the algorithm to the problem of optical multilayer design [BS95]). The reader is referred to the literature for a more detailed presentation of these variants.

6.6   OUTLOOK

Over a period of thirty years, evolution strategies have evolved and are now in a state where a variety of different variants are known and research continues into several directions, including applications, extensions of algorithms, theory, and cross-fertilization with other evolutionary algorithms. The theory of (μ,λ)-evolution strategies has reached a relatively mature state — although a lot of particular questions are still open (see the contribution "Evolution Strategies II: Theoretical Aspects" in this volume).

Concerning the algorithm itself, a contemporary evolution strategy called $(\mu, \kappa, \lambda, \varrho)$-ES was recently proposed by Schwefel and Rudolph [SR95]. The algorithm extends the (μ,λ)-ES by some new features, including:

- The life span of individuals is limited to $\kappa \geq 1$ generations (better: reproduction cycles), which allows a free variation of the selection scheme between the extreme cases $\kappa = 1$ ((μ,λ)-selection) and $\kappa = \infty$ ($(\mu+\lambda)$-selection). This is implemented by extending the individuals by a life span counter θ, which is initialized by $\theta = \kappa$ at $t = 0$ and whenever an offspring individual is born by recombination and mutation. The selection operator chooses an individual to survive only, if its remaining life span θ is larger than zero — otherwise, an individual which is worse with respect to fitness will survive. The selection condition (6.15) therefore changes to

$$\forall \tilde{a} \in \tilde{P} : \theta_{\tilde{a}} > 0 \wedge (\nexists \vec{a} \in P - \tilde{P} : \theta_{\vec{a}} > 0 \wedge f(\vec{x}) \leq f(\tilde{\vec{x}})) \quad . \qquad (6.26)$$

At the end of the selection process, the remaining life durations are decremented by one for each of the μ surviving individuals.

- Tournament selection is incorporated as an alternative to (μ,λ)-selection. This method is well suited for parallelization and works by selecting μ times the best individual from a subset B_k of size $|B_k| = \zeta$ (tournament size, $2 \leq \zeta \leq \mu+\lambda$), $k \in \{1,\ldots,\mu\}$, whose elements are chosen uniformly at random from $P = \sqcup_{i=1}^{\lambda}\{\mathbf{mut}(\mathbf{rec}(P^{(t)}))\} \sqcup P^{(t)}$ (note that duplicates are explicitly allowed to occur in the B_k). The union of the best members of these μ subsets forms the new parent population.
- The application of recombination and mutation is controlled by additional probabilities $p_m, p_r \in [0,1]$, such that the operators are applied with probability p_m and p_r, respectively.
- Further recombination types such as crossover from genetic algorithms are incorporated.

The $(\mu, \kappa, \lambda, \varrho)$-ES reflects the attempt to incorporate useful features from genetic algorithms and to benefit from modeling the natural concept of limited life spans of individuals in a more flexible way than in a (μ,λ)-ES with their one-generation restriction of life spans. The algorithm provides a generalization of the state-of-the-art evolution strategy, and certainly many experiments are necessary to investigate the effects of the additional features, especially concerning their impact on the self-adaptation of strategy parameters.

REFERENCES

[BHS93] Bäck T., Hammel U., and Schwefel H.-P. (1993) Modelloptimierung mit evolutionären Algorithmen. In Sydow A. (ed) *Simulationstechnik: 8. Symposium in Berlin*, Fortschritte in der Simulationstechnik, pages 49–57. Vieweg, Wiesbaden.

[BS95] Bäck T. and Schütz M. (1995) Evolution strategies for mixed-integer optimization of optical multilayer systems. In *Proceedings of the 4th Annual Conference on Evolutionary Programming*. (accepted for publication).

[DSM94] Davidor Y., Schwefel H.-P., and Männer R. (eds) (1994) *Parallel Problem Solving from Nature — PPSN III*, International Conference on Evolutionary Computation, volume 866 of *Lecture Notes in Computer Science*. Springer, Berlin.

[ERR95] Eiben A. E., Raué P.-E., and Ruttkay Z. (1995) Genetic algorithms with multi-parent recombination. In Davidor *et al.* [DSM94], pages 78–87.

[GBD⁺94] Geist A., Beguelin A., Dongarra J., Jiang W., Manchek R., and Sunderam V. (1994) *PVM: Parallel Virtual Machine — A User's Guide and Tutorial for Networked Parallel Computing*. The MIT Press, Cambridge, MA.

[HB92] Hoffmeister F. and Bäck T. (1992) Genetic self–learning. In Varela F. J. and Bourgine P. (eds) *Proceedings of the 1st European Conference on Artificial Life*, pages 227–235. The MIT Press, Cambridge, MA.

[Her92] Herdy M. (1992) Reproductive isolation as strategy parameter in hierarchically organized evolution strategies. In Männer and Manderick [MM92], pages 207–217.

[Hof91] Hoffmeister F. (1991) Scalable parallelism by evolutionary algorithms. In Grauer M. and Pressmar D. B. (eds) *Parallel Computing and Mathematical Optimization*, volume 367 of *Lecture Notes in Economics and Mathematical Systems*, pages 177–198. Springer, Berlin.

[KS70] Klockgether J. and Schwefel H.-P. (March 24–26, 1970) Two–phase nozzle and hollow core jet experiments. In Elliott D. (ed) *Proc. 11th Symp. Engineering Aspects of Magnetohydrodynamics*, pages 141–148. California Institute of Technology, Pasadena CA.

[Kur91] Kursawe F. (1991) A variant of Evolution Strategies for vector optimization. In Schwefel and Männer [SM91], pages 193–197.

[MM92] Männer R. and Manderick B. (eds) (1992) *Parallel Problem Solving from Nature 2*. Elsevier, Amsterdam.

[OGH94] Ostermeier A., Gawelczyk A., and Hansen N. (1994) Step-size adaptation based on non-local use of selection information. In Davidor *et al.* [DSM94], pages 189–198.

[Rec73] Rechenberg I. (1973) *Evolutionsstrategie: Optimierung technischer Systeme nach Prinzipien der biologischen Evolution*. Frommann–Holzboog, Stuttgart.

[Rec94] Rechenberg I. (1994) *Evolutionsstrategie '94*, volume 1 of *Werkstatt Bionik und Evolutionstechnik*. frommann–holzboog, Stuttgart.

[Rud91] Rudolph G. (1991) Global optimization by means of distributed evolution strategies. In Schwefel and Männer [SM91], pages 209–213.

[Rud92a] Rudolph G. (1992) On correlated mutations in evolution strategies. In Männer and Manderick [MM92], pages 105–114.

[Rud92b] Rudolph G. (1992) Parallel approaches to stochastic global optimization. In Joosen W. and Milgrom E. (eds) *Parallel Computing: From Theory to Sound Practice, Proceedings of the European Workshop on Parallel Computing*, pages 256–267. IOS Press, Amsterdam.

[Rud93] Rudolph G. (1993) Massively parallel simulated annealing and its relation to evolutionary algorithms. *Evolutionary Computation* 1(4): 361–382.

[Sch77] Schwefel H.-P. (1977) *Numerische Optimierung von Computer-Modellen mittels der Evolutionsstrategie*, volume 26 of *Interdisciplinary Systems Research*. Birkhäuser, Basel.

[Sch87] Schwefel H.-P. (June 1987) Collective phenomena in evolutionary systems. In *Preprints of the 31st Annual Meeting of the International Society for General System Research, Budapest*, volume 2, pages 1025–1033.

[Sch92] Schwefel H.-P. (1992) Imitating evolution: Collective, two-level learning processes. In Witt U. (ed) *Explaining Process and Change — Approaches to Evolutionary Economics*, pages 49–63. The University of Michigan Press, Ann Arbor, MI.

[Sch95] Schwefel H.-P. (1995) *Evolution and Optimum Seeking*. Sixth-Generation Computer Technology Series. Wiley, New York.

[SM91] Schwefel H.-P. and Männer R. (eds) (1991) *Parallel Problem Solving from Nature — Proceedings 1st Workshop PPSN I*, volume 496 of *Lecture Notes in Computer Science*. Springer, Berlin.

[Spr90] Sprave J. (December 1990) *Parallelisierung Genetischer Algorithmen zur Suche und Optimierung*. Diploma thesis, University of Dortmund.

[Spr93] Sprave J. (1993) Zelluläre evolutionäre Algorithmen zur Parameteroptimierung. In Hofestädt R., Krückeberg F., and Lengauer T. (eds) *Informatik in den Biowissenschaften*, Informatik aktuell, pages 111–120. Springer, Berlin.

[SR95] Schwefel H.-P. and Rudolph G. (1995) Contemporary evolution strategies. In Morán F., Moreno A., Merelo J. J., and Chacón P. (eds) *Advances in Artificial Life. Third International Conference on Artificial Life*, volume 929 of *Lecture Notes in Artificial Intelligence*, pages 893–907. Springer, Berlin.

[Whi89] Whitley D. (1989) The GENITOR algorithm and selection pressure: Why rank–based allocation of reproductive trials is best. In Schaffer J. D. (ed) *Proceedings of the 3rd International Conference on Genetic Algorithms*, pages 116–121. Morgan Kaufmann Publishers, San Mateo, CA.

7

Evolution Strategies II: Theoretical Aspects

Hans-Paul Schwefel[1] and Thomas Bäck[2]

7.1 SOME DEFINITIONS FROM PROBABILITY THEORY AND STATISTICS

This section shortly presents some notions and corollaries from the theory of stochastic events that will be helpful later on in discussing the convergence reliability and convergence velocity of evolution strategies (ESs). More detailed foundations may be found in textbooks on statistics and probability theory.

7.1.1 Random variables, distribution and density functions

A continuous random variable $X : \Omega \to I\!\!R$ defined over a probability space $(\Omega, \mathcal{A}, p)$ with Ω as set of elementary events, $\mathcal{A}$ as algebra of the events, and p as probability measure is characterized by the distribution function

$$
\begin{aligned}
F_X : I\!\!R &\to [0,1] \\
x &\mapsto F_X(x) = p(X \le x).
\end{aligned}
$$

The density function $f_X : I\!\!R \to I\!\!R$ of the random variable X is implicitly defined by the equation

$$
F_X(x) = \int_{-\infty}^{x} f_X(z)dz \qquad \forall x \in I\!\!R. \tag{7.1}
$$

---

[1] University of Dortmund, Department of Computer Science, 44221 Dortmund, Germany.
  E-mail: schwefel@ls11.informatik.uni-dortmund.de
[2] Informatik Centrum Dortmund, Joseph-von-Fraunhofer-Str. 20, 44227 Dortmund, Germany. E-mail: baeck@ls11.informatik.uni-dortmund.de

As *support*(X) one denotes the set $\{x \in R \mid F_X(x) > 0\}$ of all real numbers with strictly positive density.

One dimensional distributions

The expectation of a one dimensional continuous random variable X with density f_X is defined by

$$\xi \;=\; \mathrm{E}[X] \;:=\; \int_{-\infty}^{\infty} x f_X(x) dx. \tag{7.2}$$

As measures for the dispersion the variance

$$\sigma^2 \;=\; D^2[X] \;:=\; \int_{-\infty}^{\infty} (x - \mathrm{E}[X])^2 f_X(x) dx$$

and the standard deviation $\sigma = D[X]$ are used. Let $h : R \to R$ be a monotonous function of a continuous random variable with density f_X, then the expectation of h can be calculated as

$$\mathrm{E}[h(X)] \;=\; \int_{-\infty}^{\infty} h(x) f_X(x) dx.$$

Multidimensional distributions

The vector of expectations of an n-dimensional random set $X = (X_1, \ldots, X_n)^T : \Omega \to R^n$ is defined as

$$\xi \;=\; \mathrm{E}[X] \;:=\; (\mathrm{E}[X_1], \ldots, \mathrm{E}[X_n])^T .$$

The dispersion of two random variables X_i and X_j may be presented in the form of a covariance matrix $\mathrm{Cov}[X_i, X_j]$ with elements

$$\sigma_{ij} = \sigma_{ji} \;=\; \mathrm{E}\left[(X_i - \mathrm{E}[X_i])(X_j - \mathrm{E}[X_j])\right].$$

Diagonal elements $\sigma_{ii} = \sigma_i^2$ correspond to the individual one dimensional variances. The covariance matrix

$$\Sigma_X : = \begin{pmatrix} \sigma_1^2 & \sigma_{12} & \cdots & \sigma_{1n} \\ \sigma_{21} & \sigma_2^2 & \cdots & \sigma_{2n} \\ \vdots & \vdots & & \vdots \\ \sigma_{n1} & \sigma_{n2} & \cdots & \sigma_n^2 \end{pmatrix}$$

contains the complete information on the dispersion of an n-dimensional random set. If $\sigma_{ij} = 0$ for any two random variables X_i and X_j, then they are termed uncorrelated. Stochastically independent random variables are always uncorrelated; the inverse, however, does not hold, generally.

7.1.3   Special distributions

The Gaussian or normal distribution

A distribution for a random variable like

$$F_X(x) \;=\; \Phi(x) \;=\; \int_{-\infty}^{x} \frac{1}{\sqrt{2\pi}\sigma} \exp\left(-\frac{(z-\xi)^2}{2\sigma^2}\right) dz$$

is called a normal or Gaussian distribution if ξ is identical to its expectation and σ^2 identical to its variance. The corresponding density of the $N(\xi, \sigma^2)$ distribution is

$$f_X(x) \;=\; \phi(x) \;=\; \frac{1}{\sqrt{2\pi}\sigma} \exp\left(-\frac{(x-\xi)^2}{2\sigma^2}\right).$$

In the following Φ and ϕ shall always denote the distribution and density functions of a standard normal distribution $N(0,1)$. Furthermore, N, N', N'', and $N_i (i = 1, \dots, n)$ always denote standard normally distributed random variables.

A normally distributed random variable $X \sim N(\xi, \sigma^2)$ can always be transformed into a standard normally distributed random variable $Y \sim N(0,1)$ by means of the operation $Y := (X - \xi)/\sigma$.

An important and later on often used quality of normal distributions is given by their addition theorem: Let $X_1, \dots, X_n$ be stochastically independent $N_i(\xi_i, \sigma_i^2)$ distributed random variables, then

$$\sum_{i=1}^{n} X_i \;\sim\; N\left(\sum_{i=1}^{n}\xi_i, \sum_{i=1}^{n}\sigma_i^2\right) \tag{7.3}$$

is valid. The sum over normally distributed random variables is always also normally distributed, and the resulting expectation as well as the resulting variance may be calculated easily as sums of the individual items according to (7.3).

The n-dimensional normal distribution

The multidimensional normal distribution is a simple and natural extension of the one dimensional normal distribution introduced above to a vector $X = (X_1, \dots, X_n)^T$ of random variables. All marginal distributions are standard normal distributions. The density function of the vector X has the form

$$f_X(x) \;=\; \frac{1}{(2\pi)^{\frac{n}{2}}\sqrt{|\Sigma_X|}} \exp\left(-\frac{1}{2}(x-\xi)^T \Sigma_X^{-1}(x-\xi)\right)$$

where ξ is identical to the vector of the individual expectations and Σ_X is identical to the covariance matrix, $|\Sigma_X|$ being its determinant.

An important characteristic of the multidimensional normal distribution is the equivalence of the two terms 'stochastic independence' and 'uncorrelatedness' of the single components of the random vector. If the correlation matrix is a diagonal matrix, then the single components of the random vector are stochastically independent from each other, and the elements of the main diagonal are equal to the variances of the single random variables, the marginal distributions of which are all normally distributed. Helpful for all following conclusions is the next definition:

Definition 7.1.1 *An n-dimensional random vector Z is called elliptically distributed if it has a representation $Z = RX$ where the random vector X is uniformly distributed over an n-dimensional ellipsoid and R is a non-negative random variable that is stochastically independent of X. Z is called spherically distributed if X is uniformly distributed over an n-dimensional sphere in addition to the demands above.*

The multidimensional normal distribution belongs to the class of elliptical distributions. In case of stochastic independence of the components of the random vector and identical variances it is spherically distributed. The distribution of the parameter R (sometimes called total mutation step size in ESs) may be reduced to a χ^2 distribution in that case.

The χ^2 distribution

Let $X_1, \ldots, X_n$ be stochastically independent and identically standard normal distributed components, then the distribution of $Y := \sum_{i=1}^{n}(X_i)^2$ is a χ^2 distribution with n degrees of freedom. Its expectation is n, its variance $2n$. For large n the χ_n^2 distribution can be approximated by an $N(n, 2n)$ distribution, a fact that can be expressed by

$$\chi_n^2 \to n + \sqrt{2n}N. \tag{7.4}$$

$R := +\sqrt{Y}$ then may be handled as if $R \sim N(\sqrt{n}, \sqrt{\frac{1}{2}})$.

7.1.4 Order statistics

Let $X_1, \ldots, X_n$ be stochastically independent continuous random variables with distribution function F_X and density function f_X, then the distribution function F_{Y_1} of $Y_1 := \max\{X_1, \ldots, X_n\}$ can be calculated by

$$F_{Y_1}(x) \;=\; \prod_{i=1}^{n} F_{X_i}(x). \tag{7.5}$$

If in addition the X_i are identically distributed, then

$$F_{Y_1}(x) \;=\; [F_X(x)]^n \tag{7.6}$$

is valid for the distribution function of the ensemble, and after (7.1) the density of the maximum of the X_i is

$$f_{Y_1}(x) \;=\; \frac{\partial}{\partial x} F_{Y_1}(x) \;=\; n f_X(x)\,[F_X(x)]^{n-1}.$$

This is a special case only of the more general result from order statistics, which states the distribution function of the mth largest X_i to be

$$f_{Y_m}(x) \;=\; n\binom{n-1}{m-1} f_X(x)\,[F_X(x)]^{n-m}\,[1 - F_X(x)]^{m-1}.$$

7.2   ON THE CONVERGENCE BEHAVIOR OF EVOLUTION STRATEGIES

This section summarizes so far known results on the convergence properties of evolution strategies. At first, the simple (1+1) ES will be considered. The extension to $(\mu + \lambda)$ ESs is trivial since they exert results that are at least as good as those of the (1+1) ES. More difficult are assertions about (μ, λ) ESs. A (1,1) ES, for example, represents a simple random walk strategy without selection. Moreover, figure 7.1 shows that it always diverges in nonlinear cases except for $\sigma = 0$ (lowermost curve).

Criteria for the merits of an optimization technique comprise the reliability of finally reaching an ϵ-vicinity of the optimum that is sought and, even more important, the time or number of iterations necessary to reach that goal. More specifically, the convergence rate should say something about the reduction of the distance to the goal over time — may it be with respect to the objective function values or the Euclidean distance in the space of the decision variables. So far, results of this kind have been achieved only for a very limited class of situations. Nevertheless, such theoretical results are very helpful in tuning external strategy parameters to actual situations at hand.

7.2.1   Convergence reliability

In this subsection we consider a (1+1) ES starting from $x^{(0)}$ with a mutation operator $x^{(t+1)} := x^{(t)} + Z$ where $Z \sim N(0, \sigma^2 I_n)$ is a normally distributed vector with stochastically independent components of same variance σ^2. I_n denotes the n-dimensional unit matrix.

In order to guarantee the convergence of an ES, one has to restrict oneself to the class of regular optimization problems. This restriction is not stark, however, since the features to be demanded to the objective function are rather weak and moreover essential for most other optimization procedures, as well.

Definition 7.2.1 *An optimization problem*

$$f^* \;=\; f(x^*) \;=\; \min\{f(x) \mid x \in M \subseteq R^n\}$$

is called regular if and only if

- $f^* > -\infty$,
- $x^* \in \text{int}(M)$, and
- $\mu(\{x \in M \mid f(x) \in U_\epsilon(f^*)\}) > 0 \qquad \forall \epsilon > 0,$

where μ is the Lebesgue measure, $\text{int}(M)$ the set of internal points of M, and U_ϵ an ϵ-environment of its argument.

One calls f the objective function, f^* the global minimum, and x^* the solution to the optimization problem.

The necessity of the first requirement is immediately clear, whereas the second one only simplifies the analysis and is used in the proof of the following convergence theorem. The third requirement excludes optimization tasks with isolated global optima, which cannot be reached with a probability greater than zero.

Theorem 7.2.1 *Let $\epsilon > 0$ and $p_t := p(x^{(t)} \in \{x \in M \mid f(x) \in U_\epsilon(f^*)\})$ be the probability that a population of the (1+1) ES has reached the point $x^{(t)}$ at iteration t, the objective function value belonging to which is closer to the goal f^* than ϵ. Then, assuming*

$$\sum_{t=0}^{\infty} p_t = \infty \qquad (7.7)$$

implies that

$$p(\lim_{t \to \infty} (f(x^{(t)}) - f^*) = 0) = 1$$

for any starting point $x^{(0)} \in M$.

The condition (7.7) in Theorem 7.2.1 looks rather abstract, but one can show that it is the result of a more obvious condition that is fulfilled in many practical situations:

Lemma 7.2.1 *If $M \subseteq support(f_Z)$, where f_Z denotes the probability density of the random vector Z of the mutation operator, and M is bounded, then (7.7) is valid.*

Of course, this result is of academic interest only since nobody likes to wait for a result until the end of time. More interesting are results about the expected convergence velocity.

7.2.2 Convergence velocity

The following definition is useful in order to give some quantitative contents to the term convergence velocity:

Definition 7.2.2 *The value $\delta_t := E[f(x^{(t)}) - f^*]$ is called expected error at step t. An algorithm has polynomial convergence order if $\delta_t = O(t^{-\alpha})$ with $\alpha > 0$; its convergence rate is exponential if $\delta_t = O(\beta^t)$ with $\beta \in (0,1)$.*

In order to gain statements about the convergence rate, one must restrict oneself to a special class of problems, here to the class of strictly convex problems. Strict convexity implies among other features continuous differentiability and unimodality, and thus a stark restriction. Especially, strictly convex problems may easily be solved by means of gradient type strategies and are thus not the domain of evolutionary algorithms. EAs, however, should guarantee something at least in such simple situations as well, and there is some hope that it will be possible to weaken these conditions in the future in order the arrive at results for a significantly broader and more interesting class of optimization problems.

Theorem 7.2.2 *Let $f : M \to \mathbb{R}$ be strictly convex and the mutation step size of a (1+1) ES be spherically distributed with $Z = RU$ where $support(R) = (0, a) \neq \emptyset$. Then the expected error δ_t at step t for any start position $x^{(0)} \in M$ is*

$$\delta_t = \begin{cases} O(t^{-2/n}) & \text{for a constant mutation step size} \\ O(\beta^t) & \text{for an adaptive mutation step size} \end{cases}$$

with $\beta \in (0,1)$ and step size adaptation according to $R^{(t+1)} = \|\nabla f(x^{(t)})\| R^{(t)}$.

For a static mutation probability distribution one can thus state only polynomial convergence rates, whereas exponential convergence rates may be achieved in case of an appropriate step size control. Rappl [Rap84] proved, in addition, that a rule similar to the so-called success rule introduced by Rechenberg [Rec94] to adapt the variance of the mutation operator, can yield an exponential convergence rate.

Whereas the definition of the term convergence rate above only covers the order of magnitude of the convergence velocity, the factor β is of decisive interest in practical applications. That is why it is interesting to calculate optimal parameters for the step size control. For a simple strictly convex objective function such calculations have been performed [BRS93, Rec94, Schw95]. The results will be summarized in the following.

7.3   THE SPHERE MODEL

The following section contains results gained for an objective function of type $f :$ $\mathbb{R}^n \rightarrow \mathbb{R}$, $f(x) := \|x - x^*\|^2 = r^2$. For different ES versions that all use normally distributed mutation vectors with mean $\xi = 0$ and covariance matrix $\sigma^2 I_n$ (i.e., with just one strategy parameter σ) optimal values σ^* for the mutation step sizes shall be calculated. On that basis one can derive the corresponding success probability , i.e., the mean number of cases in which the successor is better than its predecessor. This figure is independent of the current distance r to the optimum and thus may be used to control the step size adaptation.

There is some hope that the same control mechanism may yield success probabilities of the same order of magnitude also for situations that differ largely from that of the sphere model. At least one completely different situation has been identified for which this is valid, i.e., the so-called corridor model, a simple linear function with constraints [Rec94, Schw95].

7.3.1   The (1+1) ES or two membered evolution strategy

Here we want to find out the optimal value of the standard deviation σ in case of a (1+1) ES with mutation operator $x^{(t+1)} := x^{(t)} + Z$ where $Z \sim N(0, \sigma^2 I_n)$ is n-dimensionally normal distributed. For that purpose we consider the relative progress during step t

$$P \quad := \quad \frac{r^{(t)} - r^{(t+1)}}{r^{(t)}} \tag{7.8}$$

that denotes the relative approximation to the goal in the space of the decision variables. Using the addition theorem of the normal distribution and relation (7.4), one gets (writing simply r instead of $r^{(t)}$):

$$r^{(t+1)} \quad = \quad \|x^{(t)} - x^* + Z\|$$

$$= \quad \sqrt{\sum_{i=1}^{n} \left((x_i^{(t)} - x_i^*) + \sigma N_i \right)^2}$$

$$= \sqrt{\sum_{i=1}^{n}(x_i^{(t)} - x_i^*)^2 + 2\sigma \sum_{i=1}^{n}(x_i^{(t)} - x_i^*)N_i + \sigma^2 \sum_{i=1}^{n} N_i^2}$$

$$\sim \sqrt{r^2 + 2\sigma r N + \sigma^2(n + \sqrt{2n}N')}$$

$$\sim \sqrt{r^2 + \sigma^2 n + \sqrt{2n\sigma^4 + 4r^2\sigma^2}N''}.$$

Inserting this result into (7.8) yields the relative progress in terms of the dimensionless item $\check\sigma := \sigma n/r$ (using a Taylor series for $\sqrt{1+x}$ and restricting to the linear term):

$$P = 1 - \sqrt{\frac{r^2 + \sigma^2 n + \sqrt{2n\sigma^4 + 4r^2\sigma^2}N''}{r^2}}$$

$$= 1 - \sqrt{1 + \frac{\check\sigma^2}{n} + \frac{\check\sigma^2}{n}\sqrt{\frac{4}{\check\sigma^2} + \frac{2}{n}}N''}$$

$$\approx 1 - \sqrt{1 + \frac{\check\sigma^2}{n} + \frac{2\check\sigma}{n}N''} \approx 1 - \left(1 + \frac{\check\sigma^2}{2n} + \frac{\check\sigma}{n}N''\right)$$

$$= -\frac{\check\sigma^2}{2n} - \frac{\check\sigma}{n}N'' \sim N\left(-\frac{\check\sigma^2}{2n}, \frac{\check\sigma^2}{n^2}\right). \tag{7.9}$$

Since the (1+1) ES only accepts improvements that in case of the sphere model occur if the distance to the optimum is diminished, the expectation of the random variable $P_{1+1} := \max\{0, P\}$ has to be considered only. This can be done in the following way

$$E[P_{1+1}] = \int_0^\infty x f_P(x)dx =: \frac{\varphi_{1+1}}{r} = \frac{\check\varphi_{1+1}}{n}$$

$$= \int_0^\infty \frac{nx}{\sqrt{2\pi}\check\sigma} \exp\left[-\frac{1}{2}\left(\frac{nx + \check\sigma^2/2}{\check\sigma}\right)^2\right] dx$$

$$= \frac{1}{n}\left\{\frac{\check\sigma}{\sqrt{2\pi}}\exp\left(-\frac{\check\sigma^2}{8}\right) - \frac{\check\sigma^2}{2}\left[1 - \Phi\left(\frac{\check\sigma}{2}\right)\right]\right\}$$

where Φ is the distribution function of a standard normal random variable and φ Rechenberg's progress velocity, $\check\varphi_{1+1} = \varphi_{1+1}n/r$ denoting the dimensionless correspondent to $\check\sigma$. The solid line in figure 7.1 shows that this mean value is maximal for $\check\sigma^* = 1.224$ at $\check\varphi_{1+1}^* = 0.2025$ and thus for

$$\sigma^* = \frac{1.224}{n}r = \frac{0.612}{n}\|\nabla f(x)\| \quad \text{at} \quad \varphi_{1+1}^* = \frac{0.2025}{n}r.$$

The corresponding success probability p_s to the optimally adjusted step size σ^* becomes

$$p_s^* = p(P > 0)|_{\sigma=\sigma^*} = p\left(-\frac{\sigma^{*2}}{2n} - \frac{\sigma^*}{n}N'' > 0\right)$$

$$= \Phi\left(-\frac{\sigma^*}{2}\right) \approx 0.270.$$

This result has been the basis for designing an algorithm that adapts its current step size automatically.

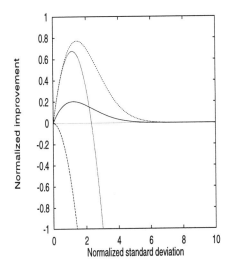 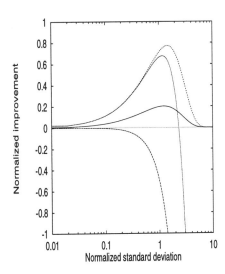

Figure 7.1: The normalized expected improvement $\check{\varphi}$ over the normalized standard deviation $\check{\sigma}$ (from top to bottom the curves correspond to $(1+5)$-, $(1,5)$-, $(1+1)$-, and $(1,1)$ evolution strategies; normal scale on the left, logarithmic scale for $\check{\sigma}$ on the right hand side)

7.3.2 The $(\mu + 1)$ evolution strategy

Rather early, the two membered evolution strategy has been the basis for expanding the population size. Three different ways of realizing this idea can be identified:

- Increasing the number of predecessors,
- Increasing the number of successors,
- Increasing both numbers.

The corresponding ES versions have been termed $(\mu + 1)$-, $(1 + \lambda)$-, and $(\mu + \lambda)$ ESs, a variant of the latter being called (μ, λ) evolution strategy.

Within the $(\mu + 1)$ ES only the number of predecessors is increased from 1 to μ. It still works in a strictly sequential way by creating just one successor at a time. Only if the successor outperforms at least one of the μ predecessors, it replaces one of the latter - generally the worst one. Therefore, one might speak of an 'extinction-of-the-worst' principle that can be found in a couple of other optimization techniques, as well, e.g., in the EVOP- or evolutionary operation method of G. E. P. Box[Box57], a deterministic factorial design technique for experimental optimization, and in all polyhedron strategies like Nelder and Mead's Simplex[NM65]- or M. J. Box's Complex[Box65] methods.

The $(\mu+1)$ evolution strategy invites straightforwardly the incorporation of another evolution principle: recombination. Any mix of two (or even more, especially when used as a computer program) parental genomes may be transferred to the successor before the mutation operator does its work. Rechenberg[Rec94] showed already in the early 1970ies that recombination may enhance the efficiency of an evolution strategy considerably. On the other hand it consumes additional storage capacity, which was relatively expensive in those days. The main reason why the $(\mu + 1)$ ES is no longer used today lies in its inability to incorporate self-adaptation of the internal strategy parameters. We therefore skip all theoretical analyses of that ES variant here.

7.3.3 The $(1 + \lambda)$ evolution strategy

Increasing the size of the predecessor's progeny per generation from just only 1 to λ improves the convergence velocity per iteration cycle considerably. The same problem as above, the sphere model, is considered again. Only the best successor becomes predecessor of the next cycle here, and if it turns out to be worse than its predecessor, then the predecessor 'survives'.

According to (7.9), the relative change in the distance to the goal for each successor is already known to be

$$P \approx -\frac{\breve{\sigma}^2}{2n} - \frac{\breve{\sigma}}{n} N''.$$

The distribution function thus is $F_P(x) = \Phi((x - \theta)/\eta)$ with expectation $\theta = -\breve{\sigma}^2/2n$ and standard deviation $\eta = \breve{\sigma}/n$. According to subsection 7.1.4 on order statistics the maximal change P_λ, i.e. that of the best successor, is, according to the stochastic independence of the successors, distributed according to

$$F_{P_\lambda}(x) = \Phi^\lambda\left(\frac{x - \theta}{\eta}\right).$$

The expectation of P_λ can be calculated by means of (7.2) as

$$\mathrm{E}[P_\lambda] = \frac{1}{\eta} \int_{-\infty}^{\infty} x f_{P_\lambda}\left(\frac{x - \theta}{\eta}\right) dx, \qquad (7.10)$$

where the density of P_λ results from (7.1) as

$$f_{P_\lambda}(x) = \frac{d}{dx} \Phi^\lambda(x) = \frac{\lambda}{\sqrt{2\pi}} \exp\left(-\frac{x^2}{2}\right) \Phi^{\lambda-1}(x).$$

The expectation of the relative improvement with respect to the random variable $P_{1+\lambda} := \max\{0, P_\lambda\}$ can be calculated in analogy to the (1+1) ES case according to

$$\mathrm{E}[P_{1+\lambda}] = \frac{1}{\eta} \int_{0}^{\infty} x f_{P_\lambda}\left(\frac{x - \theta}{\eta}\right) dx.$$

Substituting $z := (x - \theta)/\eta$ one arrives at

$$\begin{aligned}
\mathrm{E}[P_{1+\lambda}] &= \int_{-\theta/\eta}^{\infty} (z\eta + \theta) f_{P_\lambda}(z) dz \\
&= \frac{\breve{\sigma}}{n} \int_{\breve{\sigma}/2}^{\infty} z f_{P_\lambda}(z) dz - \frac{\breve{\sigma}^2}{2n}\left[1 - \Phi^\lambda\left(\frac{\breve{\sigma}}{2}\right)\right].
\end{aligned}$$

A numerical evaluation is presented in figures 7.2 and 7.3 (solid lines on the left hand sides for the $(1+\lambda)$ ES): Increasing the number λ of successors should be accompanied by enlarging the mutation step size. Since only the best successor is of importance here, the increasing number of failures is completely ignored.

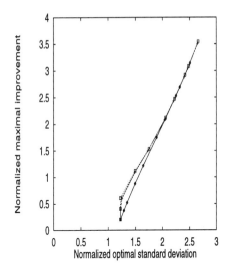
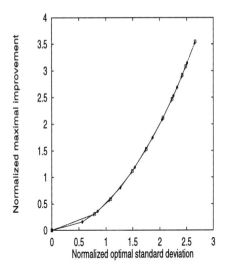

Figure 7.2: The normalized maximal improvement $\check{\varphi}^*$ over the normalized optimal standard deviation $\check{\sigma}^*$ for plus-ESs (left) and comma-ESs (right) (from bottom to top in each figure the curves correspond to the cases $\mu = 1$, $\mu = 10$, and $\mu = 50$ with varying ratios $\frac{\lambda}{\mu}$)

7.3.4   The (1,λ) evolution strategy

Each non-elitist (or comma-) version of the evolution strategies that ignores the so far achieved best intermediate result during the selection process, is prone to diverge — at least temporarily. A convergence proof thus is a difficult task. Nevertheless, this task has been solved meanwhile [Rud94]. We shall not go into the details here, however.

For certain optimization problems it is even possible to calculate mean convergence rates. We shall demonstrate that here for the above introduced objective function $f(x) := \|x - x^*\|^2 = r^2$.

In contrary to all plus-versions of the ES above, the $(1, \lambda)$ ES may lead to intermediary deterioration. The expected relative improvement is given by P_λ directly and thus may be negative, especially. The substitution $z = (x - \theta)/\eta$ in (7.10) yields

$$
\begin{aligned}
E[P_{1,\lambda}] &= \int_{-\infty}^{\infty} (z\eta + \theta) f_{P_\lambda}(z) dz \\
&= \frac{\check{\sigma}}{n} \int_{-\infty}^{\infty} z f_{P_\lambda}(z) dz - \frac{\check{\sigma}^2}{2n} \\
&= (2\check{\sigma} c_{1,\lambda} - \check{\sigma}^2)/2n,
\end{aligned} \tag{7.11}
$$

where

$$
c_{1,\lambda} := \int_{-\infty}^{\infty} z f_{P_\lambda}(z) dz
$$

denotes the expectation of the maximum of λ stochastically independent standard normally distributed random variables, a constant that depends on λ only.

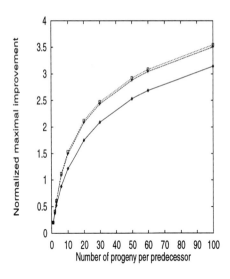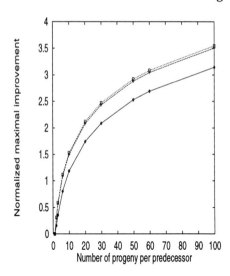

Figure 7.3: The normalized maximal improvement $\check{\varphi}^*$ over the ratio $\frac{\lambda}{\mu}$ for plus-ESs (left) and comma-ESs (right) (from bottom to top in each figure the curves correspond to the cases $\mu = 1$, $\mu = 10$, and $\mu = 50$ with varying ratios $\frac{\lambda}{\mu}$)

Differentiating equation (7.11) with respect to $\check{\sigma}$, one obtains due to

$$\partial^2 \mathrm{E}[P_{1,\lambda}]/\partial\check{\sigma}^2 \equiv -1 < 0$$

as optimal normalized standard deviation $\check{\sigma}^* = c_{1,\lambda}$. The optimal mean mutation step size thus becomes

$$\sigma^* = \frac{c_{1,\lambda}}{n} r = \frac{c_{1,\lambda}}{2n}\|\nabla f(x)\|.$$

The expectation of an improvement corresponds to $\mathrm{E}^*[P_{1,\lambda}] = c_{1,\lambda}^2/2n$, or in normalized terms, to $\check{\varphi}_{1,\lambda}^* = \frac{1}{2}c_{1,\lambda}^2 = \frac{1}{2}\check{\sigma}^{*2}$, respectively. Values for $c_{1,\lambda}$ are tabulated and may be found in [Rec94]. For large n and large λ one may make use of an approximation reported in [BRS93] saying that $c_{1,\lambda}$ is approximately proportional to $\sqrt{2\ln\lambda}$. With this result one may conclude for the expectation of the optimal relative improvement per generation

$$\mathrm{E}^*[P_{1,\lambda}] \sim \frac{\ln\lambda}{n} \quad \text{or} \quad \check{\varphi}_{1,\lambda}^* \sim \ln\lambda \tag{7.12}$$

in case of large n and large λ. This result is approximately valid also for the $(1 + \lambda)$ ES under the same assumptions since for large λ the probability that all successors are worse than their predecessor diminishes towards zero. A numerical evaluation is presented in figures 7.2 and 7.3 (solid lines on the right hand sides for the $(1, \lambda)$ ES): Differences to the $(1+\lambda)$ ES obviously appear only if λ is chosen to be small, especially for $\lambda \le 5$.

If the progeny of one generation can be generated and evaluated in parallel, then the approximation above helps to calculate the order of the speedup for a $(1, \lambda)$ ES over the merely sequential $(1 + 1)$ ES. Using

$$\beta := \mathrm{E}\left[\frac{r^{(t)} - r^{(t+1)}}{r^{(t)}}\right] = \begin{cases} \mathrm{E}[P_{1+1}] = & 0.2025/n \quad \text{for the } (1+1) \text{ ES} \\ \mathrm{E}[P_{1,\lambda}] = & \ln \lambda/n \quad \text{for the } (1, \lambda) \text{ ES.} \end{cases} \tag{7.13}$$

we find by induction from (7.13) the expected error in generation t

$$\mathrm{E}[r^{(t)}] = (1 - \beta)r^{(t-1)} = (1 - \beta)^t r^{(0)}.$$

A simple transformation yields

$$\mathrm{E}[r^{(t)}] = (1 - \beta)^t r^{(0)} < \epsilon \Leftrightarrow t > \log_{1-\beta}(\epsilon/r^{(0)}),$$

and as expected number of steps that a (1+1)- or $(1, \lambda)$ ES needs for reaching an ϵ-vicinity of the optimum we get

$$\mathrm{E}[t_\epsilon] = \log_{1-\beta}(\epsilon/r^{(0)}).$$

The speedup can be calculated by means of a Taylor series for $\log(1 - x)$ cut off after the linear term:

$$\begin{aligned} S_\lambda &:= \frac{\mathrm{E}[t_{\epsilon,1}]}{\mathrm{E}[t_{\epsilon,\lambda}]} = \frac{\log_{(1-\beta_1)}(\epsilon/r^{(0)})}{\log_{(1-\beta_\lambda)}(\epsilon/r^{(0)})} \\ &= \frac{\ln(\epsilon/r^{(0)})}{\ln(1 - 0.2025/n)} \frac{\ln(1 - \ln\lambda/n)}{\ln(\epsilon/r^{(0)})} \\ &\approx \frac{(\ln\lambda)/n}{0.2025/n} = O(\ln\lambda). \end{aligned}$$

Thus, a $(1, \lambda)$ ES provides an expected speedup that is only logarithmic in λ. The same is true for a $(1 + \lambda)$ ES since (7.12) is valid for large λ in both cases.

7.3.5   The (μ, λ) evolution strategy

In order to achieve a higher speedup one has to increase the number of survivors during the selection process. Only at the first sight this reduces the convergence velocity. Recently it has been shown, at least approximately [Bey94], that equation (7.12) has to be rewritten for a (μ, λ) ES as

$$\check{\varphi}_{\mu,\lambda}^* \sim \ln\frac{\lambda}{\mu}.$$

A numerical evaluation is presented within figures 7.2 and 7.3: Especially figure 7.3 demonstrates that the ratio $\frac{\lambda}{\mu}$ dominates the speed of convergence. Increasing the population size beyond $\mu = 10$ does not help in case of no recombination.

Nature's trick for a speedup lies in the recombination itself. We shall not dive into the rather lengthly calculation, but just provide the result from [Bey94] here. There is no difference in the expected maximum of the convergence velocity between two kinds of recombination, i.e., intermediary recombination of all μ predecessors and global discrete recombination:

$$\check{\varphi}^*_{\mu,\kappa=1,\lambda,\rho=\mu} \sim \mu \ln \frac{\lambda}{\mu}$$

The mechanisms of the two types of recombination being quite different from each other, it is not astonishing that the corresponding optimal mutation step sizes are different as well:

$$\sigma^* = \begin{cases} \sqrt{2\varphi^*} & \text{for uniform crossover} \\ \sqrt{2\mu\varphi^*} & \text{for global intermediary multirecombination.} \end{cases} \tag{7.14}$$

With respect to a theory of the autoadaptation to those optimal step sizes we refer to recent results of Beyer [Bey95].

REFERENCES

[BRS93] T. Bäck, G. Rudolph and H.-P. Schwefel, Evolutionary Programming and Evolution strategies: Similarities and Differences, in: D. B. Fogel and W. Atmar, editors, *Proceedings of the Second Annual Conference on Evolutionary Programming*, San Diego, Feb. 25-26, 1993 Evolutionary Programming Society, La Jolla CA, pp. 11–22.

[Bey94] H.-G. Beyer, *Towards a theory of 'evolution strategies'—results from the N-dependent (μ, λ) and the multi-recombinant $(\mu/\mu, \lambda)$ theory*, technical report SYS-5/94, Systems Analysis Research Group, University of Dortmund, Department of Computer Science, Oct. 1994.

[Bey95] H.-G. Beyer, *Towards a theory of 'evolution strategies': The $(1, \lambda)$-self-adaptation*, technical report SYS-1/95, Systems Analysis Research Group, University of Dortmund, Department of Computer Science, May 1995.

[Box57] G. E. P. Box, Evolutionary operation—a method for increasing industrial productivity, Appl. Stat. **6**(1957), 81–101.

[Box65] M. J. Box, A new method of constrained optimization and a comparison with other methods, Comp. J. **8**(1965), 42–52.

[NM65] J. A. Nelder and R. Mead, A simplex method for function minimization, Comp. J. **7**(1965), 308–313.

[Rap84] G. Rappl, *Konvergenzraten von Random Search Verfahren zur globalen Optimierung*, PhD Thesis, Hochschule der Bundeswehr, Munich, 1984.

[Rec94] I. Rechenberg, *Evolutionsstrategie '94*, Frommann-Holzboog, Stuttgart, 1994.

[Rud94] G. Rudolph, Convergence of non-elitist strategies, in: Z. Michalewicz, J. D. Schaffer, H.-P. Schwefel, and D. B. Fogel, editors, *Proceedings of the 1st IEEE Conference on Evolutionary Computation, IEEE World Congress on Computational Intelligence*, Orlando FL, June 27-29, 1994, vol. 1, pp. 63–66.

[Schw95] H.-P. Schwefel, *Evolution and Optimum Seeking*, Wiley, New York, 1995.

8

Genetic Algorithms in Fuzzy Control Systems

JUAN R. VELASCO and LUIS MAGDALENA[1]

8.1 INTRODUCTION

Fuzzy Logic Controllers [Lee90] (FLCs) are being widely and successfully applied in different areas. Fuzzy Logic Controllers can be considered as knowledge-based systems, incorporating human knowledge into their Knowledge Base through Fuzzy Rules and Fuzzy Membership Functions (among other information elements). The definition of these Fuzzy Rules and Fuzzy Membership Functions is actually affected by subjective decisions, having a great influence over the performance of the Fuzzy Controller. From this point of view, FLCs can be interpreted as a particular type of real time expert systems. A second interpretation more adequate for the analysis of the control properties of the FLC is to think about FLCs as non-linear, time-invariant control laws. In addition, recent works have demonstrated the ability of Fuzzy Controllers to approximate continuous functions on a compact set with an arbitrary degree of precision; different kinds of FLCs are universal approximators ([Buc93, Cas95]). Combining ideas related to these different interpretations, some efforts have been made to obtain an improvement in system performance (a better approximation to an optimal controller, with a certain performance criterion) by incorporating learning mechanisms to modify predefined rules and/or membership functions, represented as parameterized expressions. The main goal will be to combine the ability to incorporate experts' knowledge with a knowledge-based point of view (Knowledge engineering),

---

[1] Dr. Velasco (jvelasco@gsi.dit.upm.es) is with Dept. Ingeniería de Sistemas Telemáticos and Dr. Magdalena (llayos@mat.upm.es) is with Dept. Matemática Aplicada a las Tenologías de la Información.
ETSI Telecomunicación, Universidad Politécnica de Madrid, MADRID 28040 (SPAIN).

with the possibility of tuning by applying learning (Machine learning) or adaptation (Adaptive control) techniques through the analytical representation of the FLC. Ideas arising out of two main areas have been applied with this aim: ideas coming from Artificial Neural Networks (ANNs) and from Genetic Algorithms (GAs).

When applying ideas coming from ANNs ([Mag95]), the learning techniques use basically the topological properties of $\mathbb{R}^n$ (e.g., the properties of gradient), where $\mathbb{R}^n$ represents the space of parameters of the controller. On the other hand, GAs ([DeJ88, Hol75]) are probabilistic search and optimization procedures based on natural genetics, working with finite strings of bits that represent the set of parameters of the problem, and with a fitness function to evaluate each one of these strings. The finite strings used by GAs may be considered as a representation of elements of $\mathbb{R}^n$, but usually, the learning mechanisms make no use of the topological properties of this space of parameters.

The application of Genetic Algorithms to FLCs with a learning purpose, has produced some interesting works like those described in Section 8.3. This chapter presents an overview of the area and a deeper analysis of two different works applying Genetic Algorithms to Fuzzy Logic Controllers whose Rule Base is defined through a set of Fuzzy Rules. The use of a set of Fuzzy Rules (and not a Fuzzy Relational Matrix or a Fuzzy Decision Table) is adapted to the application to complex control problems containing a large number of variables, since it reduces the dimensionality of the Knowledge Base for this kind of system. The first approach uses the Knowledge Base of the system as the population of the genetic system (a single rule containing the description of the corresponding Fuzzy Sets is an individual of the population). The second one uses the Knowledge Base (containing a set of Fuzzy Rules and a set of Membership Functions) as the individual of the genetic system, working with a population of Fuzzy Controllers. Each system has its own scheme to code the information evolved through the GA, and its evolution operators.

8.2   FUZZY LOGIC CONTROLLERS

We will first present the concept of a fuzzy set and other related ideas as a previous knowledge to introduce Fuzzy Logic Controllers.

A set may be defined using different methods: enumerating their elements, defining a condition that separates the elements belonging to the set, from the remaining elements, using a characteristic function that takes value 1 for all the elements belonging to the set, and value 0 otherwise, etc. When using a characteristic function, a set might be defined as a function from the universe (U) to the set $\{0,1\}^2$. A generalization of this definition based on a characteristic function, obtained by allowing values from the whole interval $[0,1]$, will produce a new type of set that will be called *fuzzy set* [Zad65].

A *fuzzy set* F in a universe of discourse X is characterized by a membership function μ_F that takes values within $[0,1]$.

$$\mu_F : X \to [0,1] \tag{8.1}$$

---

[2] In the following this kind of set will be called *crisp* sets.

A point-wise fuzzy set or a singleton, is a fuzzy set that is reduced to a single element with membership function 1. The support of a fuzzy set F is the crisp set containing those elements of X, with a membership function μ_F greater than 0. The α-cut of a fuzzy set F is the crisp set containing those elements of X, with a membership function μ_F greater than or equal to α. A fuzzy set A is convex if and only if,

$$\forall x, y \in X, \forall \lambda \in [0,1] : \mu_A(\lambda x + (1-\lambda)y) \geq \min(\mu_A(x), \mu_A(y)). \tag{8.2}$$

The use of fuzzy sets provides a basis for a systematic way of manipulating vague and imprecise concepts.

A fuzzy set may be defined in a discrete universe of discourse or in a continuous universe of discourse, and in both cases there are different ways of representing the fuzzy set. Fuzzy sets defined in a discrete universe of discourse are usually described by pairs composed of an element ($x \in X$) and its corresponding membership function ($\mu_F(x)$), this method may be considered as the equivalent to an enumeration of a discrete crisp set. When the universe of discourse is continuous, different representations of a fuzzy set are applied. The first method is the use of parameterized membership functions[3] (triangular, trapezoidal, Gaussian, sigmoidal and other types of parameterized fuzzy sets are used), using the parameters as representation. The second method is the use of the representation of discrete fuzzy sets, after discretizing the continuous fuzzy set. The third method is the use of several α-cuts (occasionally referred to as h-levels or horizontal levels) of the fuzzy set, this method is a sort of discretization but on the values of the function not on the variables.

The basic operations defined for crisp sets (intersection, union and complementation) may be generalized for fuzzy sets. The intersection of two fuzzy sets is defined through any *t-norm*,

$$\mu_{A \cap B}(x) = \mu_A(x) \text{ T } \mu_B(x). \tag{8.3}$$

A t-norm is a function of two arguments, nondecreasing in each argument, commutative, associative, and satisfying the boundary conditions $x \text{ T } 0 = 0$ and $x \text{ T } 1 = x$. The union of fuzzy sets is defined by any *s-norm*, called *t-conorm* too,

$$\mu_{A \cup B}(x) = \mu_A(x) \text{ S } \mu_B(x). \tag{8.4}$$

An s-norm is a function of two arguments, nondecreasing in each argument, commutative, associative, and satisfying the boundary conditions $x \text{ S } 0 = x$ and $x \text{ S } 1 = 1$. The complement of a fuzzy set is usually obtained as $\mu_{\overline{A}}(x) = 1 - \mu_A(x)$ but may be defined using different expressions. A complement operation c should satisfy at least two conditions: $c(0) = 1$, $c(c(a)) = a$, and $a < b \text{ implies } c(a) > c(b)$. Other mathematical concepts may be generalized for fuzzy sets through the extension principle, that in a simple expression may be defined as follows: If U and V are two universes of discourse and f is a mapping from U to V, for a certain fuzzy set A in U the extension principle defines a fuzzy set B ($f(A)$) in V by

$$\mu_B(v) = \sup_{u \in f^{-1}(v)} \{\mu_A(u)\}. \tag{8.5}$$

This expression may be applied to generalize fuzzy sets, arithmetical operations or any other mathematical concept defined as a mapping.

---

[3] Most of the papers cited in the following Section use this method.

This Section will introduce some ideas on Fuzzy Logic Controllers that will be used later. A survey of the main aspects of the analysis and design of FLCs may be found in the previous cited references or in [Lee90, Ped93, Wan94].

A fuzzy rule based system is characterized by a set of rules and by the definitions of some concepts related to information processing: the aggregation operators, the fuzzy connectives and the inference method. Rules are defined by their antecedents and consequents. Antecedents, and frequently consequents, are associated with fuzzy concepts:

$$R_1: \text{if } x_1 \text{ is } A_{11} \text{ and } \ldots \text{ and } x_m \text{ is } A_{m1} \text{ then } y \text{ is } B_1,$$

$$\ldots$$

$$R_n: \text{if } x_1 \text{ is } A_{1n} \text{ and } \ldots \text{ and } x_m \text{ is } A_{mn} \text{ then } y \text{ is } B_n. \tag{8.6}$$

Occasionally consequents are analytical functions of the input variables:

$$R_1: \text{if } x_1 \text{ is } A_{11} \text{ and } \ldots \text{ and } x_m \text{ is } A_{m1} \text{ then } y = f_1(x_1, \ldots, x_m),$$

$$\ldots$$

$$R_n: \text{if } x_1 \text{ is } A_{1n} \text{ and } \ldots \text{ and } x_m \text{ is } A_{mn} \text{ then } y = f_n(x_1, \ldots, x_m). \tag{8.7}$$

In both cases x_i are input variables, A_{ij} are fuzzy sets related to input variables, y is the output variable, B_k are fuzzy sets related to the output variable, and f_l are functions of the input variables

$$f_l(x_1, \ldots, x_m) = a_{0l} + a_{1l}x_1 + \ldots + a_{ml}x_m. \tag{8.8}$$

The fuzzy connective *and*, between fuzzy concepts, is usually implemented through the product or the minimum operators (any t-norm may be used as the *and* connective). Systems using the first type of rule (expression 8.6) are usually called Mamdani type controllers [Mam74], while those that use the second type of rule (expression 8.7) are usually named TSK (Takagi, Sugeno and Kang) fuzzy systems [TS85].

A fuzzy relationship over the collections of fuzzy sets $A_1, \ldots, A_m, B$, is a fuzzy subset over their Cartesian product, $A_1 \times \ldots \times A_m \times B$. In a fuzzy relationship, each element of the Cartesian product $(A_{1i}, \ldots, A_{mj}, B_k)$, has a grade of membership to the Relation, represented by $\mu_R(A_{1i}, \ldots, A_{mj}, B_k)$ and taking values within the interval $[0, 1]$. Table 8.2 shows the tabular representation of a fuzzy relation for an FLC with one input (X) and one output (Y) variable, with three fuzzy sets (A_{11}, A_{12}, A_{13}) related to the input variable and four fuzzy sets related to the output variable (B_1, B_2, B_3, B_4).

Table 8.1   A fuzzy relation R.

R	B_1	B_2	B_3	B_4
A_{11}	0.5	0.8	0.2	0.0
A_{12}	0.0	0.3	1.0	0.1
A_{13}	0.0	0.0	0.3	1.0

The complete behavior of a fuzzy system may be characterized by a fuzzy relation that is the combination of the fuzzy relations defined by each element of the rule set.

This combination may be represented through the connective *also*,

$$R = \text{also} \ (R_1, \ldots, R_n), \tag{8.9}$$

where also is usually implemented with the maximum operator (any t-conorm or s-norm may be used as the *also* connective), generating a fuzzy output. This fuzzy output is a fuzzy subset of Y, from which a non-fuzzy output is usually obtained (deffuzification). Two deffuzification methods often used are the Center of Area (COA) method and the Mean of Maxima (MOM) method. The non-fuzzy output obtained with these two methods has the following expressions:

$$y = \frac{\displaystyle\int_Y \mu_o(y) \, y \, dy}{\displaystyle\int_Y \mu_o(y) \, dy} \ (\text{COA}), \quad y = \frac{\displaystyle\int_{Y^*} y \, dy}{\displaystyle\int_{Y^*} dy} \ (\text{MOM}). \tag{8.10}$$

Where Y is the universe of discourse of the output variable, $\mu_o(y)$ is the membership value of the output value y to the fuzzy output, and Y^* is the α-cut of the fuzzy output, with α equal to the maximum of $\mu_o(y)$.

When using consequences that are functions of the input variables, the *also* connective is implemented as a weighted sum of rules output, producing, then, a numerical output,

$$y = \frac{\displaystyle\sum_n \mu_i y_i}{\displaystyle\sum_n \mu_i}, \tag{8.11}$$

where μ_i is the truth value of the antecedent of rule i, and y_i is the output of rule i.

In addition to the representation of the behavior of the fuzzy system through a fuzzy relation or with a set of fuzzy rules, a third possibility is the representation using a fuzzy decision table. A fuzzy decision table represents a special case of a crisp relation (the ordinary type of relations we are familiar with) defined over the collections of fuzzy sets corresponding to the input and output variables. Table 8.2 shows a fuzzy decision table for an FLC with two input (X_1, X_2) and one output (Y) variable, with three fuzzy sets $(A_{11}, A_{12}, A_{13}; A_{21}, A_{22}, A_{23})$ related to each input variable and four fuzzy sets related to the output variable (B_1, B_2, B_3, B_4).

Table 8.2 A fuzzy decision table.

	A_{21}	A_{22}	A_{23}
A_{11}		B_1	B_2
A_{12}	B_1	B_2	B_3
A_{13}	B_1	B_3	B_4

The modular structure of an FLC is:

1. A *Fuzzification interface* that transfers the values of input variables into fuzzy information, assigning grades of membership to each fuzzy set defined for that variable.

2. A *Knowledge Base* that comprises a Data Base, and a fuzzy control Rule Base. The Data Base is used in fuzzification and defuzzification processes.

3. An *Inference Engine* that infers fuzzy control actions employing fuzzy implications and the rules of inference of fuzzy logic.

4. A *Defuzzification interface* that yields a non fuzzy control action from an inferred fuzzy control action.

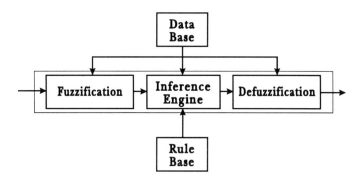

Figure 8.1   Structure of an FLC.

Figure 8.1 shows the defined structure. The fuzzy systems that will be described in the following have this structure.

8.3   GENETIC ALGORITHMS AND FUZZY LOGIC CONTROLLERS

The systems enumerated in this Section apply the general ideas of GAs to FLCs, using their own coding scheme. GAs are applied to modify the membership functions and/or the Rule Base. Some of these works use strings of numbers instead of strings of bits. The evaluation function, the composition of the first generation (initial population) and the termination condition are related to the task for which each FLC was designed.

When modifying the membership functions, these functions are parameterized with one to four coefficients (some examples are shown in Figure 8.2), and each of these coefficients will constitute a gene of the chromosome for the GA. Triangular functions are used in [Kar91, CV93, PKL94] (symmetric, Figure 8.2 center) and [LT93] (Figure 8.2 right); trapezoidal functions in [KG93] and other more complex functions with continuous derivatives in [SFH95] (radial functions) and [LM94] (Gaussian functions). Each of the coefficients will constitute a gene of the chromosome for the GA. This gene may be a binary code (representing the coefficient) [Kar91, CV93, KG93, LT93, SFH95] or a real number (the coefficient) [PKL94, LM94].

Different methods are defined to apply GAs to the Rule Base, depending on its represntation: a set of rules, a decision table or a relational matrix.

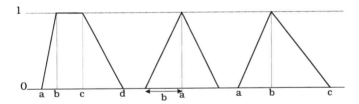

Figure 8.2   Some parameterized fuzzy membership functions.

Liska and Melsheimer [LM94] use a rule base defined as a set with a fixed number of rules. The membership functions are labeled, coding each rule with integer numbers (labels) that define the membership function related to a certain input or output variable that is applied by the rule (membership functions for every variable are ordered). In [LT93] and [SFH95] the fuzzy membership functions are included in the code of the rules, applying GAs to obtain these functions and the coefficients of the consequent for a TSK fuzzy model. One gene is used to code each coefficient of a consequent (rules for a system with two input variables need three coefficients to obtain the output value, as shown in expression 8.7), in [SFH95] a single coefficient is considered for the output. In [CV93], rules with fuzzy inputs and fuzzy outputs (like those in expression 8.6) are used.

In [Thr91], GAs are used to modify the decision table of an FLC, which is applied to control a system with two input and one output variables. A chromosome is formed from the decision table by going row-wise and coding each output fuzzy set as an integer in 0, 1, ..., n, where n is the number of membership functions defined for the output variable of the FLC. Value "0" indicates that there is no output, and value "k" indicates that the output fuzzy set has the k-th membership function. Applying this code to the fuzzy decision table represented in table 8.2, the obtained string is (0,1,2,1,2,3,1,3,4). The mutation operator changes a fuzzy code either up a level or down a level, or to zero (if it is already zero, then it chooses a non-zero code at random).

Occasionally GAs are used to modify the fuzzy relational matrix (R) of an FLC with one input and one output. The chromosome is obtained by concatenating the $m \times n$ elements of R, where m and n are the number of fuzzy sets associated with the input and output variable respectively. The elements of R that will make up the genes may be represented by binary codes [PK91] or real numbers [PKL94], e.g., (0.5,0.8,0.2,0.0,0.0,0.3,1.0,0.1,0.0,0.0,0.3,1.0) for the relation defined by table 8.2.

All these works propose really interesting ways of adding learning capabilities to an FLC by means of GAs, obtaining results that show the ability of GAs to improve the performance of predefined FLCs. The application examples of these works present FLCs with two to five variables (input and output) and up to twenty membership functions (adding those from each variable). The question is how do they work when the problem dimensions increase.

Two new approaches that use different coding strategies to avoid the problems that will probably produce larger dimension FLCs are proposed in the following sections.

8.4   THE FUZZY RULE-BASED APPROACH

As pointed out by different authors, when applying GAs to FLCs, there are two basic decisions to be made: how to code the possible solutions to the problem as a finite bit string, and how to evaluate the merit of each string.

The way we code a possible solution as a bit string constitutes the way we translate the problem defined on the space of parameters of the controller (represented by $\mathbb{R}^n$) to a certain space where GAs will operate. But this is not the first question to be answered. Previously we have to define the general characteristics of the FLC to be applied, characteristics that will affect the dimension and the properties of the space of parameters where the adaptation or learning process will occur. As a result of the dimension and properties of this space, the learning process will become faster or slower and even treatable or untreatable. The obvious conclusion is that an adequate selection of the characteristics of the FLC is crucial for the whole learning process. The following Sections will present the coding scheme used by each learning system, but previously some common characteristics to both FLCs will be described.

The general structure of the FLC will be composed of a normalizer, a fuzzifier, an inference engine, a defuzzifier and a denormalizer; the scheme of Figure 8.1 with the addition of a normalization and a denormalization elements. The input variables will be linearly normalized from its real interval to [-1,1] interval. The fuzzy sets will have trapezoidal membership functions defined through four parameters (Figure 8.2, left). The rules applied by the inference engine will have a fuzzy antecedent and a fuzzy consequent (as in expression 8.6), and will be described as a set of fuzzy rules. The output variables will be linearly denormalized from [-1,1] interval to its real interval.

The use of normalized variables in fuzzification and defuzzification processes may be interpreted with two different meanings. If the normalization limits are fixed a priori, the only effect is conceptual, producing an FLC that works with variables defined in a normalized universe of discourse. If the normalization limits are considered as adaptable parameters throughout the learning process, the obtained effect may be interpreted as that of a controller with a parameterized gain with respect to each input and output variable. The parameterized gain for each variable will be directly related to the corresponding normalization limits, and may be tuned through the learning process.

The use of trapezoidal membership functions is one of the possible selections for parameterized membership functions. The most broadly used parameterized membership functions are: triangular, trapezoidal, Gaussian, bell and sigmoidal. These parameterized functions may be classified into two main groups: the piece-wise linear functions (triangular and trapezoidal) and the differentiable functions (Gaussian, bell and sigmoidal). While the first group produces a reduced computational complexity, the second one is particularly interesting when derivatives are involved in the learning process (most of the fuzzy-neural systems). In the applications that will be presented in the following, the learning process will not involve any kind of derivative, and the computational complexity leads us to select piece-wise linear functions. From piece-wise linear functions, trapezoidal membership functions will be selected since triangular membership functions may be considered as a particular case of a trapezoidal function.

When applying GAs to FLCs, rules are generally represented as a fuzzy decision table or in some cases (systems with one input and one output) as a fuzzy relational matrix. The use of a decision table has some advantages from the point of view of an easier analysis of completeness and consistency of the knowledge base. The idea of working with sets of rules has been applied recently as a way to solve dimensionality problems ([CV93, Mag94, LM94, SFH95]) as will be commented on in the two following sections. Our systems will apply this idea, working directly with the set of fuzzy rules.

For applying these ideas, two different learning structures are possible if we consider an approach closer to the Pittsburgh or the Michigan approaches. By saying *closer to the Pittsburgh approach* we mean maintaining a population of complete solutions to the problem, while by saying *the Michigan approach* we mean using the rules as the individuals of the population, representing a complete solution with the whole population. All the previously referenced works are closer to the Pittsburgh approach. In the following, a couple of Fuzzy Genetic systems will be described, the first one closer to the Michigan approach and the second one closer to the Pittsburgh approach.

8.5   WORKING WITH A POPULATION OF RULES

This first approach starts from Artificial Intelligence theory, with a conventional expert system architecture (with fuzzy knowledge), adding some new modules to create rules, evaluate their performance, etc. Nevertheless, the work is oriented towards the control of complex processes. Figure 8.3 shows the main blocks of the system, that will be described below:

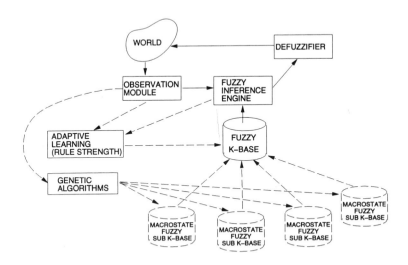

Figure 8.3   Rule System Architecture

Observation module: The facts base. This module must receive, analyze and filter the set of signals that the control system gets as inputs from the process. From this set of signals, the observation module generates the real input vector for the system.

Bad quality data are marked up, in order to avoid making inferences from them.

Fuzzy inference engine: It has been adapted to manage rules with explicit fuzzy sets and rule strength (see below).

Defuzzifier: The output of the system must be a real number. The process doesn't understand terms such as "high" or "small".

Adaptive Learning: Every rule has a strength (credibility) value that is modified by this module after every inference whether the rule has been applied or not.

Genetic Algorithms: To create new rules.

Fuzzy Knowledge bases: Split into several different KBs, each one adapted to a particular big state of the world. These *macrostates* may be obtained from experts or from a clustering process [VV94, QB94]. The reason for splitting the knowledge base into several small bases is that it will be easier to learn if the process is working in a small area of data than if it is working in the whole space. Initial knowledge bases can be obtained from three different ways:

1. Random base: it is the classical starting point for genetic learning algorithms. It may be useful if the objective is to know how the system learns, but it is unacceptable for real process control. The process cannot be controlled by a random knowledge base until it has learned some good rules.
2. Knowledge from experts: it is the artificial intelligence approach.
3. Knowledge from other learning algorithms: to obtain rules when experts are unaffordables or to complement their rules.

8.5.1 Rule Structure.

Rules have an IF <condition> THEN <action> structure, where <condition> is a conjunction of terms $< V_x \in$ [a,b,c,d]$>$, and <action> has the expression $< V_y \in$ [e,f,g,h]$>$. Each rule has a set of parameters associated to it: rule strength (credibility of the rule), rule strength average, life span, etc, used in the learning algorithms and in the fuzzy inference engine.

Each term must be read as "V_x belongs to the set defined by [a,b,c,d]", where [a,b,c,d] defines a trapezoidal fuzzy set (Figure 8.2, left). Conditions may have an indeterminate number of terms. On the other hand, actions are unique. This is because rule strength (rule credibility) is related to how good or bad the action proposed for the given conditions is: different actions should have different strengths.

The following expression shows a rule sample:

$$\textbf{If} \quad V_1 \in [0.1, 0.2, 0.4, 0.5] \text{ and } V_3 \in [0.3, 0.3, 0.4, 0.6]$$
$$\textbf{then} \quad V_7 \in [0.2, 0.3, 0.3, 0.4] \tag{8.12}$$

This rule definition gives flexibility to the system: facts are not limited to using predefined fuzzy sets. Rule generation algorithms will allow the system to find the best adapted rules for the control system.

Defuzzification is based on the COA method after pondering the consequents with rule strength, truth value and knowledge base activation value. This activation value represents the degree of membership of the actual state to the macrostate associated with the corresponding KB.

8.5.2 Rule Generation.

Genetic algorithms have been selected for creating new knowledge because they allow the systems to find new rules both near to good ones and far from them, looking for unknown good control actions. Unlike traditional systems, in this architecture GAs are not fired in every cycle of the system. In fact, the rule generation process only works from time to time, when a particular KB has been used several times. While GAs do not work, the rule evaluation algorithm adjusts rule strength. The application of the GA will have four steps.

1. Selection: It is made randomly according to rule strength: the more strength, the more the probability of being selected.
2. Uniform Crossover: Mixes genetic material for creating new individuals. (see fig 8.4). This operator has been selected over standard crossover because it allows the system to create new rules with whatever possible antecedents. Traditional crossover needs some kind of antecedents reordering (inversion operator).

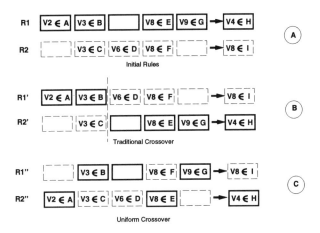

Figure 8.4   Crossover

3. Soft and Hard Mutation: Changes the fuzzy set (soft) or the variable (hard) of the fact (see fig 8.5). Soft mutation moves the (a, b, c, d) values of the trapezium to create a new fuzzy set. New values are in $([0, a+\frac{b-a}{2}]$, $[a + \frac{b-a}{2}, b + \frac{c-b}{2}]$, $[b + \frac{c-b}{2}, c + \frac{d-c}{2}]$, $[c + \frac{d-c}{2}, 1])$ ranges. This operator is introduced to create variety in fuzzy sets. On the other hand, hard mutation changes the variable of the fact (antecedent or consequent). It ensures that every variable is going to be taken into account.
4. Insertion: New rules are placed in the *Limbo*, a special place where rules are tested and evaluated without affecting the output of the controller. Only when a rule has proved a good one, it is accepted and inserted into the KB.

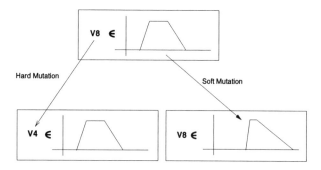

Figure 8.5   Hard and Soft Mutation

8.5.3   Evaluation.

Every genetic system must have an evaluation algorithm to decide which individuals are adapted to the environment and which are not. In a control system, we can have three different objectives, always defined from an objective variable: to maintain it at a given value, to minimize or to maximize it. These three objectives may be seen as a cost minimization problem, where the cost function is defined as:

$$C(t) = V(t) \qquad \text{if the problem is to minimize,}$$
$$C(t) = -V(t) \qquad \text{if the problem is to maximize,}$$
$$C(t) = |V(t) - \text{value}| \quad \text{if the problem is to maintain a given value.}$$

To evaluate each particular rule, we first have to calculate whether the system is near reaching its objective. But that is not enough: we have to estimate the influence that this rule has in the final action (in other words, how the output of the rule affects the system). Both combined values will give us the rule evaluation. Of course, this calculus must be made for every rule whose premises are true according to the fuzzy logic paradigm used.

System performance: Our objective will be to guide a specified variable X (the cost function) to its minimum value.

It is possible to try to minimize the absolute cost value, but if the system does that, it is almost sure that at a given moment this absolute minimum is unreachable. It is possible too that the action carried out could be the best action at that moment, with only a little decrement in the cost function. The proposed system will minimize the variable in an $N-$window, taking into account only the last N cases: the system is going to look for the local minimum at this moment.

If we look for the minimum value in the last n cases, X_{min} and X_{max} will change with time (in more detail in [MVFM92]). System performance will be:

$$SP = \begin{cases} 1 & \text{if } X_t \leq X_{min} \\ \frac{X - X_{t-1}}{X_{min} - X_{t-1}} & \text{if } X_t > X_{min} \wedge X_t \leq X_{t-1} \\ \frac{X_{t-1} - X}{X_{max} - X_{t-1}} & \text{if } X_t < X_{max} \wedge X_t > X_{t-1} \\ -1 & \text{if } X_t > X_{max} \end{cases} \qquad (8.13)$$

Applied rules: To know the influence of each applied rule, we have to analyze them rule by rule. Our data are: the fuzzy set (A) suggested for the variable V_a in the action part of the rule, and the real values for this variable (X) at times t and $t-1$. Rule influence will be:

$$RI = \begin{cases} \mu_A(X_t) & \text{if } \mu_A(X_t) > 0 \\ 0 & \text{if } \mu_A(X_t) = 0 \wedge |A - X_t| \leq |A - X_{t-1}| \\ \frac{X_t - X_{t-1}}{X_{t-1} - X_l} & \text{if } \mu_A(X_t) = 0 \wedge |A - X_t| > |A - X_{t-1}| \end{cases} \qquad (8.14)$$

where $|A - X|$ is the distance from X to the fuzzy set A (i.e., to the nearest point of the support of A), and X_l is the upper (if $X_t > X_{t-1}$) or lower (if $X_t < X_{t-1}$) limit of the Universe of Discourse of variable X.

Rule evaluation: Final rule evaluation is obtained by multiplying the rule influence estimation by system performance. If this performance is positive (the process is working well), rule evaluation depends on the rule influence: if the rule has been applied, final evaluation will be positive; if not, it will be negative. Dual considerations may be made for negative performance. And, of course, once we have the rule evaluation, it must be pondered by its overall truth value, to obtain the final value[4].

New strength for the rule: Rules increase their strength if their evaluation is positive, and decrease it otherwise. The expression to obtain this strength variation is:

$$S_{R,t} = \begin{cases} S_{R,t-1} + K \cdot T_R \cdot E \cdot (1 - S_{R,t-1}) & \text{if } E \geq 0 \\ S_{R,t-1-} + K \cdot T_R \cdot E \cdot S_{R,t-1} & \text{if } E < 0 \end{cases} \qquad (8.15)$$

where K is a constant that allows fixing of the memory of the system (K near 0 produces slow variations and vice-versa) T_R is the truth value of the rule, and E is the final evaluation of rule R (see [MVFM92]).

8.5.4   The limbo: How it works.

As has been shown, the evaluation algorithm allows the assignation of strength to every rule with positive truth value, regardless of whether they have been used by the fuzzy inference engine or not. In fact, what the algorithm uses is the *system performance* and the proposed actions of true rules. The real action made by the controller is not taken into account. This means that rule strength may be updated for those rules

---

4 The most significant aspect of this evaluation algorithm is that we can obtain values for every rule with a positive truth degree, regardless of their having been used at inference time or their being in the Limbo.

that are in the KB and for those that are in the limbo and would have been able to be fired. The rules that live in the limbo have been created by GAs but they have to prove their suitability before being used by the inference engine. In order to select or reject a rule from the limbo several parameters are used:

- Rule parameters:

  — Rule Age ($R.A.$): Number of inferences made since rule joined on to the limbo.
  — Rule Use ($R.U.$): Sum, over these inferences, of the successive truth values of the rule.
  — Rule Activations ($R.Ac.$): Number of inferences in which the rule should have taken part (the antecedent had a truth value grater than zero).
  — Equivalent Rule Evaluation ($E.R.E.$): Constant evaluation that the rule should have obtained to reach its present strength value after the same number of inferences.

- Limbo parameters

  — Limit Age ($L.A.$): Rule Age at which a rule has to leave the limbo.
  — Minimum Rule Activations ($M.R.Ac.$): Value of $R.Ac.$ that once reached allows a rule to be promoted from the limbo to the KB, before its Limit Age arrives.
  — Minimum Equivalent Evaluation ($M.E.E.$): Minimium value of the $E.R.E.$ that allows a rule to be promoted to the KB.

With these parameters in mind, a rule is killed once it has reached the Limit Age ($R.A. \geq L.A.$), if it has not been used ($R.Ac. = 0$) or it has not been a good rule ($E.R.E. < M.E.E.$). On the other hand, a rule is promoted from the limbo to the KB if $R.A. \geq L.A.$ and $E.R.E. \geq M.E.E.$; or if $R.Ac. \geq M.R.Ac.$ and $E.R.E. \geq M.E.E.$ (in this second case, the system does not wait to reach $L.A.$: if the rule is proving a good one, after a minimum number of uses it is promoted to the KB). In any other case the rule continues in the limbo.

In order to prevent the convergence of the KB (what in this case means to have a lot of identical rules in it), there is an additional filter to avoid too many copies of any rule: if a new rule is going to pass from the limbo to the KB, it should not have more than a given number of copies of it in the KB. In other situations, the rule is not inserted in the Rule Base.

8.5.5   *How to create rules from data records.*

With the proposed learning paradigm the control system can be trained with data files. The system will take a vector from a file, will determine the suggested actions after fuzzy evaluation, will read a new vector from the file, and will adjust rule strength both in the KB and in the limbo. From time to time GAs generate new rules to insert in the limbo and some rules go from the limbo to the KB.

That means that working with data files, and starting from a random rule base, the system can teach itself the initial rule base. This rule base will be adapted on line, once the program is controlling the process.

8.5.6   An application example and some learning results.

A control system with this algorithm is being installed in a fossil power plant at Velilla (Palencia, Spain). The objective of the system is to get low consumption while generated power is constant. In fact it is not a control system but a suggestion system.

An acquisition module gives 23 variables to the suggestion system. It will suggest operating 11 operation variables. The objective is to minimize the heat rate (ratio between used coal and generated power). It has been decided to use a continuous learning system because this environment is very complex and time variant (broken or dirty pipes, slow pumps, air temperature, humidity, ... are parameters that are out of our control and can affect the control system).

In order to analyze how the system learns, the following experiment has been designed: The control system will start with a 100-rule random Knowledge Base, and it will be trained with three different files as shown in the previous section:

1. A real historic data file.
2. A random data file, to compare learning results with the first case.
3. The real file, with random cost function, to assure that learning is achieved thanks to the algorithm and it has nothing to do with the data file used.

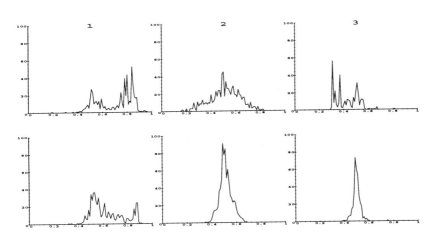

Figure 8.6   Final (upper) and average (lower) strength distribution.

The parameters of the three tests are:

1. Number of cases: 3514.
2. Max. Number of Rules in the KB: 1000.
3. Learning Constant K (for strength update): 0.25.
4. Age Limit for the limbo: 300.
5. Minimum Rule Activations: 10.
6. Minimum Equivalent Evaluation: 0.1.

The six graphics (Figure 8.6) show the strength and average strength distribution after the 3514 cases. As it is shown, in both random cases the strength distribution is around 0.5 (the initial strength), while in the real case it is distributed over the upper part of the axis. Only less than a 10% of learned rules have a strength lower than 0.5.

As final conclusion, the proposed learning system based on Genetic Algorithms applied over fuzzy rules is able to learn good fuzzy control rules and can be used in a fuzzy control system for complex time variant processes.

8.6   WORKING WITH A POPULATION OF KNOWLEDGE BASES

In this case we will use each Knowledge Base of the FLC as an individual of the population, then we will maintain a population of different Knowledge Bases to be applied by the FLC. This idea is illustrated in Figure 8.7.

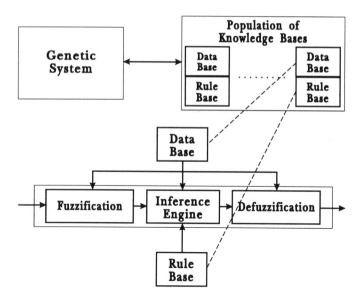

Figure 8.7   The genetic system works with a population of Knowledge Bases.

The Knowledge Base of the FLC contains the information to be coded, and is divided into a Data Base and a Rule Base. From the encoding point of view, the Data Base contains three different types of information: a set of parameters, a set of normalization limits and a set of membership functions; and the Rule Base contains a set of fuzzy control rules. All these types of information and their encoding schemes will be described in this section.

8.6.1   Encoding the Data Base Information.

Three elements must be encoded: the parameters of the FLC, the normalization limits and the membership functions.

The set of parameters defines the system dimensions, that is, the number of input variables (N) and output variables (M), and (assembled on vectors n and m) the number of linguistic terms (or the number of fuzzy sets) associated with each member of the set of input variables and output variables. The i-th component of vector n ($n = \{n_1, \ldots, n_N\}$) represents the number of linguistic terms associated with the i-th input variable. The j-th component of vector m ($m = \{m_1, \ldots, m_M\}$) is the number of linguistic terms associated with the j-th output variable.

An array of $(N + M) \times 2$ real numbers will represent the set of normalization limits. Each row in this array contains the limits of one input or output variable of the system ($\{v_{min}, v_{max}\}$).

The set of membership functions contains the trapezoidal membership functions of L fuzzy sets, where L could be obtained from n and m as shown in the following equations:

$$L_a = \sum_{i=1}^{N} n_i \ , \ L_c = \sum_{j=1}^{M} m_j \ , \ L = L_a + L_c. \qquad (8.16)$$

This set will be represented by an array of $L \times 4$ real numbers ranged in [-1,1] (as the variables are normalized, the fuzzy sets must be defined with the same range). Each row in the array contains the four parameters that describe a trapezoidal fuzzy set. We can obtain a string of reals by concatenating the rows of this array as has been done in some works presented in Section 8.3. The code of membership functions will not be included in the code of rules, as it was in the method described in the previous section.

8.6.2   Fuzzy Rules Representation.

As has been previously said, the fuzzy system will be characterized by a set of fuzzy rules.

When working with a multiple input system, decision tables become multidimensional. A system with three input variables produces three-dimensional decision tables. The number of "cells" of these decision tables is obtained by multiplying the number of linguistic terms associated with each input variable. Using the previous definitions of n and N, the number of "cells" (L_r) is:

$$L_r = \prod_{i=1}^{N} n_i. \qquad (8.17)$$

Each cell of the table describes a fuzzy rule, we will refer to these fuzzy rules as elemental fuzzy rules.

The structure of the fuzzy control rules contained in our FLC with parameters $\{N, M, n, m\}$ is:

$$\begin{aligned} &\textbf{If} \quad\quad x_i \text{ is } A_{io} \text{ and } \ldots \text{and } x_k \text{ is } A_{kp} \\ &\textbf{then} \quad y_j \text{ is } B_{jq} \text{ and } \ldots \text{and } y_l \text{ is } B_{lr} \end{aligned} \qquad (8.18)$$

where x_i is an input variable, C_{io} is a fuzzy set associated with this variable ($o \leq n_i$), y_j is an output variable and D_{jq} is a fuzzy set associated with this variable ($q \leq m_j$). All fuzzy inputs are 'connected' by the fuzzy connective 'and'. Several fuzzy sets related to the same variable could be connected with the aggregation operator 'or', appearing in a single rule, such as:

$$\textbf{If } x_i \text{ is } (A_{io} \text{ or } A_{ip}) \text{ and } \ldots \textbf{then } y_j \text{ is } (B_{jq} \text{ or } B_{jr}) \text{ and } \ldots$$

8.6.3   Encoding the Rule Base Information.

A set of rules represented by numerical values on a decision table or a relation matrix, have a direct translation into a string by means of a concatenating process. This method does not apply to a set of rules with a structure like that defined by equation 8.18.

In our system, each rule will be encoded into two strings of bits: one string of length L_a for the antecedent (a bit for each linguistic term related to each input variable) and one string of length L_c for the consequent. To encode the antecedent we will start with a string of L_a bits all of them with an initial value 0. If the antecedent of the rule contains a fuzzy input like x_i is C_{ij}, a 1 will substitute the 0 at a certain position (p) in the string:

$$p = j + \sum_{k=1}^{i-1} n_k. \tag{8.19}$$

This process will be repeated for all the fuzzy inputs of the rule. It is important to point out that using this code, an input variable for which all the corresponding bits have value 0, is an input variable whose value has no effect over the rule. The process to encode the consequent is similar to that described above, by only replacing n with m in expression 8.19. In this case, when all the bits corresponding to an output variable have value 0, the rule has no effect over that output variable.

Considering an FLC with three input and one output variables ($N = 3, M = 1$) and parameters $n = \{5, 3, 5\}$ and $m = \{7\}$ ($L_a = 13, L_c = 7$), the fuzzy rule

$$\begin{aligned} &\textbf{If} \quad x_1 \text{ is } (A_{13} \text{ or } A_{14}) \text{ and } x_3 \text{ is } (A_{31} \text{ or } A_{32}) \\ &\textbf{then} \quad y_1 \text{ is } (B_{14} \text{ or } B_{15}) \end{aligned} \tag{8.20}$$

is encoded as

$$0011000011000 - 0001100. \tag{8.21}$$

Each rule of this FLC will be represented as a string of 20 bits (fixed length). The rule base will contain an unfixed number of rules, with a maximum value of $L_r = 75$, then it will be encoded as a string of up to 75 strings (unfixed length) of 20 bits.

8.6.4   Evolving the Knowledge Base.

The process of evolution learning may be described with the following scheme:

1. Start with a first generation $G(0)$.
2. Evaluate $G(0)$: taking each member (string or other structure) of the population, decoding it, evaluating it by means of the evaluation function, and assigning a fitness value to the member.

3. While the termination condition was not met:
   (a) Create a new generation $G(t+1)$, by applying the evolution operators to the generation $G(t)$.
   (b) Evaluate $G(t+1)$.
4. Stop.

The keys of this process are: the code, the evaluation function, the termination condition and the evolution operators. The code has been widely presented before and will be summarized here. The evaluation function and the termination condition are application specific. The main questions in this section are the evolution operators, or in a more general sense, the creation of $G(t+1)$ from $G(t)$.

The code that contains the information of the knowledge base is:

1. A string of $2 + N + M$ integers containing the dimensions of the FLC.
2. A string of $(N+M) \times 2$ real numbers containing the normalization limits of the variables.
3. A string of $(L_a + L_c) \times 4$ real numbers containing the definition of the trapezoidal membership functions.
4. A string of up to L_r rules, where each rule is a string of $L_a + L_c$ bits.

It is possible to apply the evolution learning to any part of this code, according to the conditions of the learning process. From this point and in the application example we will work only with normalization limits (2) and with rule bases (4). The dimensions of the system and the membership functions will not be modified.

The effect produced by the modification of the normalization limits of a certain variable on the corresponding fuzzy sets (membership functions) are two:

- Each fuzzy set shrinks or expands in the same proportion as variable ranges do. The effect is the same as that produced when changing the gain of a controller.
- Each fuzzy set may be shifted to the right or to the left depending on its position and on the modifications of the normalization limits.

These changes are more restricted than those obtained with other methods, but the length of the employed code is reduced substantially, producing a shorter learning process; and a shorter or larger process may produce a treatable or untreatable problem.

Some evolution operators are obtained by directly adapting the classical genetic operators to the code. Others are new operators taking advantage of the code structure, or reducing its disadvantages.

Reproduction

The reproduction operator starts with an elite process that may be defined on the basis of a number of members, a percentage of members or an evaluation threshold (fixed or variable). By this process, a subset of $G(t)$, referred to as the elite of generation t ($E(t)$), will be directly reproduced (copied) on $G(t+1)$.

In a second step, individuals of $G(t)$ will be copied in the mating pool with a probability criterion based on the fitness of each one. According to the classical reproduction operator, members with a larger fitness value receive a larger number of copies. In addition to this reproduction operator, a second definition of reproduction based on a slightly modified operator has been defined. When working with this modified version of the operator, members (including those from the elite) with a larger fitness value have a higher probability of receiving a single copy (each individual will or will not receive a copy). This modified operator has been applied to the gait synthesis problem and is defined to avoid a sort of degenerative effect that produced a loss of diversity of members, when working with a small population (this is only an experimental effect, with no theoretical base).

Once the elite and the chromosomes to be reproduced have been selected, the number of members of $G(t + 1)$ must be adjusted to the maximum population. This process is performed by adding new elements to the elite (if the number of members is under the maximum population) or by extracting chromosomes from the mating pool (if the number of members is over the maximum population).

Crossover

Once a pair of parents (m_i and m_j taken from the mating pool) has been selected to be crossed (first step of the process), the crossover operator produces two new chromosomes by mixing the information provided by the parents' genes. A set of strings contains this information, two strings in our case (Normalization limits and Rules). The information contained in each string is not independent, then, if possible, the operator must work simultaneously (and not independently) on both strings. Each chromosome is composed of a pair of subchromosomes encoding the rule base (r) and the data base (d). The crossover of $m_i = (r_i, d_i)$ and $m_j = (r_j, d_j)$ will produce two new chromosomes (m_u and m_v).

Rule base subchromosomes have no fixed length, and their genes are rules:

$$r_i = \{r_{i1}, \ldots, r_{ik}\}$$
$$r_j = \{r_{j1}, \ldots, r_{jl}\}. \tag{8.22}$$

To cross r_i and r_j a cutting point must be selected for each string. Since the lengths of the strings may not be equal, cutting points (β and γ) will be obtained independently for r_i and r_j

$$r_i = \{r_{i1}, \ldots, r_{i\beta}|r_{i\beta+1}, \ldots, r_{ik}\}$$
$$r_j = \{r_{j1}, \ldots, r_{j\gamma}|r_{j\gamma+1}, \ldots, r_{jl}\}, \tag{8.23}$$

producing the new rule bases

$$r_u = \{r_{i1}, \ldots, r_{i\beta}|r_{j\gamma+1}, \ldots, r_{jl}\}$$
$$r_v = \{r_{j1}, \ldots, r_{j\gamma}|r_{i\beta+1}, \ldots, r_{ik}\}. \tag{8.24}$$

After rule bases are crossed, the process of crossing data bases will consider which rules from r_i and r_j go to r_u or r_v. An elemental fuzzy rule contains fuzzy inputs for all the variables. The rules we use contain fuzzy inputs and fuzzy outputs for only a subset of the input and output variables, then, normalization limits for the remaining variables have no influence on the meaning of the rule. A larger influence of a certain variable on rules that proceeding from r_i go to r_u, will produce a higher probability for this variable to reproduce in d_u its corresponding range from d_i. The influence is evaluated by simply counting the number of rules that, containing the variable, are reproduced from r_i to r_u. The process of selection is independent for each variable and for each descendent (m_u and m_v), consequently it is possible for both descendent to reproduce a certain range from the same antecedent.

Rules Reordering

When a fuzzy system is characterized by a set of fuzzy rules, their ordering is immaterial, the sentence connective *also* has properties of commutativity and associativity. When concatenating decision tables or relation matrices, the information of a certain gene depends on its content and its position. In our string of rules, the meaning of a gene becomes independent of the position. Therefore, rule position is arbitrarily defined for the members of the first population, it is immaterial for the output, but it biases crossover, then an operator to reorder rules will be added to the system.

This operator is applied to each set of rules produced by crossover operator, with a probability defined as a parameter of the evolution system. To reorder a rule base (r_i) a cutting point (β) is selected uniformly at random, to create a new rule base (r_j)

$$
\begin{aligned}
r_i &= \{r_{i1}, \ldots, r_{i\beta} | r_{i\beta+1}, \ldots, r_{ik}\} \\
r_j &= \{r_{i\beta+1}, \ldots, r_{ik} | r_{i1}, \ldots, r_{i\beta}\}.
\end{aligned}
\tag{8.25}
$$

The operator has no effect over the data base.

Mutation

The mutation is composed of two different processes: rule mutation and range mutation.

The rule mutation process will work at the level of bits that compose a rule. Each rule is composed of ($L_a + L_c$) bits and has the structure

$$
\begin{aligned}
p_{11} \ldots p_{1n_1} \ldots p_{N1} \ldots p_{Nn_N} \\
c_{11} \ldots c_{1m_1} \ldots c_{M1} \ldots c_{Mm_M}
\end{aligned},
\tag{8.26}
$$

where p_{ij} is the bit related to j-th fuzzy set of the i-th input variable, and c_{ij} is the bit related to j-th fuzzy set of the i-th output variable. The rule mutation operator works as the classic genetic mutation applied to the string of bits defined by equation 8.26. This operator differs from classical mutation because it does not work at the level of genes (rules), but with the simplest element (bits) that compose genes.

The range mutation operator for a variable with range $[\lambda_l, \lambda_u]$, could be described by the following equation:

$$
\begin{aligned}
\lambda_l(t+1) &= \lambda_l(t) + K P_1 S_1 (\lambda_u(t) - \lambda_l(t))/2 \\
\lambda_u(t+1) &= \lambda_u(t) + K P_2 S_2 (\lambda_u(t) - \lambda_l(t))/2,
\end{aligned}
\tag{8.27}
$$

where $K \in [0, 1]$ is a parameter of the learning system that defines the maximum variation (shift, expansion or shrinkage). P, P_1 and P_2 are random values uniformly distributed on $[0, 1]$, and S, S_1 and S_2 take values -1 or 1 by a 50% chance. The symmetry of ranges is maintained, then when a variable has symmetric ranges ($\lambda_l = -\lambda_u$) the following conditions are imposed: $P_2 = P_1$ and $S_2 = -S_1$.

8.6.5   An application example and some learning results.

This approach has been applied to the control of a simulated anthropomorphic (1.75m, 70kg) biped walking machine ([Mag94, MM95]). The model is a six-link, 2-D structure, with two legs, a hip and a trunk (Figure 8.8). Each leg has a punctual foot, a knee

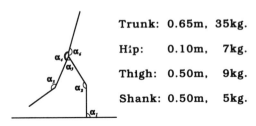

Trunk: 0.65m, 35kg.

Hip: 0.10m,   7kg.

Thigh: 0.50m,   9kg.

Shank: 0.50m,   5kg.

Figure 8.8   Variables and dimensions of the model.

with a degree of freedom and a hip with another degree of freedom. The contact of the stance-foot with the ground defines and unpowered degree of freedom(α_1) whose behavior is not directly controlled[5], but controlled through the adequate movements of the joints ($\alpha_2 - \alpha_6$). The control cycle of the FLC is 0.01 sc.

The goal of the FLC is to define joint trajectories in such a way that the biped system describes a regular walk without falling or stopping. In this case, the GAs are applied not with an optimization aim, but with a diversification aim. The idea is to use biomechanical studies ([McM84, PVB84]) to obtain the KB of an FLC controlling a regular walk with a certain speed and stride length, and then apply GAs to create other knowledge bases capable of controlling regular walks with different speeds and stride lengths.

When the information contained in a KB[6] is applied by the FLC, a sequence of movements is produced on the biped model. This sequence of movements will be evaluated ([Mag94]), based on the *stability* and *regularity* of the walk, over a ten-second simulation (a thousand control cycles). The evaluation function measures the stability as a function of the time and the number of steps before falling (if the system falls before ending the simulation), or as a fixed value if the system has not fallen at the end of the simulation. The regularity of the walk is only computed if the system does not fall or stop, and is a function of the deviation of the stride duration from the mean period along the ten-second simulation. Some chromosomes will contain valid gait patterns, that is, gait description that produces a walk simulation without falling, or stopping the biped.

Some tests have demonstrated the ability of the learning system to generate valid gaits [Mag94]. Figure 8.9 shows four sequences of walk produced by genetically generated KBs. All of them have been obtained as the result of a single learning process with a reproduction rate of 0.8, a reordering rate of 0.5, a rule mutation rate

---

[5] This problem is closely related to the simpler problem of controlling an inverted pendulum[BAC83].

[6] This KB is encoded by a chromosome that constitutes an element of the genetic population.

of 0.01, a range mutation rate of 0.05 and a mutation constant (the parameter K on expression 8.27) equal to 0.5. The maximum population was 500 individuals the number of generations is forty and the initial population contained five KBs producing valid gaits, and a set of other individuals producing the fall of the biped system. These five KBs were extracted from biomechanical studies and the generated gaits showed speeds in the $[1.05,1.15]m/sc$ interval and stride lengths in the interval $[0.67,0.68]m$. The sequences (Figure 8.9) present simulation results obtained with the biped model, containing a twenty images per second representation, except sequence 4 that is a ten images per second representation. Each sequence represents a particular case: the best, the shortest, the longest and fastest, and the slowest gait.

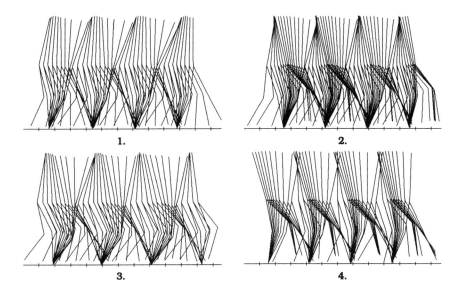

Figure 8.9 Different genetic generated gaits.

The main characteristics of each sequence are described below: the mean speed of the biped (S), the stride length (L) and the time covered by the sequence (T).

1. The most regular (best): $S = 1.21m/sc$, $L = 0.68m$ and $T = 2sc$.
2. The shortest: $S = 0.76m/sc$, $L = 0.64m$ and $T = 3sc$.
3. The longest and fastest: $S = 1.29m/sc$, $L = 0.78m$ and $T = 2sc$.
4. The slowest[7]: $S = 0.54m/sc$, $L = 0.67m$ and $T = 4.5sc$.

The genetic process has produced the evolution of the covered ranges for speed and stride length from the initial $[1.05,1.15]m/sc$ and $[0.67,0.68]m$, to $[0.54,1.29]m/sc$ and $[0.64,0.78]m$.

---

[7] This sequence contains only ten images per second.

8.7   CONCLUSIONS

Fuzzy Logic Control constitutes a growing and promising area of control theory. The main goal of this chapter was the description of a learning methodology for Fuzzy Logic Controllers, based on the evolution of their knowledge base. Two different approaches have been analyzed. Both of them use rule bases constructed as sets of rules, reducing the dimensionality of the learning space. Both methods have been successfully applied to different and complex control problems. A brief description of two of these applications has been included in the chapter.

AKNOWLEDGEMENTS

This work has been partially supported by the Commission of the European Communities and the Spanish Industry Ministry (Project PASO-095 - CORAGE) and by CICYT (Spanish Comisión Interministerial de Ciencia y Tecnología) (Project TAP93-0971-E).

REFERENCES

[BAC83] Barto A.G. S. R. and C.W. A. (Sep/Oct 1983) Neuronlike adaptive elements that can solve difficult learning control problems. *IEEE Transactions on Systems, Man and Cybernetics* 13(5): 834–846.

[Buc93] Buckley J. (1993) Sugeno type controllers are universal controllers. *Fuzzy Sets and Systems* 53: 299–304.

[Cas95] Castro J. (April 1995) Fuzzy logic controllers are universal approximators. *IEEE Transactions on Systems, Man and Cybernetics* 25(4): 629–635.

[CV93] Cooper M. and Vidal J. (October 1993) Genetic design of fuzzy controllers. In *Proceedings 2nd International Conference on Fuzzy Theory and Technology.*

[DeJ88] DeJong K. (October 1988) Learning with genetic algorithms: An overview. *Machine Learning* 3(3): 121–138.

[Hol75] Holland J. (1975) *Adaptation in Natural and Artificial Systems.* University of Michigan Press, Ann Arbor.

[Kar91] Karr C. (1991) Design of an adaptive fuzzy logic controller using a genetic algorithm. In *Proceedings 4th. International Conference on Genetic Algorithms,* pages 450–457. Morgan Kaufmann.

[KG93] Karr C. and Gentry E. (February 1993) Fuzzy control of pH using genetic algorithms. *IEEE Transactions on Fuzzy Systems* 1(1): 46–53.

[Lee90] Lee C. (Mar/Apr 1990) Fuzzy logic in control systems: Fuzzy logic controller - part I and II. *IEEE Transactions on Systems, Man and Cybernetics* 20(2): 404–435.

[LM94] Liska J. and Melsheimer S. (June 1994) Complete design of fuzzy logic systems using genetic algorithms. In *Proceedings 3rd IEEE International Conference on Fuzzy Systems, FUZZ-IEEE'94,* volume II, pages 1377–1382.

[LT93] Lee M. and Takagi H. (March 1993) Integrating design stages of fuzzy systems using genetic algorithms. In *Proceedings 2nd IEEE International Conference on Fuzzy Systems, FUZZ-IEEE'93,* volume 1, pages 612–617.

[Mag94] Magdalena L. (1994) *Estudio de la coordinación inteligente en robots bípedos: aplicación de lógica borrosa y algoritmos genéticos.* Doctoral dissertation, Universidad Politécnica de Madrid (Spain).

[Mag95] Magdalena L. (1995) A first approach to a taxonomy of fuzzy-neural systems. In *IJCAI'95 Workshop on Connectionist-Symbolic Integration: From Unified to Hybrid Approaches.*

[Mam74] Mamdani E. (1974) Application of fuzzy algorithms for simple dynamic plant. *Proceedings IEE, Part D* 121: 1585–1588.

[McM84] McMahon T. (1984) Mechanics of locomotion. *International Journal of Robotics Research* 3(2): 4–28.

[MM95] Magdalena L. and Monasterio F. (March 1995) Evolutionary-based learning applied to fuzzy controllers. In *Proceedings 4th IEEE International Conference on Fuzzy Systems and the Second International Fuzzy Engineering Symposium, FUZZ-IEEE/IFES'95*, volume III, pages 1111–1118.

[MVFM92] Magdalena L., Velasco J., Fernández G., and Monasterio F. (May 1992) A control architecture for optimal operation with inductive learning. In *Preprints IFAC Symposium on Intelligent Components and Instrument for Control Applications, SICICA'92*, pages 307–312.

[Ped93] Pedrycz W. (1993) *Fuzzy Control and Fuzzy Systems.* Research Studies Press Ltd., second, extended, edition edition.

[PK91] Pham D. and Karaboga D. (1991) Optimun design of fuzzy logic controllers using genetic algorithms. *Journal of Systems Engineering* pages 114–118.

[PKL94] Park D., Kandel A., and Langholz G. (January 1994) Genetic-based new fuzzy reasoning models with application to fuzzy control. *IEEE Transactions on Systems, Man and Cybernetics* 24(1): 39–47.

[PVB84] Plas F., Viel E., and Blanc Y. (1984) *La marche humaine.* MASSON, S.A., Paris.

[QB94] Qin S. and Borders G. (February 1994) A multiregion fuzzy logic controller for nonlinear process control. *IEEE Transactions on Fuzzy Systems* 2(1): 74–81.

[SFH95] Shimojima K., Fukuda T., and Hasegawa Y. (March 1995) RBF-fuzzy system with GA based unsupervised/supervised learning method. In *Proceedings 4th IEEE International Conference on Fuzzy Systems and the Second International Fuzzy Engineering Symposium, FUZZ-IEEE/IFES'95*, volume I, pages 253–258.

[Thr91] Thrift P. (1991) Fuzzy logic synthesis with genetic algorithms. In *Proceedings 4th. International Conference on Genetic Algorithms*, pages 509–513. Morgan Kaufmann.

[TS85] Takagi T. and Sugeno M. (January/February 1985) Fuzzy identification of systems and its applications to modeling and control. *IEEE Transactions on Systems, Man, and Cybernetics* 15(1): 116–132.

[VV94] Velasco J. and Ventero F. (1.994) Some applications of fuzzy clustering to fuzzy control systems. In *3rd Int. Conf. on Fuzzy Theory and Technology.* Durham, NC, USA.

[Wan94] Wang L.-X. (1994) *Adaptive Fuzzy Systems and Control: Design and Stability Analysis.* PTR Prentice-Hall, Englewood Cliffs, New Jersey.

[Zad65] Zadeh L. (1965) Fuzzy sets. *Information and Control* 8: 338–353.

9

Tackling Fuzzy Genetic Algorithms

F. HERRERA, M. LOZANO, J.L. VERDEGAY

9.1 INTRODUCTION

Recently, numerous papers and applications combining fuzzy concepts and genetic algorithms (GAs) have become known, and there is increasing interest in the integration of these two topics. In this paper we explore this combination from the point of view of the so-called *fuzzy genetic algorithms* (FGAs).

According to this name, FGAs, two approaches have been presented in the specialized literature:

- either, the use of fuzzy logic based techniques to model different GA components,
- or, to manage problems in an imprecise environment, where the imprecision is modeled by fuzzy sets, the GAs that represent and run these problems are also called FGAs.

Fuzzy Logic (FL) may be viewed with two different meanings as was expounded in [Zad93] by Prof. Zadeh, the pioneer of fuzzy logic and fuzzy set theory.

"(a) a narrow interpretation, FLn, in which fuzzy logic is basically a logic of approximate reasoning; and (b) a wide interpretation, FLw, in terms of which fuzzy logic is coextensive with the theory of fuzzy sets, that is, classes of objects in which the transition from membership to nonmembership is gradual rather than abrupt.

In this narrow sense, fuzzy logic may be viewed as a generalization and extension of multivalued logic. But the applicability of fuzzy logic is far than that of multivalued logic because FLn provides many concepts and techniques which are not a part of multivalued logic. Among such concepts and techniques - which play a key role in the applications of fuzzy logic - are those of the linguistic variable; the concepts of possibility and necessity; concepts of truth and usuality-qualification; fuzzy quantification and cardinality. Furthermore, the agenda of FLn is quite different from that of classical multivalued logical systems.

In its wide sense, FLw, fuzzy logic is a very broad theory with many branches, among them fuzzy sets, fuzzy arithmetic, fuzzy mathematical programming, fuzzy pattern recognition, fuzzy control, fuzzy probability theory, fuzzy topology, the calculi of fuzzy rules and fuzzy graphs, and fuzzy logic, FLn, in its narrow sense. It should be noted that there is a growing trend to interpret fuzzy logic in its wide sense since the label fuzzy logic is more euphonious and more self-explanatory than fuzzy set theory. Regardless of its interpretation, the role model for fuzzy logic is the human mind". (For more information see [WL93, YZ92, Zim90]).

The FGAs are considered in the wide sense of FL, FLw. Different fuzzy logic based techniques have been used for trying to improve the GAs' behaviour: fuzzy logic control systems for designing self-control parameter processes according to some performance measures, fuzzy operators and fuzzy connectives for designing genetic operators with different properties, etc. On the other hand, the interest for applying the GAs to problems with imprecise information represented by fuzzy sets has led to the design of alternative GAs for managing fuzzy information.

In this paper, we present a general study of the FGAs according to the two aforesaid approaches. In Section 2, we review the application of FLw to GAs. The GA approaches for managing imprecise information is dealt with in Section 3. Some final remarks are pointed out in Section 4.

9.2 FUZZY LOGIC BASED TECHNIQUES FOR GA DEVELOPMENTS

As we have already mentioned, several fuzzy logic based techniques have been used for improving GA behavior. These techniques were included in different parts of GA development:

- *Representation tasks*; for including some more complex genotype versus phenotype relationships similar to the ones occurring in nature [Voi92], and for building more flexible representations for working with fuzzy sets [San93]. Clearly, these new representation models should be accompanied by new genetic operators.
- *Parameter settings choice*; for dynamically computing apropriate settings using the exprerience and knowledge of the GA experts in order to induce suitable exploitation/exploration relationships throughout the GA's execution for avoiding the premature convergence problem [LT93, XV94].
- *Crossover operator definition*; for designing powerful crossover operators that establish adequate population diversity levels and so attacking the premature convergence problem as well [HHLV94, HLV95a, HLV95b, HLV95c].

- *Analysing the solution performance of GAs*; for handling solution predictions in order to force the GA to reach optimal solutions with a user-defined accuracy [MF94].

We shall review each one of these attempts in more detail.

9.2.1   GAs as a Fuzzy Decision Process

The classical evolutionary algorithms, GAs and evolution strategies (ESs), do not take into account the *development* of an individual or organism from the gene level to the mature phenotype. There are no one-gene, one-trait relationships in natural evolved systems. The phenotype varies as a complex, non-linear function of the interaction between underlying genetic structures and current environmental conditions [Fog94]. An attempt to deal with more complex genotype/phenotype relations in evolutionary algorithms was presented by Voight in [Voi92]. A fuzzy representation and fuzzy operations were proposed to model the developmental process based on fuzzy decisions. This type of evolutionary algorithms, which were called *Fuzzy Evolutionary Algorithms*, shall be described below, but first we shall review, by using an example, the mappings from genotype to phenotype under which GAs and ESs work [Voi92], this shall be done by analysing optimization problems of the form

$$\min\{f(x)|x \in X\},\ X \subseteq \Re^n,\ f : \Re^n \to \Re$$

GENETIC ALGORITHMS

Let us consider that the genotype, g, of an individual belonging to a population is the following string of binary values:

$$g = (0, 1, 0, 0, 0, 1).$$

By decoding this genotype with a fixed point integer decoding scheme, where each integer is represented by three bits, the intermediate phenotype h for this genotype shall be:

$$h = (2, 1).$$

This intermediate phenotype shall be mapped to the feasible region X to form the phenotype. For example, if $X = (5, 10) \times (3, 6)$ then the phenotype p is:

$$p = (6.43, 3.43).$$

The genetic representation by a binary string corresponds via the transformation

$$g \to h \to p$$

to a one-to-one mapping of the genotype to the phenotype and vice versa. Therefore, there may not be a developmental process for individuals.

For Voight the GAs are mainly based on binary decisions because of the binary representation of the genotype.

EVOLUTION STRATEGIES

In the ESs, the genetic representation is given by an array of floating-point values and this representation is at the same time as the phenotypic representation, i.e.

$$g = p$$

Again, we may conclude that there is no individual developmental process. Decisions are made in the space of phenotypes on the basis of continuous variables in the range of the machine accuracy.

FUZZY EVOLUTIONARY ALGORITHMS

The modelling of the developmental process was attempted as a fuzzy decision process to emphasize the complexity of development from genotype to a mature phenotype. The genotypic representation is done by associating a number of n_d decision genes belonging to the interval $[0, 1]$ to each problem parameter. The chromosomes are formed by the link of the values of the decision genes in each parameter. This coding scheme is similar to the one in GAs, however, each gene may have a real number in $[0,1]$ instead of a binary digit. Clearly, the idea here is to generalize the concept of *binary decision gene* (with values of zero or one) into *fuzzy decision gene* (with values between zero and one). We shall describe an example.

Let us assume that an individual is coded by the following array of fuzzy decision values:

$$g = (0.22, 0.86, 0.14, 0.03, 0.18, 0.92).$$

Now we may apply, as in the case of the classical GA, a fuzzy fixed point integer decoding scheme where each fuzzy integer is represented by three fuzzy decision variables ($n_d = 3$). The intermediate phenotype, h, for this genotype is

$$h = (0.22 \cdot 2^2 + 0.86 \cdot 2^1 + 0.14 \cdot 2^0, 0.03 \cdot 2^2 + 0.18 \cdot 2^1 + 0.92 \cdot 2^0) = (2.74, 1.40).$$

This intermediate phenotype shall be mapped to the feasible region X to form the phenotype. Again, if $X = (5, 10) \times (3, 6)$ then the phenotype p is:

$$p = (6.96, 3.60).$$

When $n_d > 1$, this coding type breaks the one-to-one correspondence between genotype and phenotype, since two different genotypes may induce the same phenotype. So, it is impossible to find inferences from phenotype to genotype, i.e., the mapping from genotype to phenotype is not *isomorphic*.

The recombination in this model is done using union and intersection operators of fuzzy sets for defining all possible children. Given two arrays of fuzzy decision values:

$$F_1 = (d_1^1, ..., d_n^1) \text{ and } F_2 = (d_1^2, ..., d_n^2),$$

first, the min- and max-rules are applied for computing the fuzzy subsets

$$F_{min} = (\min(d_1^1, d_1^2), ..., \min(d_n^1, d_n^2)) \text{ and } F_{max} = (\max(d_1^1, d_1^2), ..., \max(d_n^1, d_n^2)),$$

finally, the child shall be defined at random by using a uniform probability distribution applied to the range of all fuzzy min and max values.

The mutation of a fuzzy decision gene is carried out by choosing a random value from the interval $[0, 1]$.

9.2.2 Fuzzy Crossover Using Fuzzy Templates

In [San93] a fuzzy crossover was proposed for working with chromosomes consisting of strings of real numbers in the interval $[0, 1]$, which were used for problems where fuzzy sets need to be represented. For presenting such a crossover operator first it was suggested that the classical one-point crossover operators under binary coding may be defined by means of *templates*. A binary chromosome has associated a set whose elements are the loci of the genes with value one. For example, the chromosome $C = (0, 1, 1, 1, 0, 1)$ has associated the set $SC = \{2, 3, 4, 6\}$. Supposing two binary chromosomes, $C_1 = (b_1^1, ..., b_n^1)$ and $C_2 = (b_1^2, ..., b_n^2)$ to be crossed, with S_1 and S_2 being their associated sets, an instance of the application of the one-point crossover operator on these chromosomes, e.g. $H_1 = (b_1^1, ..., b_i^1, b_{i+1}^2, ..., b_n^2)$ and $H_2 = (b_1^2, ..., b_i^2, b_{i+1}^1, ..., b_n^1)$, $i \in \{1, ..., n - 1\}$, may be expressed in terms of the associated sets by means of another set, $T = \{i + 1, ..., n\}$, which was called template, as follows:

$$SH_1 = (S_1 \cap \bar{T}) \cup (S_2 \cap T) \text{ and } SH_2 = (S_1 \cap T) \cup (S_2 \cap \bar{T}),$$

with $\bar{T}$ being the complement of T in all the possible loci set.

The idea is to generalize these issues for the case in which all sets considered are fuzzy, i.e., the chromosomes C_1 and C_2 are vectors of real number in $[0, 1]$ and the templates are also fuzzy sets of loci, i.e., *fuzzy templates*. Under this condition the previous computation of SH_1 and SH_2, considering that operators $\cup$ and $\cap$ are fuzzy ones, becomes a fuzzy crossover operator. The effect of using fuzzy templates, instead of crisp ones, is to smooth the fuzzy set represented in the fuzzy chromosomes generated through crossover. Other ways to define fuzzy operators may be considered, such as $S_1 \cup S_2$ and $S_1 \cap S_2$.

We should point out that this attempt is very similar to one presented by Voight (previous subsection); Sanchez considers chromosomes as sets and Voight as binary decisions; both generalize these ideas to fuzzy tools, i.e, fuzzy sets and fuzzy decisions, respectively.

9.2.3 Fuzzy Logic Based Systems for Dynamic Control Parameters

An important problem in the use of GAs is the premature convergence problem; the search becomes trapped in a local optimum before the global optimum is found. Their main causes are the lack of diversity in the population and a disproportionate exploitation/exploration relationship; e.g., the proper balance between a broad search and a sufficient refinement is not established [LLK92]. Premature convergence causes a drop in the GA's efficiency; the genetic operators don't produce the feasible diversity to tackle new search space zones and thus the algorithm reiterates over the known zones producing a slowing-down in the search process.

The GA control parameter settings such as mutation rate, crossover rate and population size are key factors in the determination of the exploitation versus exploration tradeoff. It has long been acknowledged that they have a significant impact on GA performance [Gre86]. If poor settings are used, the exploitation/exploration balance may not be reached in a profitable way; the GA's performance shall be severely affected [Dav89].

Finding settings that work well on any problem is not a trivial task, since the interaction of GA control parameters with GA performance is complex and the optimal settings are problem dependent. One of the first attempts for determining robust parameter settings was made by Grefenstette [Gre86]. He used a meta-GA for finding the appropriate settings for obtaining the best Online and Offline measures behaviour in the five De Jong functions [Jon75]. However, the optimal value for each operator probability may vary during the course of a run. For that reason, adaptive techniques that change the control parameter based on performance measures are required. In [Dav89] one of such solutions was offered, the technique involves adapting the crossover and mutation probabilities based on their performance observed as the run takes place. Another approach was tackled in [LLK92], diversity functions were used that may describe the behaviour of a GA and may be used for handling the control parameters in order to induce an ideal exploitation/exploration relationship.

Currently, a certain body of expertise, experience and knowledge on GAs has become available as a result of empirical studies conducted over a number of years. This human expertise and knowledge would be very useful for increasing the capabilities of GAs to avoid premature convergence. However, generally too much of this information is vague, incomplete, or ill-structured to a certain degree which causes it to be rarely applied. This latter feature suggests the use of a fuzzy tool for handling this type of knowledge. One application of the FL that is useful for controlling GA parameters following control strategies underlying in the human expertise and knowledge are the fuzzy logic controllers (FLCs). FLCs implement an expert operator's approximate reasoning process in the selection of a control action. An FLC allows one to qualitatively express the control strategies based on experience as well as intuition. These control strategies may be expressed in a form that permits both computers and humans to efficiently share them.

Next, we study an application of FLC for the dynamic control of GA parameters [LT93], before we describe briefly the generic structure of FLCs.

DESCRIPTION OF THE FLCs

An FLC is composed by a *Knowledge Base*, that comprises the information given by the expert in the form of linguistic control rules, a *Fuzzification Interface*, which has the effect of transforming crisp data into fuzzy sets, an *Inference System*, that uses them together with the Knowledge Base to make inference by means of a reasoning method, and a *Defuzzification Interface*, that translates the fuzzy control action so obtained to a real control action using a defuzzification method. The generic structure of an FLC is shown in Figure 9.1 [Lee90].

The Knowledge Base encodes the expert knowledge by means of a set of fuzzy control rules. A fuzzy control rule is a conditional statement with the form **IF** *a set of conditions are satisfied* **THEN** *a set of consequences can be inferred* in which the antecedent is a condition in its application domain, the consequent is a control action

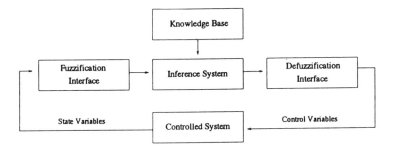

Figure 9.1 Generic structure of an FLC

Dynamic Parametric GA

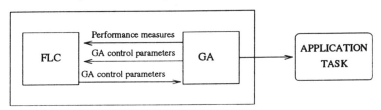

Figure 9.2 Structure of the Dynamic Parametric GA

to be applied in the controlled system (notion of control rule) and both antecedent and consequent are associated with fuzzy concepts, that is, linguistic terms (notion of fuzzy rule).

Thus the Knowledge Base is composed of two components, a *Data Base*, containing the definitions of the fuzzy control rules linguistic labels, that is, the membership functions of the fuzzy sets specifying the meaning of the linguistic terms, and a *Rule Base*, constituted by the collection of fuzzy control rules representing the expert knowledge (a more complete explanation may be seen in chapter 3 of this book).

DYNAMIC CONTROL OF GA USING FLCs

In [LT93] the *Dynamic Parametric GA* was proposed, a GA that uses an FLC for controlling GA parameters. The inputs to the FLC are any combination of GA performance measures or current control settings, and outputs may be any of the GA control parameters (see Figure 9.2).

Below we describe an example of the application of this technique. First we consider the inputs (state variables) and output (control variable) of the FLC by establishing their ranges and linguistic labels associated, then we define the Data Base and the Rule Base.

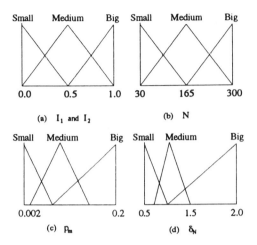

Figure 9.3   Meaning of the linguistic terms associated with the inputs and the

Inputs.   Four inputs were considered; two that describe the convergence stage of the
population:

- $I_1 = \dfrac{\text{average fitness}}{\text{best fitness}}$, $I_1 \in [0,1]$.
- $I_2 = \dfrac{\text{worst fitness}}{\text{average fitness}}$, $I_2 \in [0,1]$.

and two others that are GA control parameters: the mutation probability, p_m, with
$p_m \in [0.02, 0.2]$ and the population size, N, with possible values $\{30, ..., 300\}$. The
linguistic labels set of all these inputs are $\{Small, Medium, Big\}$.

Output.   The output is a variable that controls a variation on the current population
size. This variable, noted as δ_N, represents the degree in which the current population
size should vary. The population size variation shall be carried out by multiplying the
δ_N value, obtained by the FLC, by the current population size.
    The action interval of δ_N shall be $[0.5, 2.0]$ and its linguistic labels associated shall be
$\{Small, Medium, Big\}$ as well. A value of 0.5 for δ_N implies that the new population
size shall be half the current one, and a value of 2.0 implies doubling it.

Data Base.   The Data Base for this example is shown in Figure 9.3. The meaning of
the linguistic terms associated with inputs I_1, I_2 are depicted in (a), the ones for N in
(b), the ones for p_m in (c), and finally, the ones for δ_N in (d). For each linguistic term,
there is a *triangular fuzzy set* that defines its semantic, i.e, its meaning [Lee90, Zim90].

Rule Base.   The rules describe the relation between the inputs and outputs, for
example, the following three rules present a possible relation between these variables:

IF I_1 is *Big* **THEN** δ_N is *Big*.

IF I_2 is *Small* **THEN** δ_N is *Small*.

IF p_m is *Small* **AND** N is *Small* **THEN** δ_N is *Big*.

In [Lee93], an automatic technique was proposed for obtaining both the Data Base and the Rule Base for those cases where knowledge or expertise are not available. This technique is very similar to the meta-GA of Grefentette [Gre86]. By using an automatic technique, relevant relations and membership functions may be automatically determined and may offer insight for understanding the complex interaction between GA control parameters and GA performance.

The experiment carried out by the authors using the Dynamic Parametric GA exhibited better performance than a simple static GA. Finally, we should point out that another attempt, very similar to the previous one, may be referred to [XV94].

9.2.4 Fuzzy Connectives Based Crossover Operators

The crossover operator plays a central role in the GA's performance. It exploits the available information from the population about the search space. This operator has been highlighted as another key point for solving the premature convergence problem [Boo87]. Numerous investigations have been directed to finding optimal crossover rates and more powerful alternative crossovers which would allow suitable levels of exploration and exploitation to be established [ECS89].

In [HHLV94, HLV95b] the development of such crossover operators was attempted for the case of *Real-Coded GAs* (RCGAs) [Gol91, Mic92]. It was presented the fuzzy connective-based crossovers (FCB-crossovers) operators for RCGAs based on the use of fuzzy connectives: t-norms, t-conorms, average functions and generalized compensation operators [Miz89a, Miz89b]. In [HLV95b] a set of offspring selection mechanisms (OSMs) was proposed which choose the chromosomes (produced by the crossover) that will be the population members. Different exploration and exploitation degrees may be introduced with the FCB-crossover operators. The OSMs establish a relationship between these properties so that they induce different diversity levels in the population and therefore the premature convergence problem may be erradicated.

To describe the FCB-crossover operators, we follow three steps: define gene combination functions, use these functions to define crossover operators between two chromosomes and apply the crossover operators to the individuals in the population, establishing the number and type of operators along with the OSM to be used.

GENE COMBINATION FUNCTIONS

Let's consider $c_i^1, c_i^2 \in [a_i, b_i]$ two real genes to be combined and $\alpha_i = \min\{c_i^1, c_i^2\}$ and $\beta_i = \max\{c_i^1, c_i^2\}$. The action interval $[a_i, b_i]$ of these genes can be divided into three intervals $[a_i, \alpha_i]$, $[\alpha_i, \beta_i]$ and $[\beta_i, b_i]$. These intervals bound three regions to which the resultant genes of some combination of the former may belong. Moreover, considering a region $[\alpha_i', \beta_i']$ with $\alpha_i' \leq \alpha_i$ e $\beta_i' \geq \beta_i$ would seem reasonable. This is shown in Figure 9.4

The intervals described above could be classified as exploration or exploitation zones. The interval with both genes being the extremes is an exploitation zone, the two

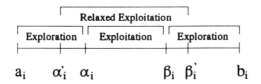

Figure 9.4   Action interval for a gene

intervals that remain on both sides are exploration zones and the region with extremes α_i' and β_i' could be considered as a relaxed exploitation zone.

With regards to these intervals, four functions were proposed: F, S, M and L defined from $[a, b] \times [a, b]$ in $[a, b]$, $a, b \in \Re$, which fulfill:

(P1) $\forall c, c' \in [a, b]$ $F(c, c') \leq \min\{c, c'\}$
(P2) $\forall c, c' \in [a, b]$ $S(c, c') \geq \max\{c, c'\}$
(P3) $\forall c, c' \in [a, b]$ $\min\{c, c'\} \leq M(c, c') \leq \max\{c, c'\}$
(P4) $\forall c, c' \in [a, b]$ $F(c, c') \leq L(c, c') \leq S(c, c')$
(P5) F, S, M, y L are monotone and non-decreasing.

Each one of these functions allows us to combine two genes giving results belonging to each one of the aforementioned intervals. Therefore, each function will have different exploration or exploitation properties depending on the range being covered by it.

T-norms, t-conorms, averaging functions and generalized compensation operators [Miz89a, Miz89b] may be used to build these functions; we may associate F to a t-norm, S to a t-conorm, M to an averaging operator and L to a generalized compensation operator. In order to do so, we need to transform the genes, that will be combined, from the interval $[a, b]$ into $[0, 1]$ and later, the results into $[a, b]$.

Complying with a set of fuzzy connectives, $\{T, G, P, \hat{C}\}$, a set of functions $\{F, S, M, L\}$, associated with it, is built as described below:

If $c, c' \in [a, b]$ then

$$F(c, c') = a + (b - a) \cdot T(s, s')$$
$$S(c, c') = a + (b - a) \cdot G(s, s')$$
$$M(c, c') = a + (b - a) \cdot P(s, s')$$
$$L(c, c') = a + (b - a) \cdot \hat{C}(s, s')$$

where $s = \frac{c-a}{b-a}$ and $s' = \frac{c'-a}{b-a}$.

These operators have the properties of being continuous and non-decreasing, and satisfy the respective properties (P1)-(P5). The families of fuzzy connectives used are shown in Table 9.1.

Fuzzy connectives families in Table 9.1 fulfill the following property:

(P6) $T_4 \leq T_3 \leq T_2 \leq T_1 \leq P_j(j = 1, ...4) \leq G_1 \leq G_2 \leq G_3 \leq G_4$

FCB-CROSSOVERS

Let us assume that $O \in \{F, S, M, L\}$ and $C_1 = (c_1^1 \ldots c_n^1)$ and $C_2 = (c_1^2 \ldots c_n^2)$ are two chromosomes that have been selected to apply the crossover operator to them. We can

Table 9.1 Families of fuzzy connectives. From top to bottom, the Logical, Hamacher, Algebraic and Einstein families are shown

t-norm	t-conorm	Averaging Fun. $(0 \leq \lambda \leq 1)$	Gen. Comp. Op.
$T_1(x,y) = \min(x,y)$	$G_1(x,y) = \max(x,y)$	$P_1(x,y) = (1-\lambda)x + \lambda y$	$\hat{C}_1 = T_1^{1-\lambda} \cdot G_1^{\lambda}$
$T_2(x,y) = \frac{xy}{x+y-xy}$	$G_2(x,y) = \frac{x+y-2xy}{1-xy}$	$P_2(x,y) = \frac{1}{\frac{y-y\lambda-xy+x\lambda}{xy}+1}$	$\hat{C}_2 = P_2(T_2,G_2)$
$T_3(x,y) = xy$	$G_3(x,y) = x+y-xy$	$P_3(x,y) = x^{1-\lambda}y^{\lambda}$	$\hat{C}_3 = P_3(T_3,G_3)$
$T_4(x,y) = \frac{xy}{1+(1-x)(1-y)}$	$G_4(x,y) = \frac{x+y}{1+xy}$	$P_4(x,y) = \frac{2}{1+(\frac{2-x}{x})^{1-\lambda}(\frac{2-y}{y})^{\lambda}}$	$\hat{C}_4 = P_4(T_4,G_4)$

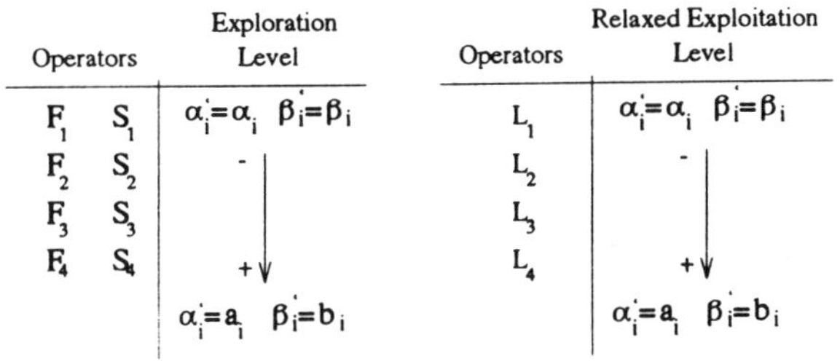

Figure 9.5   Exploration and relaxed exploitation levels

generate the chromosome $H = (h_1, ..., h_i, ..., h_n)$ as

$$h_i = O(c_i^1, c_i^2), \quad i = 1, ..., n.$$

This operator applies the same F, S, M or L function for all the genes in the chromosomes to crossover. For this reason, they will be called F-crossover, S-crossover, M-crossover and L-crossover when the F, S, M and L functions are applied respectively. It should be emphasized that these crossover operators have different properties: the F-crossover and S-crossover operators show exploration, the M-crossover operators show exploitation and the L-crossover operator shows relaxed exploitation.

According to the property (P6) of the families of fuzzy connectives in Table 9.1, we can see that the degree, in which each crossover operator shows its related property, will depend on the fuzzy connective on which it is based. Thus, we dispose of F_j-crossover and S_j-crossover operators with different exploration levels; the F_4-crossover and S_4-crossover show the maximum exploration, whereas the F_1-crossover and the S_1-crossover represent the minimum exploration. These operators give results between the extremes of the exploration domain. With respect to the L_j-crossover, the level of relaxed exploitation directly depends on the T_j, G_j and P_j used for its definition. Figure 9.5 shows the behaviour of the interval definition for the crossover operators.

Table 9.2 Application strategies (N_c is the number of chromosomes that should undergo crossover, N is the population size and p_c the crossover probability).

Strategy	FCB-crossovers	N_c	OSM	Diversity
ST1	F-, S- and M-crossover	$\frac{2}{3} \cdot p_c \cdot N$	All three offspring	strong
ST2	F-, S- and two M-crossovers	$p_c \cdot N$	The two most promising	weak
ST3	F-, S-, M- and L-crossover	$p_c \cdot N$	The two most promising	weak
ST4	F-, S-, M- and L-crossover	$\frac{1}{2} \cdot p_c \cdot N$	All four offspring	high

APPLICATION OF THE FCB-CROSSOVERS

Making use of the previously proposed crossover operators, different application strategies can be built, which are differentiated according to how they carry out the following two steps:

1. Generation of offspring using the different operators defined.
2. Selection of offspring resulting from the crossover which will form part of the population.

Four proposals were presented in [HLV95b], which are shown in Table 9.2, where we specify the strategy name, the FCB-crossover operators applied, the number of chromosomes in the population that should undergo crossover, and finally the type of OSM and the way in which it introduces diversity (measured in three diversity levels: *strong, high* and *weak*).

The experiments developed in [HLV95b] on different test functions, showed, in general, the suitability of using the Logical FCB-crossovers and the OSM that choose the two best elements from a set of four where the exploitation and exploration properties are equitably assigned. Another important conclusion was that by using F-crossover and S-crossover with F and S functions distant from the minimum and maximum (Logical F and S functions) respectively, then the population diversity levels obtained are greater, which agrees with the properties shown in Figure 9.5. This result suggested the definition of the dynamic FCB-crossovers in [HLV95a, HLV95c], which extend the use of the FCB-crossover operators in order to follow the principle *"to protect the exploration in the initial stages and the exploitation later"*. The dynamic FCB-crossovers are based on the use of parameterized t-norms, t-conorms and averaging functions [Miz89a, Miz89b]. These operators allow F-crossover operators and S-crossover operators with F and S functions distant from the minimum and maximum respectively to be considered in the first stages, and so high diversity is induced (as may be observed in Figure 9.5). Later, F-crossover operators with F functions close to the minimum and S-crossover operators with S functions close to the maximum are considered. Therefore, convergence is caused and the good behaviour of the Logical FCB-crossover operators shall be kept.

Another approach to the use of fuzzy connectives for designing crossover operators is presented in [HLV95a], the fuzzy connectives were used for building heuristic FCB-crossovers. These operators complement the FCB-crossover operators in order to include the goodness in the offspring generation for inducing heuristic exploration or heuristic exploitation. These operators were combined with the dynamic ones for designing heuristic dynamic FCB-crossover operators, which showed a robust behaviour in the experiments carried out [HLV95a].

9.2.5 Fuzzy Stop Criterion

Due the possibility of premuture convergence, GAs do not guarantee that the optimal solution shall be found. Therefore, if the optimal solution is not known, GA performance is difficult to measure accurately. In [MF94] a fuzzy stop criterion mechanism (FSCM) is developed to provide a useful evaluation of the GA's real-time performance. FSCM is based on achieving a user-defined level of performance for the given problem. In order to do so, it includes a predicting process based on statistics for estimating the value of the GA's optimal solution, then it compares the current solution to this optimal one by checking if an acceptable percentage (specified by the user) of the latter is reached. If so, the GA stops and returns belief and uncertainty measures [KF88] that provide a reliability measure for the GA's chosen solution. The whole process is outlined in Figure 9.6.

The acceptable percentage optimal solution defined by the user represents a fuzzy stop criterion for stopping GA if an appropriate solution is reached. The predicting process is invoked every 40 iterations and uses performance values such as the minimum solution value, average solution value and belief and plausibility values, all obtained during these iterations. For computing the reliability measure of the final solution, the following heuristic rules are taken into account:

1. As the number of iterations increases, the best solution value improves, the approach of the optimal solution indirectly indicates increasing reliability in the performance estimation.
2. As iterations increase, the average solution value improves, again, this is an indication of improving belief in the performance estimate.
3. As iterations increase, the accuracy of calculating the optimal solution increases. This is based on the general GA rule which states that whilst a GA may not guarantee the optimal solution shall be found, it does guarantee finding an equal or better solution than its initial one.

The underlying idea for the FSCM is that the user doesn't need to find the global solution, but rather an *approximate* solution that is close to the optimal one, i.e. the GA is used for solving a *fuzzy goal* instead of a crisp one because of the vagueness of the term *approximate*. This term is quantitatively measured by the user through the acceptable percentage optimal solution he requires in the final solution.

9.3 GENETIC ALGORITHMS IN FUZZY ENVIRONMENT

In the following, we shall study the use of GAs in an imprecise environment, managing fuzzy information. In this way FGAs will be GAs for producing approximate solutions to fuzzy optimization problems where the problem's variables translate imprecision or ambiguity into the measurement of a variable for a given point of view with the use of

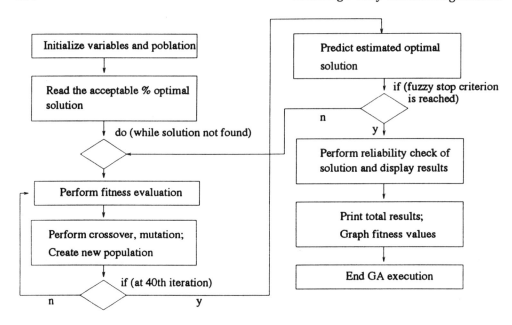

Figure 9.6   Structure of the Fuzzy Stop Criterion Mechanism

a fuzzy set in $\Re$, which is a set of ordered pairs $\{x, A(x)\}$ where $A(x)$ is termed "the grade of membership of x" for the fuzzy set $\overline{A}$.

The grade of membership can be interpreted as the degree of compatibility (degree of possibility) of the measurement of A with x. The fuzzy set will be assumed to be normal $(sup_x A(x) = 1)$ and convex $(\forall x, y, z \in \Re, y \in [x, z], A(y) \leq min\{A(x), A(z)\})$. A particular example corresponds to a trapezoidal fuzzy number defined with parameters $(a_i, b_i, \alpha_i, \beta_i)$. Figure 9.7 shows the membership function of this fuzzy number.

The two following subsections present two proposals for applying GAs in a fuzzy environment [BH94, HLV94]. The first presents a GA that manages fuzzy valued variables, and a representation method for fuzzy sets is shown. The second approaches the problem when the variables have associated fuzzy sets and then we have fuzzy valued fitness.

9.3.1   Fuzzy genetic algorithms: Proposal 1

In this subsection we present the approach proposed in [BH94].

Suppose there is a function F so that

$$\overline{Y} = F(\overline{X}).$$

$\overline{X}$ is the input and it will be a fuzzy subset of some interval $[0, M], M > 0$. $\overline{Y}$ is the output from F given $\overline{X}$. So F is a well defined mapping from fuzzy subsets of $[0, M]$ into fuzzy subsets of the real numbers. $X(r)$ represents the grade of membership of r for the fuzzy set $\overline{X}$.

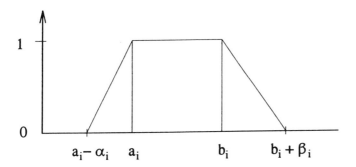

Figure 9.7 Membership functions of a trapezoidal fuzzy number

The objective is to find $\overline{X}$ in $[0, M]$ to "maximize" $\overline{Y}$. However, it is not able to maximize $\overline{Y}$ since it is a fuzzy set. So, let

$$m(\overline{Y}) = \theta$$

denote some measure (m) of how big $\overline{Y}$ is. For example, m could calculate the centroid of $\overline{Y}$, so that $\theta = centroid(\overline{Y})$. In general, m maps fuzzy sets into real numbers. Now the wish is to get an $\overline{X}$ in $[0, M]$ to maximize θ.

For the FGA the fuzzy set $\overline{X}$ is discretized. Let N be a positive integer and choose

$$z_o = 0, \ z_i = i \cdot \frac{M}{N} \text{ for } 1 \leq i \leq N,$$

and for $\overline{X}$, it is now used

$$\overline{X} = (x_0, ..., x_N),$$

where $x_i = X(z_i)$, $0 \leq i \leq N$. We input the discrete version of $\overline{X}$ to F and obtain θ the m measure of $F(x_0, ..., x_N)$. So F maps $[0, 1]^{N+1}$ to maximize θ. This is a general approach where it does not need to be fuzzy convex, normal, a fuzzy number, ...

An FGA will be applied for finding a ν on $[0, 1]^{N+1}$ to maximize θ. Making $\overline{X}$ discrete is no major drawback to the fuzzy optimization problem, because:

1. in computer applications it would probably have to make the fuzzy sets discrete; and
2. theoretically it can make the discretization of $[0, M]$ as fine as we wish (chosen N arbitrarily large).

The FGA is:

1. Randomly generate an initial population of size n from $[0, 1]^{N+1}$. Let the population be

$$\overline{X}_1 = (x_{10}, ..., x_{1N}), ..., \overline{X}_n = (x_{n0}, ..., x_{nN})$$

where $x_{ij} \in [0, 1]$.

2. Compute θ_i from each $\overline{X}_i$, $1 \le i \le n$. Let $T = \theta_1 + \cdots + \theta_n$. Let the partial sums be $s_k = \theta_1 + \cdots + \theta_k$, $1 \le k \le n$. Construct intervals $I_1 = [0, s_1)$, $I_i = [s_{i-1}, s_i)$, $2 \le i \le n - 1$, $I_n = [s_{n-1}, T]$.

3. Create a new population. Let w_i be a random number in $[0, T]$, $1 \le i \le n$. If w_i is in I_i, then choose $\overline{X}_i$ to be in the new population. Duplications (choosing the same $\overline{X}_i$ more than once) are expected.

4. The genetic operator crossover and mutation are defined in the classical way taking into account that x_{ij} values are in $[0, 1]$ for fuzzy sets, therefore operators for real coding information are needed. In particular, the FCB-crossover operators studied in the above section can be used.

9.3.2 Fuzzy genetic algorithms: Proposal 2

In this subsection we present the approach proposed in [HLV94].

When we work in a fuzzy environment, the feasible solutions of an optimization problem can hold associated fuzzy objectives [DKVV94] and the fitness associated with an individual is defined by a fuzzy number. Then, *it would be necessary to define a method to obtain the selection probabilities of the individuals.* As follows a method to do this is described.

Probability selection from fuzzy fitness

Consider the fuzzy fitness associated with each individual in the population, S, obtained by means of the evaluation function, $E(\cdot)$, and represented as a fuzzy number, $E : S \longrightarrow F(R)$, $s_i \in S$, $E(s_i) \in F(R)$, where $F(R)$ is the set of real fuzzy numbers.

From these fuzzy numbers it is necessary to define the selection probability associated with each individual to obtain a gene pool. We can consider a function $f : F(R) \longrightarrow \Re$ that allows the selection probability from the fuzzy fitness to be defined.

In fact, for each individual s_i in the population S, P_{S_i} can be defined as

$$P_{S_i} = \frac{f(E(s_i))}{f(\sum_{j=1}^{N} E(s_j))},$$

where the function $f : F(R) \longrightarrow \Re$ must be such that the set $\{P_{S_i}\}_{i=1,\ldots,N}$ verifies it to be a distribution of probability and also a selection probability, which allows the best strings to be reproduced, that is, the strongest individuals. In order for the function f to verify these desired properties, we introduce the next conditions for the function f:

1) Preserve the addition operation

$$f(\sum_{i=1}^{N} E(s_j)) = \sum_{i=1}^{N} f(E(s_j)),$$

necessary condition in order to verify $\sum_{i=1}^{N} P_{S_i} = 1$.

2) Preserve the order

$$E(s_k) \leq E(s_j) \iff f(E(s_k)) \leq f(E(s_j)),$$

necessary condition for reproducing the strongest strings by means of the selection probability. Because of this condition, the strongest individuals have greater selection probability,

$$E(s_k) \leq E(s_j) \iff P_{S_k} \leq P_{S_j}.$$

An important class of functions which verify these two conditions are the linear ranking functions [CGV92, ZL92].

Linear ranking functions. Fuzzy numbers [DP80] are fuzzy subsets of R whose α-cuts are closed and bounded intervals on $\Re$ when $\alpha > 0$. If $\alpha = 0$, A^0 will denote the closure of $supp(A)$. Hence, $\forall \alpha \in [0, 1]$ the α-cuts of $\overline{A}$ will be represented by

$$\overline{A}^\alpha = [a_1(\alpha), a_2(\alpha)] = \{x \in \Re | A(x) \geq \alpha\}$$

Extended sum and product by positive real numbers are considered to be defined in $F(R)$ by means of Extension Principle. Hence, given any two fuzzy numbers $\overline{A}, \overline{B} \in F(R)$, $\forall \alpha \in [0, 1]$, the following results regarding their α-cuts will be used,

$$(\overline{A} + \overline{B})^\alpha = [a_1(\alpha) + b_1(\alpha), a_2(\alpha) + b_2(\alpha)]$$

$$(r\overline{A})^\alpha = [ra_1(\alpha), ra_2(\alpha)] \quad \forall r \in \Re \; r > 0.$$

The problem of the comparison of fuzzy numbers has been widely investigated in the literature. Many fuzzy number ranking procedures (FNRPs) can be found, for instance in [ZL92]. Here we will focus on those FNRP that verify the two aforementioned conditions, *linear ranking functions*, which is not too restrictive because many well known FNRPs may be formulated by using linear ranking functions in some way.

Consider $\overline{A}, \overline{B} \in F(R)$, a simple method of comparison between them consists in the definition of a certain function $f : F(R) \longrightarrow \Re$. If this function $f(\cdot)$ is known, then $f(\overline{A}) > f(\overline{B})$, $f(\overline{A}) = f(\overline{B})$, $f(\overline{A}) < f(\overline{B})$ are equivalent to $\overline{A} > \overline{B}$, $\overline{A} = \overline{B}$, $\overline{A} < \overline{B}$ respectively. Usually, f is called a Linear Ranking Function (LRF) if

$$\forall \overline{A}, \overline{B} \in F(R); \; \forall r \in \Re \; r > 0; \; f(\overline{A} + \overline{B}) = f(\overline{A}) + f(\overline{B}) \; \text{and} \; f(r\overline{A}) = rf(\overline{A})$$

As is well known, from this definition several FNRPs may be considered. In particular, the two results regarding the α-cuts suggest that FNRPs using linear functions of the α-cuts could be expressed by LRFs. A good study of these LRFs is found in [CGV92]. Here, some of them are shown.

Linear functions based on only one α-cut. Consider $\alpha, \lambda \in [0,1]$ and $A \in F(R)$. One defines

$$f(\overline{A}) = \lambda a_1(\alpha) + (1 - \lambda)a_2(\alpha)$$

The definition contains two parameters which depend on the decision-maker: the first, α, is an accomplishment degree of the comparison, and the second, λ, is an optimism-pessimism level. Particular cases of this definition are:

(a) If $\alpha = 1$ and $\overline{A} \in F(R)$ is any unimodal fuzzy number, then the first index of Yager, [Yag78], is obtained.
(b) If $\alpha \in [0,1]$ is any fixed value and $\lambda = 0$, then the index of Adamo [Ada80] is deduced.

This definition is a particular case of the discrete method proposed in [GV91]. It is very easy to show that $f(\cdot)$ is linear in the sense of the definition above.

Linear functions based on all the α-cuts. Let $\overline{A} \in F(R)$, $\lambda \in [0,1]$ and $P(\cdot)$ be any additive measure in $[0,1]$. One defines

$$f(\overline{A}) = \int_0^1 (\lambda a_1(\alpha) + (1 - \lambda)a_2(\alpha)) \, dP(\alpha).$$

As may be seen, now one has only one parameter, λ, which acts as an optimism-pessimism degree. Particular cases of this definition are:

(a) If $P(\cdot)$ is Lebesgue's measure and $\lambda = \frac{1}{2}$, then the fourth index of Yager, [Yag81], is obtained.
(b) If $P(\cdot)$ is given by $P([a,b]) = b^2 - a^2$ and $\lambda = \frac{1}{2}$, then the index of Tsumura, [TTS81], is deduced.

This general definition of $f(\overline{A})$ was proposed in [Gon90], where it was also shown that this ranking function is linear. In particular for triangular fuzzy numbers, $f(\overline{A})$ takes the following expression:

$$f(\overline{A}) = a_1(1) - \frac{(a_1(1) - a_1(0))}{(r+1)} + \lambda \frac{(a_2(0) - a_1(0))}{(r+1)}$$

where r is a parameter according to which $f(\overline{A})$ can take values either close to the modal values ($r > 1$) or close to the values on the support ($r < 1$).

9.3.3   Some considerations and applications

First, to point out that these two ways to manage fuzzy information have a similar root because the measure m in the first method may be viewed as an LRF, in fact the root is to consider a measure that permits a real value for defining the selection process to be obtained. The differences are in the conception of the variables, the first proposal considers variables with fuzzy values representing these in a chromosome and the second considers non fuzzy value variables but with a fuzzy evaluation, having a fuzzy fitness.

A lot of fuzzy optimization problems can be considered for applying these two FGA approaches [DKVV94]. As follows we consider an example that was used in [BH94, HLV94] where we can observe the application of the two proposals, this is the problem of *maximum flow in a network with fuzzy capacities*, [BJ77, CK82, CK84, Cha87].

Let $S = (N, A)$ be a network with m vertices and n arcs where a liquid or a gas goes through it from a source node n_1 until a sink node n_m with fuzzy arc capacities. N denotes the set of nodes and $A \subset N \times N$ the set of arcs. The problem consists of finding the maximum flow from n_1 to n_m where, associated to each arc (i, j), there is a capacity $l_{ij} = 0$ giving a lower bound for the flow, and a fuzzy capacity $\overline{U}_{ij} \in F(R)$ giving an upper bound for the flow on the arc flows. If the arc (i, j) does not exist, then we assign $\overline{U}_{ij} = 0$.

Resolution with fuzzy values. Let $\overline{X}_{ij}$ denote the fuzzy flow from node i to node j, $\overline{X}_{ij} \leq \overline{U}_{ij}$. The mathematical problem is to maximize the variable $\overline{X}$, subject to the constraints:

$$\overline{X} = \sum_{i=2}^{m} \overline{X}_{1i}$$

$$\sum_{i=1, i \neq j}^{m} \overline{X}_{ij} = \sum_{k=1, k \neq j}^{m} \overline{X}_{jk}, \ j = 2, ..., m-1$$

$$\overline{X} = \sum_{i=1}^{m-1} \overline{X}_{im},$$

solving the final problem

$$max(\overline{X} - \sum_{j=1}^{m} \psi_j \overline{\Omega}_j),$$

for penalty constants ψ_j and penalty expressions $\overline{\Omega}_j$ in order to carry out the constraints. Every variable $\overline{X}_{ij}$ is a fuzzy subset of $[0, M_{ij}]$ verifying $\overline{X}_{ij} \leq \overline{U}_{ij}$.

Now, for applying the FGA to the fuzzy optimization problem, it is necessary to deduce on θ (possible centroid), discretize the intervals $[0, M_{ij}], i, j = 1, ..., m$, and concatenate the discrete $\overline{X}_{ij}, i, j = 1, ..., m$ for running the algorithm.

For example, considering the example shown in Figure 9.8 [BH94], the problem is:
max $\overline{X}$
subject to:

$$\overline{X} = \overline{X}_{12} + \overline{X}_{13},$$
$$\overline{X}_{24} = \overline{X}_{12},$$
$$\overline{X}_{13} = \overline{X}_{34} + \overline{X}_{35},$$

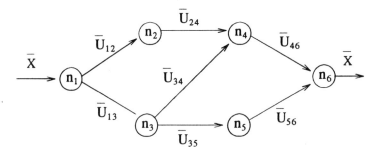

Figure 9.8   Fuzzy maximum flow problem with fuzzy capacities

$\overline{X}_{46} = \overline{X}_{24} + \overline{X}_{34}$,
$\overline{X}_{56} = \overline{X}_{35}$,
$\overline{X} = \overline{X}_{46} + \overline{X}_{56}$,
$0 \leq \overline{X}_{ij} \leq \overline{U}_{ij}$ all arcs.
The penalty expressions (after simplification) are:
$\overline{\Omega}_1 = \overline{U}_{34} - (\overline{X}_{13} - \overline{X}_{35})$.
$\overline{\Omega}_2 = \overline{U}_{46} - (\overline{X}_{12} + \overline{X}_{13} - \overline{X}_{35})$,
$\overline{\Omega}_2 = \overline{X}_{13} - \overline{X}_{35}$,

and the problem becomes

$$max(\overline{X} + \psi_1\overline{\Omega}_1 + \psi_2\overline{\Omega}_2 + \psi_3\overline{\Omega}_2$$

for penalty constants ψ_i and $\overline{X}_{13}, \overline{X}_{12}, \overline{X}_{35}$ fuzzy subsets of their respective intervals (for further explanations see [BH94]).

Resolution with fuzzy fitness.   Next the concept of cut in a network (basic in the development of this approach) as well as the concept of fuzzy capacity of a cut are introduced.

Definition. (Separation of the nodes n_1 and n_m [BJ77])
    Let N_1 be any set of nodes in the network such that n_1 belongs to N_1 and n_m do not belong to it, and let N_2 be $N_2 = N - N_1$. Then $(N_1, N_2) \equiv \{(i, j) : n_i \in N_1, n_j \in N_2\}$ is called a cut which separates the nodes, n_1 and n_m.
Definition. (Fuzzy capacity of a cut [CK84])
    The fuzzy capacity of the cut (N_1, N_2) is the fuzzy number defined as

$$\overline{C}(N_1, N_2) = \sum_{(N_1,N_2)} \overline{U}_{ij} = \sum_{n_i \in N_1, n_j \in N_2} \overline{U}_{ij}$$

extending the addition operation to fuzzy numbers.
Theorem. (Maximum flow-minimum cut theorem [FF56])
    The value of the maximum flow from n_1 to n_m is equal to the value of the minimum cut-set (N_1, N_2) separating n_1 from n_m.

The value of such a cut-set is the sum of the capacities of all arcs whose initial vertices are in N_1 and the final vertices in N_2. Then the minimum cut-set is the cut-set with the smallest value $\overline{C}(N_1, N_2)$.

Thus, the problem is reduced to obtain a partition of the set N into N_1, N_2, which separates the two vertices n_1 and n_m with minimum $\overline{C}(N_1, N_2)$. It is possible to formulate the minimum cut problem as a 0-1 programming problem as follows: Let x_i be a 0-1 variable associated with the node $n_i \in N$ defined as

$$x_i = \begin{cases} 1 & \text{if } n_i \in N_1 \\ 0 & \text{if } n_i \in N_2 \end{cases}$$

then the minimum cut problem can be formulated in the same way as minimizing

$$\overline{g}(x) = \sum_{i=1}^{m} \sum_{j=i+1}^{m} \overline{U}_{ij}[x_i(1 - x_j) + (1 - x_i)x_j]$$

where every possible solution x has associated the fuzzy fitness $g(x)$.

Then with an adequate genetic representation of the solutions, a vector of 0-1 values, the selection probability for each individual in the population will be obtained using an LRF, $f(\cdot)$, as was described earlier.

9.4   FINAL REMARKS

In this chapter, we have reviewed the combination of FL and GAs from the point of view of the so-called fuzzy genetic algorithms, GAs with components based on fuzzy logic techniques in order to improve their behaviour or to manage imprecise information represented by the fuzzy set theory.

The use of fuzzy logic techniques permits GA behaviour to be improved in different ways, such as: representation models for dealing with more complex genotype/phenotype relationships, fuzzy logic control systems for introducing a self-control parameter process according to performance measures, fuzzy operators and fuzzy connectives for designing genetic operators that introduce different population diversity levels and mechanisms that allow GA performance to be analysed from a more human point of view.

On the other hand, it should be remarked that the use of GA may offer a great potential for the fuzzy logic based optimization approach for representing uncertainty and approximation in relationships between system variables, given the potential of GAs in fuzzy environments as a flexibility tool for optimization and search.

REFERENCES

[Ada80] Adamo J. M. (1980) Fuzzy decision trees. *Fuzzy Sets and Systems* 4: 207–219.

[BH94] Buckley J. J. and Hayashi Y. (1994) Fuzzy genetic algorithm and applications. *Fuzzy Sets and Systems* 61: 129–136.

[BJ77] Bazaraa M. S. and Jarvis J. J. (1977) *Linear Programming and Network Flows.* John Wiley & Sons.

[Boo87] Booker L. (1987) Improving search in genetic algorithms. In *Genetic Algorithms and Simulated Annealing*, pages 61–73.

[CGV92] Campos L., Gonzalez A., and Vila M. A. (1992) On the use of the ranking function approach to solve fuzzy matrix games in a direct way. *Fuzzy Sets and Systems* 49: 193–203.

[Cha87] Chanas S. (1987) Fuzzy optimization in networks. In *Optimization Models using Fuzzy Sets and Possibility Theory*, pages 303–327. Reidel P. Co.

[CK82] Chanas S. and Kolodziejczyk W. (1982) Maximum flow in a network with fuzzy arc capacities. *Fuzzy Sets and Systems* 8: 165–173.

[CK84] Chanas S. and Kolodziejczyk W. (1984) Real-valued flows in a network with fuzzy arc capacities. *Fuzzy Sets and Systems* 13: 139–151.

[Dav89] Davis L. (1989) Adapting operator probabilities in genetic algorithms. In *Proc. of the Third Int. Conf. on Genetic Algorithms (ICGA'89)*, pages 61–69. George Mason University.

[DKVV94] Delgado M., Kacprzyk J., Verdegay J. L., and Vila M. A. (1994) *Fuzzy Optimization: Recent Advances.* Physica-Verlag.

[DP80] Dubois D. and Prade H. (1980) *Fuzzy Sets and Systems. Theory and Applications.* Academic Press.

[ECS89] Eshelman L. J., Caruana R. A., and Schaffer J. D. (1989) Biases in the crossover landscape. In *Proc. of the Third Int. Conf. on Genetic Algorithms (ICGA'89)*, pages 10–19. George Mason University.

[FF56] Ford L. and Fulkerson D. (1956) Maximal flow through a network. *Canad. J. Math* 8: 399–404.

[Fog94] Fogel D. B. (1994) An introduction to simulated evolutionary optimization. *IEEE Trans. on Neural Networks* 5(1): 3–14.

[Gol91] Goldberg D. E. (1991) Real-coded genetic algorithms, virtual alphabets, and blocking. *Complex Systems* 5: 139–167.

[Gon90] Gonzalez A. (1990) A study of the ranking function approach through mean values. *Fuzzy Sets and Systems* 35: 29–41.

[Gre86] Grefenstette J. J. (1986) Optimization of control parameters for genetic algorithms. *IEEE Trans. on Systems, Man, and Cybernetics* SMC-16: 122–128.

[GV91] Gonzalez A. and Vila A. (1991) A discrete method to study indifference and order relation between fuzzy numbers. *Information Sciences* 56: 245–288.

[HHLV94] Herrera F., Herrera E., Lozano M., and Verdegay J. L. (September 1994) Fuzzy tools to improve genetic algorithms. In *Proc. First European Congress on Fuzzy and Intelligent Technologies (EUFIT'94)*, pages 1532–1539. Aachen.

[HLV94] Herrera F., Lozano M., and Verdegay J. L. (1994) Applying genetic algorithms in fuzzy optimization problems. *Fuzzy Sets & Artificial Intelligence* 3: 39–52.

[HLV95a] Herrera F., Lozano M., and Verdegay J. L. (April 1995) Dynamic and heuristic crossover operators for controlling the diversity and convergence of real-coded genetic algorithms. Technical Report DECSAI-95113, University of Granada, Department of Computer Science and Artificial Intelligence, University of Granada, Granada, Spain.

[HLV95b] Herrera F., Lozano M., and Verdegay J. L. (February 1995) Fuzzy connectives based crossover operators to model genetic algorithms population diversity. Technical Report DECSAI-95110, University of Granada, Department of Computer Science and Artificial Intelligence, University of Granada, Granada, Spain.

[HLV95c] Herrera F., Lozano M., and Verdegay J. L. (1995) The use of fuzzy connectives to design real-coded genetic algorithms. To appear in Mathware & Soft Computing.

[Jon75] Jong K. A. D. (1975) An analysis of the behavior of a class of genetic adaptive systems. Master's thesis, University of Michigan.

[KF88] Klir G. and Folger T. (1988) *Fuzzy Sets, Uncertainly, and Information.* Prentice Hall.

[Lee90] Lee C. C. (March 1990) Fuzzy logic in control systems: Fuzzy logic controller - parts i and ii. *IEEE Transactions on Systems, Man and Cybernetics* 20(2): 404–435.

[LLK92] Li T.-H., Lucasius C. B., and Kateman G. (1992) Optimization of calibration data with the dynamic genetic algorithm. *Analytica Chimica Acta* 268: 123–134.

[LT93] Lee M. A. and Takagi H. (1993) Dynamic control of genetic algorithms using fuzzy logic techniques. In *Proc. Fifth International Conference on Genetic Algorithms (ICGA '93)*, pages 76–83. Urbana-Champaign.

[MF94] Meyer L. and Feng X. (1994) A fuzzy stop criterion for genetic algorithms using performance estimation. In *Proc. Third IEEE International Conference on Fuzzy Systems*, pages 1990–1995. Orlando.

[Mic92] Michalewicz Z. (1992) *Genetic Algorithms + Data Structures = Evolution Programs.* Springer-Verlag.

[Miz89a] Mizumoto M. (1989) Pictorial representations of fuzzy connectives, part i: Cases of t-norms, t-conorms and averaging operators. *Fuzzy Sets and Systems* 31: 217–242.

[Miz89b] Mizumoto M. (1989) Pictorial representations of fuzzy connectives, part ii: Cases of compensatory operators and seft-dual operators. *Fuzzy Sets and Systems* 32: 45–79.

[San93] Sanchez E. (1993) Fuzzy genetic algorithms in soft computing environment. In *Fifth IFSA World Congress.* Seoul. Invited Plenary Lecture.

[TTS81] Tsumura Y., Terano T., and Sugeno M. (1981) Fuzzy fault tree analysis. summary ofpapers on general fuzzy problems. Technical Report 7.

[Voi92] Voight H. (June 1992) Fuzzy evolutionary algorithms. Technical Report tr-92-038, International Computer Science Institute (ICSI), Berkeley.

[WL93] Wang P. Z. and Lee K. F. (1993) *Between Mind and Computer: Fuzzy Science and Engineering.* World Scientific.

[XV94] Xu H. Y. and Vukovich G. (1994) Fuzzy evolutionary algorithms and automatic robot trajectory generation. In *Proc. of The First IEEE Conference on Evolutionary Computation*, pages 595–600. Orlando.

[Yag78] Yager R. R. (1978) Ranking fuzzy subset over the unit interval. In *Proc. CDC*, pages 1735–1737.

[Yag81] Yager R. R. (1981) A procedure for ordering fuzzy subsets of the unit interval. *Information Sciences* 24: 143–161.

[YZ92] Yager R. R. and Zadeh L. A. (1992) *An Introduction to Fuzzy Logic Applications in Intelligent Systems.* Kluwer.

[Zad93] Zadeh L. A. (1993) Foreword. In *Between Mind and Computer: Fuzzy Science and Engineering.* World Scientific.

[Zim90] Zimmermann H. J. (1990) *Fuzzy Set Theory and its Applications.* Kluwer.

[ZL92] Zhu Q. and Lee E. S. (1992) *Comparison and Ranking of Fuzzy Numbers.* Omnitech Press Warsaw and Physica-Verlag.

10

Modeling Hybrid Genetic Algorithms

D. WHITLEY

10.1 INTRODUCTION

A "hybrid genetic algorithm" combines local search with a more traditional genetic algorithm. The most common form of hybrid genetic algorithm uses local search to improve the initial population as well as the strings produced by genetic recombination. The resulting improvements are then coded onto the strings processed by the genetic algorithm (e.g. Mühlenbein, 1991; Davis, 1991). These hybrid genetic algorithms have also called "Memetic Algorithms". The term Lamarckian Evolution is also sometimes used, where local search in this context is considered to be analogous to acquired or learned behavior that is subsequently coded back onto the "genotype" representation and passed on to offspring during reproduction.

This paper looks at how one form of hybrid genetic algorithm can be modeled in the context of the existing models for the simple genetic algorithm; it should be possible to model the integration of other types of local search with genetic algorithms using the same basic approach. A secondary goal of this paper is to review the existing models for finite and infinite population genetic algorithms and to show the relationship between the two models. The "Lamarckian" updates associated with a hybrid genetic algorithm can be represented as a matrix update to the infinite population vector used by the infinite population model of the simple genetic algorithm. In turn, the infinite population vector is also equivalent to the sampling distribution probabilities used by finite Markov models of the simple genetic algorithm. Thus, the extension of existing models to hybrid genetic algorithms is simple and straight forward.

This paper also builds on earlier work by Whitley, Gordon and Mathias (1994) by exploring the notion that the use of a hybrid genetic algorithm induces equivalence classes over the set of all possible discrete functions, such that functions in an equivalence class are in expectation processed in an identical fashion under hybrid genetic search.

The results presented here specifically focus on parameter optimization problems that have a binary encoding. The models discussed in this paper can be extended to other representations, and in particular, Whitley and Yoo (1995) show how models of simple genetic algorithms can be adapted for permutation representations.

For binary encodings, local search involves changing each of the L bits in a string encoding; if an improvement is found, the search is said to "move" to the improved string. Initially local search will be limited to one step in the search space. The arguments presented in this paper assume that each step of local search results in a deterministic choice of moves in the search space; for modeling hybrid genetic algorithms this requirement is not rigid however and the results can be reformulated to consider probabilistic choices of moves under local search. In order to make moves deterministic we use steepest ascent as our local search. For steepest ascent, all L bit changes are tested and the best improvement is taken. An alternative is to use next ascent, where the first improvement found among the L neighbors is taken. Note that if an improvement is found before checking all L neighbors then next ascent has less evaluation cost per move than steepest ascent. Also note that if next ascent checks neighbors in a fixed order, the moves it generates are also deterministic; however, in practice it is often effective to test the neighbors of a string in random order, thus making the next move a probabilistic decision.

10.1.1 Equivalent Classes Under Hybrid Genetic Search

Consider the following Function One:

e(00000) = 105	e(01000) = 73	e(10000) = 36	e(11000) = 60
e(00001) = 69	e(01001) = 71	e(10001) = 93	e(11001) = 64
e(00010) = 43	e(01010) = 89	e(10010) = 37	e(11010) = 96
e(00011) = 47	e(01011) = 43	e(10011) = 25	e(11011) = 19
e(00100) = 60	e(01100) = 44	e(10100) = 15	e(11100) = 54
e(00101) = 40	e(01101) = 29	e(10101) = 68	e(11101) = 38
e(00110) = 18	e(01110) = 48	e(10110) = 44	e(11110) = 31
e(00111) = 65	e(01111) = 91	e(10111) = 23	e(11111) = 100

where $e(x)$ is the evaluation of string x. This function was created by assigning the evaluation 100 to point 11111 and evaluation 105 to the point 00000 and by assigning a random value between 10 and 99 to all other points in the space. The problem is posed as a maximization problem. The basins of attraction for this function under steepest ascent are illustrate in Figure 10.1.

Note that there are five local optima under steepest ascent: 00000, 11111, 11010, 10001, and 10110.

In the current paper, we assume that 1-step of steepest ascent is applied to the initial population processed by the genetic algorithm, and that all offspring produced by recombination and/or mutation are also improved with 1-step of steepest ascent. As already noted, each improvement will change one bit in the string being processed. The

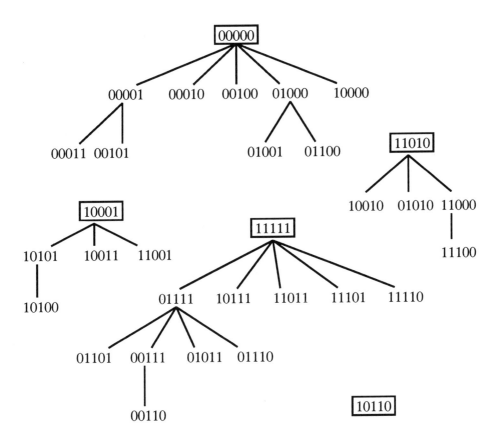

Figure 10.1   Basins of attraction under steepest ascent. The "boxed" strings
are local opitma.

first observation which can be made is that the genetic algorithm (i.e., not considering the steepest ascent component) will never process any of the saddle points in the search space. This is true in the sense that local search could be built into the evaluation function: the evaluation function is passed a string, evaluates the string, does 1-step of steepest ascent, returns the improved string (assuming the current string is not an optima) and also returns the evaluation of the improved string. In this sense the genetic algorithm can be separated from the Lamarckian updates. As will be shown, this separation carries through to the infinite and finite population models of genetic search.

A related observation is that the actual values of the saddle points are irrelevant to the genetic search process except in so much as the evaluations of the saddle points conforms to the tree structures associated with the various basins of attraction. Since the saddle point are not actually processed during genetic search except inside the evaluation function, the actual evaluations of the saddle have no impact on the search process. For Function One 22 of the 32 points are saddle points. Since Function One is posed as a maximization problem, it follows that decreasing the evaluation associated with any of the 22 saddle points would not change the tree structures of the basins of attraction in any way. The evaluations of the saddle points can also be increased as long as the tree structure remains unaltered. It thus follows that there are many functions which have exactly the same basins of attraction as Function One and which in expectation would be processed in an identical fashion by a hybrid genetic algorithm using 1 step of steepest ascent.

Figure 10.2 shows the strings in Function One that are actually evaluated by the simple genetic algorithm after 1 and 2 steps of steepest ascent. Also note that for two functions to be processed in an identical fashion by a genetic algorithm after applying steepest ascent to all members of the population every generation, all that is required is that the reduced tree structures be equivalent. It is not required that the strings which have been removed from the full function space map onto the remaining strings in the same way, or even that strings be in the same basin of attraction in the full function space. All that matters is that the reduces tree structures be identical with the same number of strings moving to the remaining nodes.

10.1.2 Models of Simple Genetic Algorithms

Goldberg (1987, 1989) and Bridges and Goldberg (1987) were the first to model critical details of how the genetic algorithm processes infinitely large populations under recombination. Vose (1990; Vose and Liepins 1991) extended and generalized this model. Whitley et al. (1992; Whitley 1993) introduce another version of the infinite population model that relates the work of Goldberg and Vose. Here, the Vose model is reviewed and it is shown how the effect of various operators, including local search, fit into this model.

The following notation is based on Vose and Liepins. The vector $p^t \, \epsilon \, \Re$ is such that the kth component of the vector is equal to the proportional representation of string k at generation t. It is assumed that $n = 2^L$ is the number of points in the search space defined over strings of length L and that the vector p is indexed 0 to $n - 1$. The vector $s^t \, \epsilon \, \Re$ represents the tth generation of the genetic algorithm after selection and the ith component of s^t is proportional representation of string i in

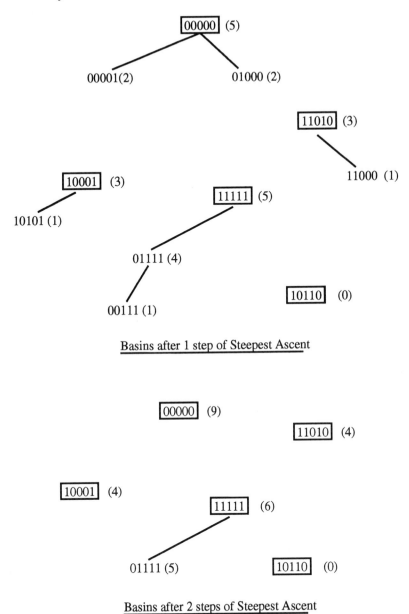

Basins after 1 step of Steepest Ascent

Basins after 2 steps of Steepest Ascent

Figure 10.2 Structure of the basins of attraction for the same problem as they appear to a Hybrid Genetic Algorithm using 1 and 2 steps of steepest ascent. The numbers in parentheses beside each string indicates the number of strings that move to that position after steepest ascent.

the population after selection but before any operators (e.g. recombination, mutation, local search) are applied. Likewise, p_i^t represents the proportional representation of string i at generation t *before* selection occurs.

The function $r_{i,j}(k)$ yields the probability that string k results from the recombination of strings i and j. (For now assume that r only yields the results for recombination; the effect of other operators could also be included in r.) Now, using $\mathcal{E}$ to denote expectation,

$$\mathcal{E}\, p_k^{t+1} = \sum_{i,j} s_i^t\, s_j^t\, r_{i,j}(k) \tag{10.1}$$

To begin the construction of a general model, we first consider how to calculate the proportional representation of string 0 (i.e., the string composed of all zeros) at generation $t + 1$; in other words, we compute p_0^{t+1}. A mixing matrix M is constructed where the $(i, j)^{th}$ entry $m_{i,j} = r_{i,j}(0)$. Here M is built by assuming that each recombination generates a single offspring. The calculation of the change in representation for string $k = 0$ is now given by

$$\mathcal{E}\, p_0^{t+1} \;=\; \sum_{i,j} s_i^t s_j^t r_{i,j}(0) \;=\; s^T M s \tag{10.2}$$

where T denotes transpose. Note that this computation gives the expected representation of a single string, 0, in the next genetic population. Vose and Liepins formalize the notion that bitwise exclusive-or can be used to access various probabilities from the recombination function r. Specifically,

$$r_{i,j}(k) = r_{i,j}(k \oplus 0) = r_{i \oplus k, j \oplus k}(0). \tag{10.3}$$

This implies that the mixing matrix M, which was defined such that entry $m_{i,j} = r_{i,j}(0)$, can provide mixing information for any string k just by changing how M is accessed. By reorganizing the components of the vector s the mixing matrix M can yield information about the probability $r_{i,j}(k)$. A permutation function, σ, is defined as follows:

$$\sigma_j < s_0, ..., s_{n-1} >^T \; = \; < s_{j \oplus 0}, ..., s_{j \oplus (n-1)} >^T \tag{10.4}$$

where the vectors are treated as columns and n is the size of the search space. The computation

$$(\sigma_q\, s^t)^T M (\sigma_q\, s^t) = p_q^{t+1} \tag{10.5}$$

thus reorganizes s with respect to string q and produces the expected representation of string q at generation $t + 1$. A general operator $\mathcal{M}$ can now be defined over s which remaps $s^T M s$ to cover all strings in the search space.

$$\mathcal{M}(s) = < (\sigma_0\, s)^T M (\sigma_0\, s), ..., (\sigma_{n-1}\, s)^T M (\sigma_{n-1}\, s) >^T \tag{10.6}$$

This model has not yet addressed how to generate the vector s^t given p^t. A fitness matrix F is defined such that fitness information is stored along the diagonal; the

$(i, i)^{th}$ element is given by $f(i)$ where f is the fitness function. Following Vose and Wright (1994)

$$s^t = Fp^t / 1^T Fp^t \tag{10.7}$$

since $Fp^t = < f_0 p_0^t, f_1 p_1^t, \ldots, f_{n-1}, p_{n-1}^t >$ and the population average is given by $1^T Fp^t = f_0 p_0^t + f_1 p_1^t + \ldots + f_{n-1}, p_{n-1}^t$.

Vose (1993) refers to this complete model as the $\mathcal{G}$ function. Given any population distribution p, $\mathcal{G}(p)$ can be interpreted in two way. If the population is infinitely large, then $\mathcal{G}(p)$ is the next exact distribution of the next population. However given any (finite or infinite) population p, if strings in the next population are chosen one at a time, then $G(p)$ defines a vector such that element i is chosen for the next generation with probability $\mathcal{G}(p)_i$. This is because $\mathcal{G}(p)$ defines the exact sampling distribution for the next generation. This is very useful when constructing finite Markov models of the simple genetic algorithm as well as the hybrid genetic algorithm as defined in this paper.

To motivate the inclusion of local search updates into the $\mathcal{G}$ function, we first look at how mutation can be added to this model in a modular fashion. Recall that M is the recombination matrix, which we initially defined to cover only crossover. Define $\mathcal{Q}$ as the mutation matrix. Assuming mutation is independently applied to each bit with the same probability, $\mathcal{Q}$ can be constructed by defining a mutation vector Γ such that component Γ_i is the probability of mutating string i and producing string 0. The vector Γ is the first column of the mutation matrix and in general Γ can be reordered to yield column j of the matrix by reordering such that element $q_{i,j} = \Gamma_{i \oplus j}$.

Having defined a mutation matrix, mutation can now be applied after recombination in the following fashion:

$$p^{t+1,m} = (p^{t+1})^T \mathcal{Q},$$

where $p^{t+1,m}$ is just the p vector at time t+1 after mutation. Mutation also can be done before crossover; the effect of mutation on the vector s immediately after selection produces the following change: $s^T \mathcal{Q}$, or equivalently, $\mathcal{Q}^T s$.

We now drop the $p^{t+1,m}$ notation and assume the original p^{t+1} vector is defined to include mutation, such that

$$p_0{}^{t+1} = (\mathcal{Q}^T s)^T M (\mathcal{Q}^T s)$$

$$p_0{}^{t+1} = s^T (\mathcal{Q} M \mathcal{Q}^T) s$$

$$p_0{}^{t+1} = s^T (\mathcal{Q} M \mathcal{Q}^T) s$$

$$\text{thus} \quad p_0{}^{t+1} = s^T M_2 s \quad \text{where} \quad M_2 = (\mathcal{Q} M \mathcal{Q}^T)$$

The matrix M as used here included only recombination, while M in the Vose model is a "mixing matrix" that includes both mutation and recombination. As long as the mutation rate is independently applied to each bit in the string, it makes no difference whether mutation is applied before or after recombination. Also, this view of mutation makes it clear how the general mixing matrix can be build by combining matrices for mutation and crossover.

10.1.3   Modeling Hybrid Simple Genetic Algorithms

The following function will be denoted Function Two:

e(0000) = 28	e(0100) = 20	e(1000) = 12	e(1100) = 4
e(0001) = 26	e(0101) = 18	e(1001) = 10	e(1101) = 2
e(0010) = 24	e(0110) = 16	e(1010) = 8	e(1110) = 0
e(0011) = 22	e(0111) = 14	e(1011) = 6	e(1111) = 30

where the function is posed as a maximization problem.

Recall that p_i^t, is the representation of string i in the population at time t. Let $p_i^{t,k}$ be the representation of string i in the population at time t after some k steps of local search. Function Two has two basins of attraction at 0000 and 1111. We now calculate how local search changes the distributions in the p vector. Note that the redistribution of string representations will be different for functions that are in different equivalence classes. For Function Two, the redistribution of points in the space occurs as follows. (The binary representation of the indices are used here to aid the reader in following the update process due to local search.)

$$
\begin{aligned}
p_{0000}^{t,k} &= p_{0000}^t + p_{0001}^t + p_{0010}^t + p_{0100}^t + p_{1000}^t \\
p_{0001}^{t,k} &= p_{0011}^t + p_{0101}^t + p_{1001}^t \\
p_{0010}^{t,k} &= p_{0110}^t + p_{1010}^t \\
p_{0100}^{t,k} &= p_{1100}^t \\
p_{1111}^{t,k} &= p_{1111}^t + p_{1110}^t + p_{1101}^t + p_{1011}^t + p_{0111}^{t,l} \\
p_{0011}^{t,k} &= 0 \\
p_{0101}^{t,k} &= 0 \\
p_{0110}^{t,k} &= 0 \\
p_{0111}^{t,k} &= 0 \\
p_{1000}^{t,k} &= 0 \\
p_{1001}^{t,k} &= 0 \\
p_{1010}^{t,k} &= 0 \\
p_{1011}^{t,k} &= 0 \\
p_{1100}^{t,k} &= 0 \\
p_{1101}^{t,k} &= 0 \\
p_{1110}^{t,k} &= 0
\end{aligned}
$$

$$(10.8)$$

In this case, k = 1. Note that the local search updates to the p vector can also be expressed in the form of a matrix update. For the same single step of steepest ascent, those updates are as follows:

$$p^{t,k} = (p^t)^T \begin{bmatrix} 1\,0\,0\,0\,0 & 0 \\ 1\,0\,0\,0\,0 & 0 \\ 1\,0\,0\,0\,0 & 0 \\ 0\,1\,0\,0\,0 & 0 \\ 1\,0\,0\,0\,0 & 0 \\ 0\,1\,0\,0\,0 & 0 \\ 0\,0\,1\,0\,0 & 0 \\ 0\,0\,0\,0\,0 & \cdots & 0 \\ 1\,0\,0\,0\,0 & 0 \\ 0\,1\,0\,0\,0 & 0 \\ 0\,0\,1\,0\,0 & 0 \\ 0\,0\,0\,0\,0 & 0 \\ 0\,0\,0\,0\,1 & 0 \\ 0\,0\,0\,0\,0 & 0 \\ 0\,0\,0\,0\,0 & 0 \\ 0\,0\,0\,0\,0 & 0 \end{bmatrix}$$

Note that the first column of the matrix has a 1 bit in those positions that directly "move" to string 0000 and in general column j of the matrix flags those strings that directly move to string j under 1 step of steepest ascent. Let L denote the update matrix for 1 step of local search; the matrix notation also has the advantage that for for k steps of local search $p^{t,k} = (p^t)^T L^k$. Thus, define $\mathcal{G}^k(p)$ to be the simple genetic algorithm function where k steps of local search are applied to each population (i.e., each generation).

10.2 THE MARKOV MODEL

Nix and Vose show how to structure the finite population for a simple genetic algorithm. Briefly, the Markov model is an N X N transition matrix Q, where N is the number of finite populations of K strings and $Q_{i,j}$ is the probability that the kth generation will be population $\mathcal{P}_j$ given that the $(k-1)^{th}$ population is $\mathcal{P}_i$.
  Let
$$< Z_{0,j}, Z_{1,j}, Z_{2,j}, ..., Z_{r-1,j} >$$
represent a population, where $Z_{x,j}$ represents the number of copies of string x in population j, and $r = 2^L$.
  We will build the population incrementally. How may ways are there to place the $Z_{0,j}$ copies of string 0 in the population?
$$\binom{K}{Z_{0,j}}$$
How many ways can $Z_{1,j}$ now be placed in the population.
$$\binom{K - Z_{0,j}}{Z_{1,j}}$$

Continuing for all strings

$$\binom{K}{Z_{0,j}}\binom{K-Z_{0,j}}{Z_{1,j}}\binom{K-Z_{0,j}-Z_{1,j}}{Z_{2,j}}\cdots\binom{K-Z_{0,j}-Z_{1,j}-\ldots-Z_{r-2,j}}{Z_{r-1,j}}$$

which yields

$$\frac{K!}{(K-Z_{0,j})!Z_{0,j}!}\ \frac{(K-Z_{0,j})!}{(K-Z_{0,j}-Z_{1,j})!Z_{1,j}!}\ \frac{(K-Z_{0,j}-Z_{1,j})!}{(K-Z_{0,j}-Z_{1,j}-Z_{2,j})!Z_{2,j}!}$$

$$\cdots\frac{(K-Z_{0,j}-Z_{1,j}-\ldots-Z_{r-2,j})!}{Z_{r-1,j}!}$$

which in turn reduces to

$$\frac{K!}{Z_{0,j}!Z_{1,j}!Z_{2,j}!\ldots Z_{r-1,j}!}$$

Let $C_i(y)$ be the probability of generating string y from the finite population P_i. Then

$$Q_{i,j} = \frac{K!}{Z_{0,j}!Z_{1,j}!Z_{2,j}!\ldots Z_{r-1,j}!}\prod_{y=0}^{r-1}C_i(y)^{Z_{y,j}}$$

and

$$Q_{i,j} = K!\prod_{y=0}^{r-1}\frac{C_i(y)^{Z_{y,j}}}{Z_{y,j}!}$$

So, how do we compute $C_i(y)$? Note that the finite population P_i can be described by a vector p. Also note that the sampling distribution from which P_j is constructed is given by the infinite population model $\mathcal{G}(p)$. Thus, replacing $C_i(y)$ by $\mathcal{G}(p)_y$ yields (Vose, In Press):

$$Q_{i,j} = K!\prod_{y=0}^{r-1}\frac{(\mathcal{G}(p)_y)^{Z_{y,j}}}{Z_{y,j}!}$$

To build the Markov model for the hybrid genetic algorithm simply replace $C_i(y)$ by $\mathcal{G}^k(p)_y$.

10.3   CONCLUSIONS

It has been shown that a simple matrix representation can be used to characterize how local search updates can be integrated into the existing models of simple genetic algorithms. The local search updates can be modeled as simple updates to the p vector representing the proportional distribution of strings in an infinitely large population. In the current paper, the use of binary representations has been assumed, but models for the simple genetic algorithm have been generalized to include other representations by Vose (Vose, In Press; Whitley and Yoo, 1995) and generalizations to other forms of hybrid genetic algorithms should be straight forward in most cases.

The vector p not only tracks the representation of strings in a infinitely large population, but can also be used to represented the sampling distribution probabilities of a finite population genetic algorithm. Thus it follows that the use that the local search updates to the p vector carry over to the use of the p vector in the construction of finite Markov models for simple genetic algorithms.

REFERENCES

Bridges C., and Goldberg D. (1987) An Analysis of Reproduction and Crossover in a Binary-Coded Genetic Algorithm. In *Proc. 2nd International Conf. on Genetic Algorithms and Their Applications.* pp: 9–13. J. Grefenstette, ed. Lawrence Erlbaum.

Davis L. (1991) *Handbook of Genetic Algorithms.* Van Nostrand Reinhold, NY.

Goldberg D. (1987) Simple Genetic Algorithms and the Minimal, Deceptive Problem. In *Genetic Algorithms and Simulated Annealing*, pp: 74–88. L. Davis, ed., Pitman.

Goldberg D. (1989) *Genetic Algorithms in Search, Optimization and Machine Learning.* Reading, MA: Addison-Wesley.

Mühlenbein H. (1991) Evolution in Time and Space - The Parallel Genetic Algorithm. In *Foundations of Genetic Algorithms*, pp 316–337. G. Rawlins, ed. Morgan-Kaufmann.

Vose M. (1990) Formalizing Genetic Algorithms. In *Proc. of Genetic Algorithms, Neural Networks and Simulating Annealing Applied to Problems in Signal Processing and Image Processing*, IEEE. Glasgow, U.K.

Vose M. (1993) Modeling Simple Genetic Algorithms. *Foundations of Genetic Algorithms -2-,* D. Whitley, ed., Morgan Kaufmann.

Vose M. (In Press) The Simple Genetic Algorithms: Foundations and Theory. Cambridge, MA: MIT Press.

Vose M. and Liepins G., (1991) Punctuated Equilibria in Genetic Search. *Complex Systems* 5:31-44.

Nix A. and Vose M. (1992) Modeling Genetic Algorithms with Markov Chains. *Annals of Mathematics and Artificial Intelligence*, 5: 79–88.

Vose M. and Wright G. (1994) Simple Genetic Algorithms with Linear Fitness. *Evolutionary Computation*, 2(4): 347–368.

Whitley D. (1993) An Executable Model of a Simple Genetic Algorithm. In *Foundations of Genetic Algorithms -2-.* D. Whitley, ed. Morgan Kaufmann.

Whitley D., Das R. and Crabb, C. (1992) Tracking Primary Hyperplane Competitors During Genetic Search. *Annals of Mathematics and Artificial Intelligence.* 6:367-388.

Whitley D., Gordon V.S. and Mathias, K.E. (1994) Lamarckian Evolution, The Baldwin Effect and Function Optimization, In *International Conference on Evolutionary Computation: PPSN 3* H-P Schwefel and R. Maenner, eds. Springer-Verlag.

Whitley D. and Yoo N.W. (1995) Modeling Simple Genetic Algorithms for Permutation Problems. *Foundations of Genetic Algorithms -3-,* D. Whitley and M. Vose, eds., Morgan Kaufmann.

11

Genetic Algorithms and Neural Networks

D. WHITLEY

11.1 INTRODUCTION

Genetic algorithms and neural networks are both inspired by computation in biological systems. A good deal of biological neural architecture is determined genetically. It is therefore not surprising that as some neural network researchers explored how neural systems are organized that the idea of evolving neural architectures should arise.

Genetic algorithms have been used in conjunction with neural networks in three major ways. First, they have been used to set the weights in fixed architectures. This includes both supervised learning applications and reinforcement learning applications. In related work, a genetic algorithm has been used to set the learning rates which in turn are used by other types of learning algorithms. Genetic algorithms have also been combined with more traditional forms of gradient based search.

Second, genetic algorithms have been used to learn neural network topologies. When evolving neural networks topologies for function approximation, this includes the problem of specifying how many hidden units a neural network should have and how the nodes are connected.

A third major application is the use of genetic algorithms to select training data and to interpret the output behavior of neural networks.

Schaffer, Whitley and Eshelman (1992) survey these various areas in an introduction to the proceeding of a 1992 workshop on *Combinations of Genetic algorithms and Neural Networks.* The current paper is tutorial in nature and highlights select cases and briefly references some of the work that has been introduced in the last 3 years.

11.2   GENETIC ALGORITHMS FOR PREPROCESSING AND INTERPRETING DATA

Two examples of using genetic algorithms for preprocessing data is given in the work of Chang and Lippmann (1991) and the work of Brill, Brown and Martin (1992). In both cases, a large number of inputs were available as input to a K nearest neighbor (KNN) classifier. In this case the coding can be a simple binary string indicating whether a particular input or combination of inputs can be deleted from the input set without significantly changing the classification behavior. In the Chang and Lippmann application, the genetic algorithm was able to reduce the input set from 153 to 33 input features. Brill, Brown and Martin also were able to reduce the input set, but their goal was not just to reduce the set of inputs to the nearest neighbor classifier, but to also identify inputs that would also work well for a counterpropagation network. The nearest neighbor classifier was used for feature selection since the evalutation of a feature set is much faster with the KNN classifier than the counterpropagation network. Nevertheless, the reduced input set for the KNN classifier also worked well for the counterpropagation network.

Genetic algorithms have not only been used to reduce the input data set but also to interpret outputs of a neural network. Eberhart and Dobbins (1991; Eberhart 1992) used a genetic algorithm to search for the decision surface that identified boundary cases of appendicitis as predicted by a neural network. For example, what inputs lead to a classification of 0.5, where 0.5 indicates a borderline case, i.e., a case that lies on the boundary between the decision regions that classify cases as positive or negative examples of appendicitis? It can also be useful to determine what are considered to be what Eberhart calls 'quintessential' examples of appendicitis as predicted by a neural network. In this case, what inputs lead to a classification of 1.0, where 1.0 corresponds to a classic case of appendicitis?

Asking for an input that yields an output of 1.0 or 0.5 is really a form of network inversion; in other words, this is analogous to running the neural network backwards. One can literally attempt to run a neural network backwards by using backpropagation to look for hidden node and input node activations that yield a particular output, but the process can be time consuming and does not always work well since the classification of a neural network is often many-to-one and not an invertible function. Eberhart and Dobbins simply searched the input space for strings that produced the desired output. By running a genetic algorithm multiple times they were able to obtain multiple patterns that mapped to a particular output.

Such information can be used in two ways. First, it can be used as an explanation tool. Knowing quintessential examples as well as borderline cases can help explain how a network classifies novel inputs. Second, it can also be used to assess what a neural network has learned and whether the cases that it considers to be quintessential and borderline are reasonable.

11.3   GENETIC ALGORITHMS FOR TRAINING NEURAL NETWORKS

The idea of training neural networks with genetic algorithms can be found in Holland's 1975 book *Adaptation in Natural and Artificial Systems*. Most of the actual work in this area is far more recent. Belew, McInernery and Schraudolph (1990), Harp, Samad and Guha (1989;1990) and Schaffer, Caruana and Eshelman (1990) all used genetic algorithm to set the learning and momentum rates for feedforward neural networks. Mühlenbein also contributed to the early efforts in this area (1990; Mühlenbein and Kindermann, 1989).

This tuning was often done in conjunction with other changes to the network, such as weight initialization or changing the network topology. In addition, there have also been several researchers that attempted to train feedforward neural networks for decision problems using genetic algorithms (Whitley and Hanson, 1989; Montana and Davis, 1989; Whitley et al. 1990). Related to this is the use of genetic search in the optimization of Kanerva's (1988) sparse distributed memories by Rogers (1990) and Wilson's work (1990) which learned predicates over input features to construct new higher order inputs to a perceptron.

Rogers (1990) has used genetic algorithms to optimize the "location addresses" (i.e. the layer mapping inputs to hidden units) of a sparse distributed memory. Das and Whitley (1992) extend the work of Rogers by using a genetic algorithm for "location address" optimization that actively extracts information about multiple local minima based on relative global competitiveness. Each local optimum in this particular definition of the search space represents a different and distinct data pattern that correlates with some output or event of interest. This allows multiple data patterns to be tracked simultaneously, where each pattern corresponds to a different local optimum in location address space.

The application of genetic algorithms to simple weight training for neural networks has been hampered by two factors. First, gradient methods have been developed that are highly effective for weight training in supervised learning applications where input-output training examples are available and where the target network is a simple feed forward network. Second, the problem of training a feed forward Artificial Neural Network (ANN) may represent an application that is inherently not a good match for genetic algorithms that rely heavily on recombination. Some researchers do not use recombination (e.g. Porto and Fogel, 1990) while other have used small populations and high mutation rates in conjunction with recombination. We first look at why optimizing the weights in a neural network may cause problem for algorithms that rely heavily on simple recombination schemes.

11.3.1   The Problem With ANN

One reason that genetic algorithms may not yield a good approach to optimizing neural network weights is the *Competing Conventions Problem*. Nick Radcliffe (1990; 1991) has also named this the *Permutations Problem*. The source of the problem is that there can be numerous equivalent symmetric solutions to a neural network weight optimization problem.

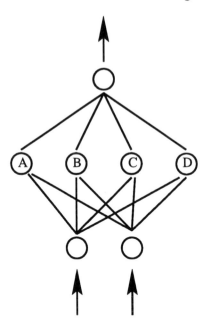

Figure 11.1 A simple feedforward neural network. Note that rearranging the positions of the hidden units does not change the functionality of the network.

Figure 11.1 illustrates a simple feedforward network. Assume that the vector

$$w_{a,1}, w_{a,2}, w_{a,3}, w_{b,1}, w_{b,2}, w_{b,3}, w_{c,1}, w_{c,2}, w_{c,3}, w_{d,1}, w_{d,2}, w_{d,3}$$

is an arbitrary assignment of weights to this neural network, where $w_{\alpha,i}$ passes through hidden node α and $i = 1, i = 2$ are input connections and $i = 3$ is the output connection. Note that for every vector of this form there are $4! = 24$ equivalent vectors representing exactly the same solution. All permutations over the set of hidden unit indices, $\{a, b, c, d\}$, are equivalent vectors in terms of neural network functionality and in terms of the resulting evaluation function. This is because rearranging the order of the hidden units has no effect on the functionality of the network. Thus, given H hidden units in a simple fully connected feedforward network, there are $H!$ symmetries and up to $H!$ equivalent solutions.

The problem this creates for a genetic algorithm that uses simple recombination is as follows. If one does simple crossover on a permutation such as [A B C D] and [D A C B] then the offspring will duplicate some elements of the permutation and will omit others. Similarly, if different strings try to map functionality of hidden nodes in different ways, then recombining these strings will result in duplication of some hidden units and omission of other hidden units. In this case, using a population-based form of search can be a disadvantage, since different strings in the population may not map functionality to the different hidden units in the same way.

Various solutions have been proposed to the Competing Conventions Problems.

Early on, Montana and Davis (1989) attempted to identify functional aspects of hidden units during recombination in order to perform a type of intelligent crossover. Radcliffe (1991) also suggested a solution whereby hidden units are treated as a multiset: hidden units with the same connectivity are considered to be the same, but hidden units might have different connectivities. During recombination, one can search through the hidden units to determine which are identical and use this information to guide crossover. Hancock (1992) has implement this idea as well as extensions to consider how similar hidden units are; he concludes the permutation problem is not as bad as has often been suggested. More recently, Korning (1994) has suggested that the traditional use of the standard quadratic error measurement, $(target - observed)^2$, is part of the problem and suggests the use of other fitness measurements. Korning also suggests "killing off" any offspring that do not meet minimal fitness requirements, which might filter out offspring from incompatible parents. Overall however, it is very difficult to find cases where genetic algorithms have been shown to yield results better than gradient based methods for supervised learning applications.

One recent report returns to a theme initially put forward by Belew et al. (1990). Part of the traditional wisdom (folklore?) which has grown up around genetic algorithms is that a genetic algorithm is good at roughly characterizing the structure of a search space and finding regions of good average fitness, but not adept at exploiting local features of the search space. One way to use a genetic algorithm then is to use it to find an initial set of good weights and then to turn the search over to a gradient based method. Skinner and Broughton (1995) have reported good results with this kind of approach and suggests this method is better than using gradient methods alone for complex problems involving large weight vectors.

11.3.2   Genetic and Evolutionarly Algorithms for Reinforcement Learning

Another stratgey is to use genetic and evolutionary algorithms for weight optimization in domains where gradient methods cannot be directly applied, or where gradient methods are less effective than in simple supervised learning applications. One such application is the use of evolutionary algorithms to train neural networks for reinforcement learning problems and neurocontrol applications. Some results suggest that evolutionary algorithms can be quite competitive against other algorithms that are applicable to reinforcement learning problems. For reinforcement learning applications the set of target outputs that correspond to some set of inputs used to train the net are not known a priori. Rather, the evaluation of the network is performance based. Most existing algorithms attempt to convert the reinforcement learning problem to a supervised learning problem by indirectly or heuristically generating a target output for each input. Some approaches compute an inverse of a system model. The system model maps inputs (current state and control actions) to outputs (the subsequent state). Given a target state, the inverse of the system model can be used to generate actions, which can then be used as a output target (the appropriate action) for a separate controller. This general description is applicable to methods such as "Back propagation through time." "Adaptive critic" methods use a separate evaluation net that learns to predict or evaluate performance at each time step. The prediction can then be used to heuristically generate output targets for an "action" net which controls system behavior. Note, however, that both of these methods compute target outputs and the resulting gradients either indirectly or heuristically.

Genetic algorithms can be directly applied to reinforcement learning problems because genetic algorithms do not use gradient information, but rather only a relative measure of performance for each set of weight vectors that is evaluated. Genetic algorithms and evolutionary algorithms have been successfully applied to training neural nets to controlling an inverted pendulum. Weiland (1990; 1991) for example trained recurrent networks to balance two inverted pendulums of different lengths at the same time, as well as a jointed pendulum. These algorithms often use smaller population sizes and higher mutation rates to cope with the "Competing Conventions Problems." Whitley et al. (1991; 1993) compared a genetic hill-climber to the well known work of Anderson (1989) which uses the "temporal difference method" (Sutton 1988) to train an "Adaptive Heuristic Critic" (AHC) which in turn is used to generate target outputs for doing reinforcement backpropagation. The results suggests that that the genetic algorithms produced training times comparable to the AHC with reinforcement backpropagation, while generalization was better for the genetic algorithm. .

Whitley et al. (1991; 1993) have argued that comparisons of algorithms for reinforcement learning (and other decision problems) should not only consider learning time but also generalization. Algorithms that learn very quickly can potentially fail to produce an adequate generalized model of the process being learned. Thus, fast learning is not in and of itself a good measurement for evaluating a training algorithm. Generalization is also effected by how the evaluation function is constructed. In reinforcement learning and control problems, the number of possible initial states can be intractable. Thus, evaluation involves sampling the set of possible start states. Evaluation based on a single fixed start state can result in fast learning, but very poor generalization. Evaluation based on a single random start state is somewhat better, but the resulting evaluation is noisy and it difficult to compare the evaluation of one string against another. Evaluation based on a set of start states that uniformly samples the input space would appear to be the best strategy.

11.4   GENETIC ALGORITHMS FOR CONSTRUCTION NEURAL NETWORKS

Some of the early efforts to encode neural network architectures assume that the number of hidden units was bounded; the genetic algorithm could then be used to determine what combinations of weights or hidden units yield improved computational behavior within a finite range of architectures. These directly coded network architectures have usually been trained using back propagation. A common fitness measurement is the training time. Miller and Todd (1989) have explored these ideas, as have Belew, McInerney and Schraudolf (1990). Whitley, Starkweather and Bogart (1990) show that the genetic algorithm can be used to find network topologies that consistently display improved learning speeds over the typical fully connected feed forward network. They also explore how to create selective pressure toward smaller nets and to reduce training time by initializing the reduced networks using weights that have already been optimized for larger fully connected networks.

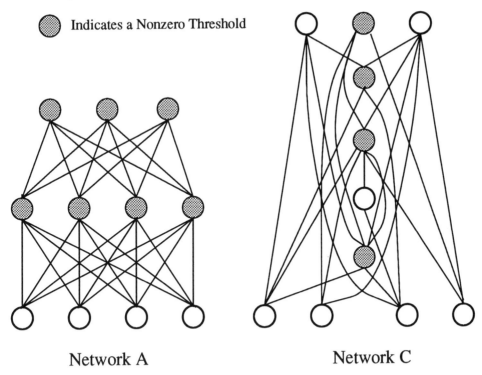

Figure 11.2   A standard feedforward networks for adding two 2-bit numbers
and an architecture evolved using a genetic algorithm. The special architecture
learns much faster.

An example of the effort to reduce the network topology for a 2-bit adder is given
in Figure 11.2. Network C was evolved by a genetic algorithm and learned to add in
between 8,000 and 9,000 training epochs on 50 out of 50 tests. Network A failed to
converge on 5 of the 50 tests, and over half of the networks required more than 50,000
training epochs to train. A Network B was created by adding direct connections to
the input-output nodes of Network A. Network B learned the training set in between
10,000 and 50,000 training epochs on 46 out of 50 tests.

Such early results were encouraging, but the difficulty with directly optimizing
a network architecture is the high cost of each evaluation. If we must run a backpropagation algorithm (or some faster, improved form of gradient descent) for each
evaluation, the number of evaluations needed to find improved network architectures
quickly becomes computationally prohibitive. The computation cost is typically so
high as to make genetic algorithms impractical except for optimizing small topologies.
For example, if an evaluation function for a modest-sized neural network architecture
on a complex problem involves one hour of computation time, then it requires *one
year* to do only 9,000 architecture evaluations. If the architecture is complex then
9,000 evaluations is most likely inadequate for genetic search to be effective.

Also, one trend in neural networks that partially addresses the architecture issue is constructive algorithms such as the Cascade Correlation Learning Architecture, which incrementally adds hidden units to the neural network as it learns. Thus, the basis for comparison is not just simple fully connected feedforward networks.

Another more recent effort to evolve neural networks is the work of Angeline et al (1994). The *GeNeralized Acquistion of Recurrent Links*, or GNARL system, uses selection and mutation to search the space of possible recurrent neural network architectures. GNARL attempts to learn weights and topology at the same time. This type of approach differs from constructive algorithms such as Cascade Correlation in that the space of possible architectures is explored in a nonmonotonic fashion.

11.4.1   Neurogenesis: Growing Neural Networks

In the last 5 years some of the most advanced work for using genetic algorithms to develop neural network have focused on growing neural network. Weights and architectures are often developed together. This can include systems such as GNARL. Other researchers have also looked at genetic programming as a way of developing architectures and weights together (Koza and Rice 1991).

Grammar based architecture descriptions have been explored by Kitano (1990), Mjolsness et al. (1988) and by Gruau (1992). Nolfi et al. (1990) have also looked at grammar based systems that retain may of the characteristics of L-systems. By optimizing grammar structures that generate network architectures instead of directly optimizing architectures this research hopes to achieve better scalability, and in some sense, reusability of network architectures. In other words, the goal of this research is to find rules for generating networks which will be useful for defining architectures for some general class of problems. In particular, this would allow developers to define neural structures for smaller problems that could reused as as building blocks for solving larger problems.

One of the earliest efforts to look at network growth was by Mjolsness et al., (1988) which defined a recursive equation for a matrix from which a family of integer matrices could be derived, and then a family of weighted neural nets. The search space is defined over the set of equation coefficients. Mjolsness uses simulated annealing instead of the genetic algorithm to search this space.

Kitano (1990) uses a grammar to generate a family of matrices of size 2^k. The element of the matrix are characters in a finite alphabet. In order to develop matrix M_{k+1} each character of the matrix M_k is replaced by a 2×2 matrix. This connectivity matrix describes the architecture of a neural net. To produce an acyclic graph for a feed forward neural network, only the upper right triangle of the matrix is used.

More recently, Kitano has presented a simple model of neurogenesis that is more biological in nature. In this approach, "axons grow while cell metabolism are being computed." (1995:81). Cell membranes are also modeled that are capable of chemical transport and diffusion. This work appears to be focused on understanding the emergent properties of this type of system.

Gruau (1992) directly develops a cellular development model for growing neural nets called *cellular encoding*. Each cell has a duplicate copy of the "genetic code." Each cell reads the code at a different position. Depending on what is read, a cell can divide, change internal parameters, and finally become a neuron. Arguably, the resulting language can describe networks in a more elegant and compact way than matrix representations, and the representation can be readily recombined by the genetic algorithm. Gruau used a genetic algorithm to recombine grammar trees representing cellular encodings and has showed that neural networks for the parity problem and symmetry problem could be found. More recently Gruau (1995) has also evolved controls for a 6 legged robot and Whitley, Gruau and Pyeatt (1995) evolve recurrent neurocontrollers for balancing 1 and 2 poles without velocity input information.

11.4.2   A Review of Cellular Development.

Each cell carries a copy of the genetic code in the form of a grammar tree. Each cell also has a pointer which points to a node into the grammar tree. Each node is a program instruction. Development starts with a single *ancestor cell* with connections to input cells and output cells.

In a *Sequential* divide, denoted by S, the parent cell splits into two cells such that the first child inherits all of the input connections of the parent and the second child inherits all of the output connections of the parent; the first child is also connected by a single connection to the second child. In Figure 11.3, during a Sequential divide the second child is placed under the first child. An S node is also a branch point, with the top child cell moving its pointer to the left branch node below S and the bottom child moving its pointer to the right branch node below S.

In a *Parallel* divide, denoted by P, the parent cell splits into two cells that inherit all of the input and output connects of the parent. In Figure 11.3, during a Parallel divide the two child cells are place side by side. A P node is also a branch point, with the left hand child cell moving its pointer to the left branch node below P and the right hand child moving its pointer to the right branch node below P.

The next symbol encountered in Figure 11.3 is the E, which is the *end* or termination symbol. A cell terminates development after reading the E symbol.

The program-symbol **A** increments the threshold of the hidden unit. The program-symbol denoted "-" sets the weight of the input link pointed by the link register to -1. In this example the link register has not been reset and so has its original default setting such that it points to the leftmost fan-in connection.

Figure 11.3 shows an example of a simple grammar tree that generates a XOR networks.

In order to reuse subcomponents of the neural network, cellular encoding uses a special recurrent program-symbol denoted **R**. Associated with **R** is a counter than controls the number of recursive jumps that can be made. When **R** is encountered by a cell, the cell moves its reading head back to the root cell of the grammar tree. The associated counter decrements each time the recursive jump is made. When the counter equals 0 the cell does not reset it pointer, but rather moves forward in the grammar tree, or gives up its reading head and terminates development. Gruau and Whitley (1993) provide an example of how the solution to the XOR net can be generalized to cover all parity problems by placing an **R** symbol in the leftmost leaf node of the grammar tree in Figure 11.3. On parity and symmetry problems, after the genetic algorithm has generated a family of recursively developed networks that handle the lower order cases (3 to 6 inputs), the recursive network encoding represents a general relation and automatically generalizes to handle arbitrarily large problems.

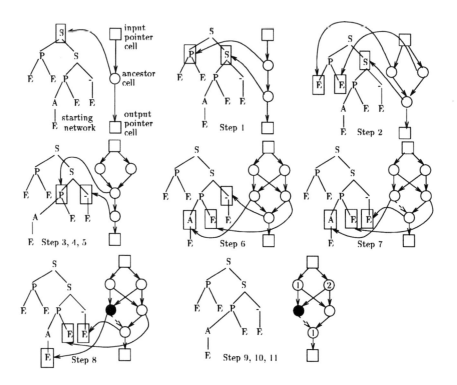

Figure 11.3 The cellular development process. In step 1 the ancestor cell does a sequential divide into 2 cells. In step 2 the uppermost cell from the previous step does a parallel divide. The two cells that are created both read termination symbols in steps 3 and 4; in step 5 the sequential divide is executed. In step 6 a parallel divide is executed. In step 7 the "-" symbol has been executed and a negative weight is introduced feeding into the output node. In step 8 the black cell has changed its threshold. In the final steps, the remaining cells just read termination symbols. (This figure is taken from Gruau and Whitley, 1993).

Another way to reuse development code is to use a form of Automatic Function Definition like that used in Genetic Programming. Subtrees are created, such that the main tree can jump to a subtree, execute the subtree, then return to the associated program-symbol in the main grammar tree. Subtrees thus function like program subroutines. Gruau has used Automatic Function Definition to evolve a mechanism to control the gait of a 6-legged robot. The use of Automatic Function Definition results in simpler, more modular and well structured neural network (Gruau 1995).

11.5   Evolution, Learning and the Baldwin Effect

There has been considerable interest recently in the idea that learning can impact evolution even if learned behaviors are not coded back on the chromosome, as in Lamarckian evolution. The work of Hinton and Nowlan (1987) explains how learning can reshape the fitness landscape, since an individual's fitness is made up of both their genetically determined behavior and learned behavior. If learned behavior has a significant impact on fitness and if the contribution of the learned behavior is stable over time, there can be a selective advantage to having a genetic predisposition that makes it easier to acquire this learned behavior, and eventually, perhaps even to add the behavior to the individual's genetically determined behaviors. Note that this can occur without Lamarckian mechanisms, since there is selection presssure for the learned behavior which can be exploited by Darwinian selection. This idea dates back to Baldwin (1896) and hence is known as the Baldwin Effect.

Such interactions in learning and evolution have been observed when training neural networks using genetic algorithms. Also, the idea of using learning on top of genetic search to speed up the search process has also been explored. Some researchers that explore the interaction of learning and evolution in neurogenetic systems include Ackley and Littman (1991), Gruau and Whitley (1991) and Belew (1989).

11.6   CONCLUSIONS

The challenge facing researchers interested in combinations of genetic algorithms and neural networks is to show how genetic algorithms can make a positive and competitive contribution in the neural networks arena. Currently, it appears that using genetic algorithms to find a set of initial weights before applying gradient based methods may be advantageous for supervised learning classification problems. The application of genetic methods to the development of neural networks for reinforcement learning application also appears to be a worthwhile area for future work. Combinations of genetic algorithms and neural networks are likely to also continue to impact the field of artificial life.

REFERENCES

Ackley D.H. and Littman M. (1991) Interactions between learning and evolution. In *Proc. of the 2nd Conf. on Artificial Life*, C.G. Langton, ed., Addison-Wesley, 1991.

Anderson C. W. (1989) Learning to Control an Inverted Pendulum Using Neural Networks. *IEEE Control Systems Magazine*, 9, 31-37.

Angeline P.J., Saunders G. M. and Pollack J.B. (1994) An evolutionary algorithm that constructs recurrent neural networks. *IEEE Transactions on Neural Networks* 5(1):54-64.

Baldwin J.M. (1896) A new factor in evolution. *American Naturalist*, 30:441-451, 1896.

Belew R. (1989) When both individuals and populations search: Adding simple learning to the genetic algorithm. In J.D. Schaffer (Ed.), *Third international conference on genetic algorithms* (pp. 34-41). San Mateo, CA: Morgan Kaufmann.

Belew R., McInerney J. and Schraudolph N. (1990) Evolving Networks: Using the Genetic Algorithms with Connectionist Learning. CSE Technical Report CS90-174, Computer Science, UCSD.

Brill F.Z., Brown D.E. and Martin W.N. (1992) Fast genetic selection of features for neural network classifiers. *IEEE Transactions on Neural Networks*, *3* (2), 324-328.

Chang E.J. and Lippmann R.P. (1991) Using genetic algorithms to improve pattern classification performance. In R.P. Lippmann, J.E. Moody and D.S. Touretsky (Eds.), *Advances in neural information processing 3* (pp. 797-803). San Mateo, CA: Morgan Kaufmann.

Das R. and Whitley D. (1992) Genetic Sparse Distributed Memories. *Combinations of Genetic Algorithms and Neural Networks*. D. Whitley and J.D. Schaffer (eds.) IEEE Computer Society Press.

Eberhart R.C. and Dobbins R.W. (1991) Designing neural network explanation facilities using genetic algorithms. *IEEE international joint conference on neural networks* (pp. 1758-1763). Singapore: IEEE.

Eberhart R.C. (1992) The role of genetic algorithms in neural network query-based learning and explanation facilities. In *Combinations of Genetic Algorithms and Neural Networks*. D. Whitley and J.D. Schaffer (eds.) IEEE Computer Society Press. Fahlman S. and Lebiere C. (1990). The Cascade Correlation Learning Architecture. In D. Touretzky (Ed), *Advances in Neural Information Processing Systems 2*, Morgan Kaufmann.

Gruau F. (1992) Genetic synthesis of Boolean neural networks with a cell rewriting developmental process. In, *Combination of Genetic Algorithms and Neural Networks*, D. Whitley and J.D. Schaffer, eds, IEEE Computer Society Press, 1992.

Gruau F. and Whitley D. (1993) Adding Learning to the Cellular Development of Neural Networks: Evolution and the Baldwin Effect. *Evolutionary Computation* 1(3): 213–233.

Gruau F. (1995). Automatic Definition of Modular Neural Networks, *Adaptive Behavior*,3(2):151-183.

Hancock P.J.B. (1992) Genetic algorithms and permutation problems: a comparison of recombination operators for neural structure specification. In *Combinations of Genetic Algorithms and Neural Networks*. D. Whitley and J.D. Schaffer (eds.) IEEE Computer Society Press.

Harp S.A., Samad T. and Guha A. (1989) Towards the genetic synthesis of neural networks. In J.D. Schaffer (Ed.), *Third international conference on genetic algorithms* (pp. 360-369). San Mateo, CA: Morgan Kaufmann.

Harp S.A., Samad T. and Guha A. (1990) Designing application-specific neural networks using the genetic algorithm. In D.S. Touretsky (Ed.), *Advances in neural information processing 2* (pp. 447-454). San Mateo, CA: Morgan Kaufmann.

Hinton G.E. and Nowlan S.J.(1987) How learning can guide evolution. *Complex Systems*, 1:495-502.

Holland J. (1975) *Adaptation in Natural and Artificial Systems*. Ann Arbor, Univ. of Michigan Press.

Kanerva Pentti (1988). *Sparse Distributed Memory*. Cambridge, Mass: MIT Press.

Kitano H. (1990) Designing neural network using genetic algorithm with graph generation system. *Complex Systems*, 4:461–476.

Kitano H. (1995) A simple model of neurogenesis and cell differentiation based on evolutionary large-scale chaos. *Artificial Life*, 2:79–99.

Korning P.G. (1994) Training of neural networks by means of genetic algorithm working on very long chromosomes. Technical Report, Computer Science Department, Aarhus C, Denmark.

Koza J.R. and Rice J.P. (1991) Genetic generation of both the weights and architecture for a neural network. In, *Intern. Joint Conf. on Neural Networks, Seattle 92.*

Miller G., Todd P. and Hedge S. (1989) Designing Neural Networks using Genetic Algorithm, In, *3rd Intern. Conf. on Genetic Algorithms*, D.J. Schaffer, ed., Morgan Kaufmann.

Mjolsness E., Sharp D.H. and Alpert B.K. (1989) Scaling, machine learning, and genetic neural nets. *Advances in Applied Mathematics, 10*, 137-163.

Montana D.J. and Davis L. (1989) Training feedforward neural networks using genetic algorithms. In *Proceedings of eleventh international joint conference on artificial intelligence* (pp. 762-767). San Mateo, CA: Morgan Kaufmann.

Mühlenbein H. (1990) Limitations of multi-layer perceptrons networks - steps towards genetic neural networks. *Parallel Computing*, 14:249–260.

Mühlenbein H. & Kindermann J. (1989). The dynamics of evolution and learning – Towards genetic neural networks. In R. Pfeifer, Z. Schreter, F. Fogelman-Soulie & L. Steels (Eds.), *Connectionism in perspective* (pp. 173-197). Amsterdam: Elsevier Science Publishers B.V. (North-Holland).

Nolfi S., Elman J.L. and Parisi D. (1990) *Learning and evolution in neural networks*. CRL Technical Report 9019, La Jolla, CA: University of California at San Diego.

Porto V.W. and Fogel D.B. (1990) Neural network techniques for navigation of AUVs. *Proceedings of the IEEE Symposium on Autonomous Underwater Vehicle Technology* (pp. 137-141). Washington, DC: IEEE.

Radcliffe N.J. (1990) *Genetic neural networks on MIMD computers*. Doctoral dissertation, University of Edinburgh, Edinburgh, Scotland.

Radcliffe N.J. (1991) *Genetic set recombination and its application to neural network topology optimization*. Technical report EPCC-TR-91-21, University of Edinburgh, Edinburgh, Scotland.

Rogers D. (1990) Predicting Weather Using a Genetic Memory: a Combination of Kanerva's Sparse Distributed Memory with Holland's Genetic Algorithm; *Advances in Neural Information Processing 2.*

Sutton R. (1988) Learning to Predict by the Methods of Temporal Differences, *Machine Learning*, 3:9-44.

Skinner A. and Broughton J.Q. (1995) Neural Networks in Computational Materials Science: Training Algorithms *Modelling and Simulation in Materials Science and Engineering*, 3:371—390.

Schaffer J.D., Whitley D. and Eshelman L. (1992) Combination of Genetic Algorithms and Neural Networks: The state of the art. *Combination of Genetic Algorithms and Neural Networks*, IEEE Computer Society, 1992.

Schaffer J.D., Caruana R.A. and Eshelman L.J. (1990) Using genetic search to exploit the emergent behavior of neural networks. In S. Forrest (Ed.), *Emergent computation* (pp. 244-248). Amsterdam: North Holland.

Weiland A.P. (1990) Evolving controls for unstable systems. In D.S. Touretsky, J.L. Elman, T.J Sejnowski & G.E. Hinton (Eds.) *Proceedings of the 1990 connectionist models summer school* (pp. 91-102). San Mateo, CA: Morgan Kaufmann.

Weiland A.P. (1991) Evolving neural network controllers for unstable systems. *IEEE international joint conference on neural networks* (pp. II-667 - II-673). Seattle, WA: IEEE.

Wilson S.W. (1990) Perceptron redux: Emergence of structure. In S. Forrest (Ed.), *Emergent Computation* (pp. 249-256). Amsterdam: North Holland.

Whitley D. and Hanson T. (1989) Optimizing neural networks using faster, more accurate genetic search. In J.D. Schaffer (Ed.), *Third international conference on genetic algorithms* (pp. 391-396). San Mateo, CA: Morgan Kaufmann.

Whitley D., Starkweather T. and Bogart C. (1990) Genetic Algorithms and Neural Networks: Optimizing Connections and Connectivity. *Parallel Computing.* 14:347-361.

Whitley, D., Dominic, S. & Das, R. (1991). Genetic Reinforcement Learning with Multilayered Neural Networks. *Proc. 4th International Conf. on Genetic Algorithms,* Morgan Kaufmann.

Whitley D., Dominic S., Das R. and Anderson C. (1993) Genetic Reinforcement Learning for Neurocontrol Problems. *Machine Learning* 13:259-284.

Whitley D., Gruau F. and Pyeatt L. (1995) Cellular Encoding Applied to Neurocontrol. In, *5th Intern. Conf. on Genetic Algorithms,* L. Eshelman, ed., Morgan Kaufmann.

12

Genetic Algorithms: A Stochastic Improvement Technique. Tools, Skills, Pitfalls and Examples

G. Winter[1] , M. Galán[1], P. Cuesta[1], D. Greiner[1]

12.1 Summary

In this chapter we try to show an introductory study addressed, mainly, to the postgraduate student.

Because of the extention and multiple branches of this stochastic technique, our proposal has been to provide a simplified panoramic view of the Genetic Algorithms field (GAs).

To begin with, we intent to mostrate some results, tools, and references in order to give an approximation to the "state of the art"; therefore we will start with a description of the main aspects and operators of GAs.

Creativity and innovation are strong features in the field of GAs which are, sometimes, more an art than a science as they strongly depend on several 'fine tune' parameters which play a decissive role in the algorithm success.

On the other hand we present several examples of the application of GAs in some simple test cases, including some new proposals, and finally we comment on our experiences with this "improvement" algorithm applied to several problems.

[1] CEANI, Centro de Aplicaciones Numéricas en Ingeniería de la U.L.P.G.C.
Edificio de Ingenierías I, Campus Universitario de Tafira
Tafira Baja, 35017 Las Palmas de Gran Canaria, España
Tf: +34 28 45 19 17, Fax: +34 28 45 19 21, E-mail: gabriel@hptitan.ulpgc.es

12.2   Introduction and approximation towards the state of the art

The initial context of GA by Holland (75) was the design of robust adaptive systems trying to imitate the mechanism of natural evolution. GAs are an iterative adaptive search algorithms, (somewhat heuristic), that differ from more standard search algorithms (e.g., gradient methods, controlled random search, hill-climbing, simulated annealing, etc) in that the search is conducted using the information of a population of structures instead of a direction or a single structure. GAs consider many structures as potential candidate solutions and work with more global sampling of the search space than traditional kinds of search algorithms, increasing the possibility of getting convergence to the global optimum, however but does not implying necessarily convergence to a global optimum in finite population. **GAs are quite robust in producing near-optimal sequences of trials** for many applications also including problems with high levels of uncertainty and problems which cannot easily be reduced to closed form. A good textbook on genetic algorithms can be found in Goldberg (89) and other textbooks of interest can be found in Davis L.(Ed.) (91) and Michalewicz Z.(94), too.

Given some finite discrete domain D and a function $\tilde{f} : D \longrightarrow R$ **the search problem or function optimisation problem** consists in finding the best or near best in D under $\tilde{f}$. R is ordered and there is a natural induced total ordering on D even if D is multi-dimensional. As a search algorithm, GAs examines some discret subset of a search space S^l and returns some individual or string s_i (which corresponds to a point of the domain function D by means of a mapping called decoding function) whose objective value $f(s_i)$ is an estimator of $Max_{x \in D}\tilde{f}(x)$, being $Max_{x \in D}\tilde{f}(x) = Max_{s \in S^l}f(s)$, where f is the information of the indiviual's environment.

In case of minimization problems for $\tilde{f}$ we can consider $g = -\tilde{f}$ and

$$Min\tilde{f}(x) = Max(-\tilde{f}(x)) = Maxg(x).$$

More specifically the process in GAs is sequential; successive generations are produced with new solutions replacing some the older ones, and each generation must be produced before it can be used as the basis for the following generation. During iteration t, a GA maintains a population of potential solutions $P(t) = \{s_1^t, s_2^t, ..., s_i^t, ..., s_n^t\}$ of individuals or chromosomes s_i (s_i^t), that are represented as strings s $(a_1, a_2, ..., a_l)$, of length l, the allele of the ith gene (or ith "locus" position) in a string is denoted by $a_i = b$, $b \in S$, where S is alphabet set of symbols, being the more usual $S = \{0, 1\}$ for binary strings.

Let S^l be the set of the strings of length l. A **decoding (encoding)** is a function $\mathbf{d} : S^l \longrightarrow D$, such that for each string $s \in S^l$ correspond a value $x \in D$, being S^l the **search space**, s_i $(a_1, ..., a_l) \in S^l$.

The **fitness function f** is a function $f : S^l \longrightarrow R$ which can be seen as the composition of $\tilde{f}$ and d, thus can be defined as $f(s) = \tilde{f}(d(s))$; to the extent of GAs, f is "blackbox" function. The only procedure on f that Genetic Algorithms (GAs) require is its evaluation at a point of the search space (GAs are blind search algorithms). Thus being the rank $\mathbf{r}$ of a string $\mathbf{s}$ the number of strings whose objective value is smaller under the induced total ordering.

So, being $r(s) = Card\{s^* \in S^l : f(s) \geq f(s^*)\}$, we can consider the fitness function

f as a set of triples $\{(s, x, r)/s \in S^l, x \in D, r \in R\}$, with $d(s) = x$, $\tilde{f}(x) = r$, thus $f(s) = r$, then any permutation in any component of the tiplet will produce new domain function and correspondingly a new encoding (Rawling 91).

The objective function or fitness function f(s) plays the role of the environment; each individual **s** is evaluated according to its fitness. In this way a new population (iteration t+1) is formed by selection of the better individuals of the former population, as they will form a new solution by means of applying selection procedure and crossover and mutation operators. It should be noted that diversity of individuals is required to find good solutions with GA.

The mapping from binary string is for $s_i \in S^l$ and $x_i \in D$, being $A \leq x_i \leq B$, is given by the expressions:

$$\left(\sum_{n=1}^{l} a_n 2^{n-1}\right)_{10} = M$$

$$x_i = A + M \frac{B-A}{2^l - 1}$$

and required precision, i.e. 'm' decimal places for the variables values x_i when f is a function of k variables $f(x_1, x_2, ..., x_k)$ and being $A_k \leq x_k \leq B_k$ given by the expresion

$$2^{l-1} \leq (B_i - A_i)10^m \leq 2^l$$

where in this case a vector string $(a_{11}, a_{12}, ..., a_{1l_1}, a_{21}, a_{22}, ..., a_{2l_2}, ..., a_{k1}, a_{k2}, ...,$ $a_{kl_k})$ is composed by k substrings of length l_j, $j = 1, 2, ..., k$ and l is given by $l = \sum_{j=1}^{k} l_j$.

Bit string encoding have several advantages over other encodings, as they are simple to mate and manipulate. On the other side performance theorems have been proved for bit string chromosomes (theorems for other encoding techniques have been proved from the former ones). There are well known advantages in the use of low cardinality alphabets ((binary representation)=2).

However, real (floating point) codings are claimed to work well and more faster in a number of practical problems (Michalewicz 94), even if the use of real-coded or floating-point genes has risen some controversial from a theoretical point of view.

Caruana and Schaffer (88) have empirically found that **Gray coding**, later described in this chapter, is often better than 'naïve' binary coding for GAs.

The encoding is very important for the success and performance of a GA. One simple, yet important, observation in practical implementation of GA, to facilitate a fast decoding from the binary string representation is that the choice of the length of the strings should be a multiple of the size of the 'natural' variable in the specific hardware/O.S.

Simple GAs operate on a finite population size n and fixed-length binary strings l for each chromosome using selection, crossover and mutation operators. The initial population should include individuals with very high diversity whose 'allele' are randomly generated. **A selection operator** identifies the fittest individuals of the current population to be chosen parents of the next generation. The fitness function provides the environmental feedback for selection procedure.

Let $P(0)$ the randomly generated initial population and $P(t)$ the population at the time t, $P(0)$, $P(t) \in (S^l)^n$. Let $p_s : S^l \longrightarrow [0, 1]$ be the function such that for all

$P(t)$: $\sum_{i=1}^{n} p_s(s_i^t) = 1$, determining the selection probabilities of the individuals in a population.

Selection operator is an operator that produces an intermediate population $P'(t)$ from the population $P(t)$.

Genetic operators (including selection) generate new individuals with potentially higher fitness values. Some different selection schemes are :

- **proportional selection schemes** (Holland 75) which are selection schemes that choose individuals for birth according to their fitness function values :

$$p_s(s_i^t) = \frac{f(s_i^t)}{\sum_{j=1}^{n} f(s_j^t)}$$

where $p_s(s_i^t)$ is the probability of individual in the tth generation to be sampled. It leads to the expectation of the individual s_i^t to occur $\eta_i^t = np_s(s_i^t)$ times in generation $t+1$, η_i^t is called generally the expected value of s_i^t.

After few generation the population average fitness may be close to the population best fitness leading to a premature convergence to a mediocre generation. This particular scheme can lead to a "superindividual" (individual with very high fitness value compared to the rest of the population) which, in a few generations, will take over the population, leading, sometimes, to a premature misconvergence. In order to solve this problem, one possible alternative is to use **scaled fitness values** by means of an additional mapping $f' : R \times (S^l)^n \longrightarrow R$. There are several proposed scaling types i.e., see : Goldberg 89 and Bäck and Hoffmeister 91.

Also several procedures have been suggested for sampling the probability distribution: roulette wheel selection (De Jong 75), Stochastic remainder selection (Booker 82, Brindle 81), Stochastic universal selection (Baker 87, Grefenstette and Baker,89),etc.

Another alternative can be the use of the following selection scheme:

- **linear ranking selection** (Baker 85); it is based on the notion of rank. So, after a reordering (according to the fittness value) of the population, and being s_1 the fittest individual, the selection probabilities are constant values :

$$p_s(s_i^t) = \frac{1}{\lambda}(\eta_{max} - (\eta_{max} - \eta_{min})\frac{i-1}{\lambda-1})$$

where λ is the number of individuals representing one generation (in standard GAs $\lambda = n$) and $\eta_{min} = 2 - \eta_{max}$, $(1 \leq \eta_{max} \leq 2)$.

It assigns the number of copies that each individual should receive according to a non-increasing function, and then performs proportional selection based on the assignment. Baker suggest a maximum expected value of $\eta_{max} = 1.1$.

Experimental results give indications relative to the advantage of ranking selection in case of multimodal functions (Baker85, Whitley 89).

- (μ, λ)-**uniform ranking selection** (Schwefel 81):

$$p_s(s_i^t) = \begin{cases} \dfrac{1}{\mu} & 1 \leq i \leq \mu \\[2ex] 0 & \mu \leq i \leq \lambda \end{cases}$$

These last schemes are generalized by Bäck and Hoffmeister (91):
(μ, λ)-Proportional Selection given by the following expression,

$$
p_s(s_i^t) = \begin{cases} \dfrac{f(s_i^t)}{\displaystyle\sum_{j=1}^{\mu} f(s_j^t)} & 1 \le i \le \mu \\[4mm] 0 & \mu \le i \le \lambda \end{cases}
$$

and (μ, λ)-linear ranking given by,

$$
p_s(s_i^t) = \begin{cases} \dfrac{1}{\mu}(\alpha_{max} - 2(\alpha_{max} - 1)\dfrac{i-1}{\mu-1}) & 1 \le i \le \mu \\[4mm] 0 & \mu \le i \le \lambda \end{cases}
$$

being schemes where only μ individuals are allowed to be selected, thus μ guides the selective pressure.

The fraction $\mu/\lambda = 1/5$ seems to give near rate of convergence (speed) for unimodal problems and much higher value must be considered to explore of the search space for multimodal problems. Always $\sum_{i=1}^{\lambda} p_s(s_i^t) = 1$ must be satisfied.

These last two selection operators are selection mechanisms borrowed from Evolution Strategies (ESs), which can be used in GAs ($\lambda = n$). It being tried and generalized to incorporate differences among different selection operators by Bäck, T. and Hoffmeister,F. (91), thus these researchers define extinctive versus preservative selection schemes, elitist or k-elitist selection schemes versus pure selection and generational versus steady-state selection, etc., classifying and experiments these last schemes and other schemes.

Using genetic algorithm the population slowly converges from the initial population until a state in which all individuals are almost similar (individuals well adapted to the environment). Many factors influence in the convergence rate, such as the population size, the parent selection scheme, crossover operator applying, mutation rate, forbidding of replicates, scaling procedures,..., etc. One way to avoid loss in diversity is changing the selection procedure overtime. A good strategy is using a less aggressive selection when the diversity of the population decreases. Another, possible, alternative is ranking based (Davis 91).

Another possibilities of interest are elitist selection strategies and/or tournament selection. **Elitist strategies** are based on copying the best individual of each generation into the next one, where some/all parents are allowed to reproduce, thus reducing the possibility that the best individual of the population may fail to produce offspring in the next generation. Further performance improvements can be obtained using this approach and replacing the worst individuals of the population with newly generated individuals (De Jong 75, Whitley 89).

When the objective is to balance towards more 'exploitation' and less 'exploration', appears **tournament selection (Wetzel, unpub.).** It is an important procedure of selection based on randomly choosing some number of individuals from a population (with or without replacement), selecting the best individual from this set as result of the tournament among several competitors for further genetic processing and repeat as often as desired, for example until the mating pool is filled (Brindle 81).

The difference between average fitness of the mating pool and the average population fitness gives a measure of the selection pressure. Tournament selection seems simple to code and efficient procedure for both non-parallel and parallel architectures and can adjust the selection pressure by varying the tournament size.

Progress about the convergence rates in several selection schemes can be found in Goldberg and Deb. 91, and more recently in Mülenbein and Schlierkamp-Voosen 93 and Thierens and Goldberg 94. Based on order statistic, one model to predict the selection pressure for tournament selection using noisy fitness functions (noisy environments) is recently presented in the very interesting paper by Miller and Goldberg (95).

To continue it is very important understanding the behavior of GAs to know about the following well-known definitions: A **Schema** H is a hyperplane in the metric space (S^l, d), being d the Hamming distance of one string to another, $x(x_1, x_2, ..., x_l)$, $y(y_1, y_2, ..., y_l) \in S^l$

$$d : S^l \times S^l \longrightarrow \{0, 1, ..., l\}$$

where,

$$d(x, y) = \|i \in \{1, ..., l\} : x_i \neq y_i\|.$$

This metric space is represented as strings by $(S \cup \{*\})^l$, being $* \notin S$ a wildcard symbol which stands for any symbol in S (Holland 75), thus unspecified position , i.e. positions that "don't matter" or "don't care" are filled with $*$'s. A schema describes a subset of strings with same values at certain string positions, thus three strings in to be (011) , (111), (010) may be represented by the schema ($* 1 *$) and for example on the other hand the schema $(01* 0 * 1)$ describes to the following subset of strings: (010011), (010001), (010011), (011001), (011011).

In a population of size n, between 2^l and $n2^l$ different schema may be represented.

An interest situation that occurs many times is when a GA can give a population dominated by individuals very close to the optimum in the phenotype space (D), but far in terms of Hamming distance applied to the genotype space (S^l). In this case one good choice can be consider Gray codes.

Order of a schema, denoted by $O(H)$ is in binary coding the number of 0 and 1 fixed positions in the schema. The notion of the $O(H)$ is important in calculating survival probabilities of the schema for mutations.

Length of the schema $L(H)$ is the distance between the first and the last fixed string positions.

For example,

$$H_1 = (* * *0110 * *01 * 0 * **)$$

$$H_2 = (* * 10010 * 0 * 0 * 101)$$

$$O(H_1) = 7 \quad 0(H_2) = 10 \quad L(H_1) = 13 - 4 = 9 \quad L(H_2) = 16 - 3 = 13.$$

The length of the schemata defines the compactness of information contained in a schema.

Low order and short length schema are known as building blocks.

Under proportional selection, simple crossover and mutation, the expected number of copies of a schema H contained in $(t+1)$th iteration is bounded by the following inequality (Goldberg 89):

$$m(H, t+1) \geq m(H, t)\frac{f(H)}{\overline{f}}[1 - p_c\frac{L(H)}{l-1} - p_m O(H)]$$

being $f(H)$ the average of the fitness values of all strings which are represented by the schema H,

$$f(H) = \frac{\displaystyle\sum_{s_i^t \in H} f(s_i)}{m(H, t)}.$$

and being $\overline{f}$ the average fitness of the entire population,

$$\overline{f} = \frac{\displaystyle\sum_{j=1}^{n} f(s_j^t)}{n}.$$

p_c, p_m are crossover and mutation probabilities.

GAs are different from other search methods in the property known as **implicit parallelism**: many schemas may be sampled according to their fitness by the combined actions of survival-of-the-fittest selection and recombination, this property may be stated at large number of the schema in a population being sampled in future trials exponentially overtime and the grow rate is their fitness ratio. Thus **Holland's Schema Theorem** says that "*above average, short, low-order schemata are given exponentially increasing numbers of trials in successive generations*".

The estimation for a GA with an optimal population size n about the number of short schemas sampled is on the order of n^3 (Holland 75, Goldberg 89) and the number of low order schemas sampled is also on the order of n^3, being an important property of GA respect other search algorithms. The cost required to converge by GA when the function is unimodal is given with high probability to be $O(n \log n)$ functions evaluations (Goldberg 91,92).

One case of special interest is when "the environment has no preference for one allele over another" then there is no selective advantage for either of two competing alternatives, and the finite population will converge to one alternative or the other in finite time (De Jong 75).

This problem of finite populations it has received a special name "**genetic drift**". One source of convergence difficulty in GAs results from the stochastic errors in sampling and selection working with small populations. Then it is necessary to take special strategies like **sizing populations**, methods using **sharing functions**, etc., provide a payoff incentive to maintain selective pressure among competing individuals (Goldberg 89, 91, 92).

The basis for the Schema Theorem was that if the average fitness of the population didn't change much over the lifetime of an individual then the expected number of offspring generated by individual i is proportional to the ratio between the distance of the fitness value to the average fitness and the range of the fitness values.

Assuming fixed size population and being at each time a single individual selected probabilistically and using proportional selection via a uniform random distribution, then the expected lifetime in a population of size n is n (one strategy possible).

Holland introduced another strategy which consists at each time step to deterministically select all individuals in order to produce their expected number of offspring in a temporary storage location and when that process is completed the offspring generated replaces the expected number of offspring within probabilistic roundoff.

Let G be the fraction of the population to be replaced in each generation (De Jong and Serma 93), then the number of offspring generated each generation is $K = Gn$. In the first strategy $G = 1/n$ and in the second strategies $G = 1$. The **steady state GAs** correspond to a constant value for K, thus few individuals are replaced each generation from a fixed size population, they seem to require larger populations (Whitley 88).

Whitley and Syswerda have proposed the reproduction technique "steady-state without duplicates" that discards children that are duplicates of current chromosomes, then every individual of the population will be different. In this selection an offspring replaces a parent if it performs better and the set of parents may change for every reproduction step (Witley 89). This type of selection seems not to work well when the "evaluation function is noisy" (each time the same chromosome is evaluated the evaluation function return different values).

In natural systems different species have found different niches. Individuals of the same species demanding same resources (and the increased competition) cause them to share with the others in the case that there are enough for all, when resources are scarce, fight (competition) appears. Goldberg and Richardson 87 proposed a way of imposing niche and specification on strings based on some measure of their distance from each other.

A **sharing function** is a way of determining the degradation of an individual's payoff due to a neighbor at some distance. Many sharing functions are possible, is not limited to one-dimensional problems, also may be evaluated using phenotype (x) or the genotype (the strings) where in this last form the Hamming distance is a useful metric. Methods using sharing functions improve the performance of genetic algorithms in many practical search and optimization problems like multiple local optimum.

Another alternative is to use rank-proportional selection proposed by Whitley 89 where a constant selection differential is maintained between the best and worst individual in the population and could depend on appropriate choices of rank-based selection functions (linear, quadratic, exponential, etc.).

On the other hand a source of loss of diversity results from poor performance of recombination operators in terms of sampling new structures: **crossover and mutation operators modifies this selective pressure providing diversity and play a critical role in the exploration/exploitation balance.** Crossover recombines features of the best chromosomes by combining building blocks of the good ones. GA search process does not form individual chromosome one by one, rather it builds high utility schema with many fixed bits from high utility schema with few fixed bits (Schema Theorem).

The **crossover operator** randomly chooses a pair of individuals among those previously selected to breed and exchanges or concatenates arbitrary substrings between them, thus crossover recombines the genetic material from two parent into their children. There are different types of crossover operators:

One-point crossover proposed by Holland. It consists of a random selection of the place to cut the parent chromosomes in two substrings. The children will have their chromosomes composed by these two substrings.

Two point crossover operator selects two positions to cut. It seems to loose less schemata than traditional one-point crossover operator, see i.e. Schaffer, Caruana, Eshelman and Das 89 and Eshelman,L.J., Caruana, R.A. and Schaffer, J.D. 89.

Uniform crossover operator is a generalization of one-point, two-point and multipoint crossover, see i.e. Syswerda (89).

parent1	1 0 0 1 0 1 1
parent2	0 1 0 0 1 1 0
random position	↑
offspring1	1 0 0 0 1 1 0
offspring2	0 1 0 1 0 1 1

Table 1. 1-point crossover

parent1	1 0 0 1 0 1 1
parent2	0 1 0 0 1 1 0
random positions	↑ ↑
offspring1	1 0 0 0 1 1 1
offspring2	0 1 0 1 0 1 0

Table 2. 2-point crossover

parent1	1 0 0 1 0 1 1
parent2	0 1 0 0 1 1 0
random positions	0 1 1 0 1 0 1
offspring1	1 1 0 1 1 1 0
offspring2	0 0 0 0 0 1 1

Table 3. uniform crossover

Other. There are other crossover procedures like segmented, shuffle, etc. Also some recombination operators of interest are proposed by Eshelman and Schaffer (92) and Mülenbein and Chlierkamp-Voosen (93).

Mutation is an operator used after applying crossover operator to increase diversity, that is to say, to take new points in the search space to evaluate. A chromosome's locus in randomly chosen for mutation. In a binary-coded the corresponding bits are "flipped" from 0 to 1 or from 1 to 0. A good property of Gray coding is that changing a 0 to a 1 produces a perturbation in either direction (Wright 91). The probability of mutation (p_m) must be low and must be appropriated according the encoding used. Mutation is a secondary operator in GAs, but seems clear that mutation depends on the length of the string l, thus it seems sensible to consider the probability of mutation depending of l. In the context of binary encoding, Bäck proposes a formula for the approximate evaluation of mutation rate. He also proposes a kind of self-adaptation under certain circunstances (Bäck, T. 92).

For **extinctive selection** (a selection scheme is called extinctive by Bäck, T. and

Hoffmeister 91), i.e. (μ, λ)-proportional or (μ, λ)-ranking selection, if

$$\forall t \geq 0, \forall P(t) = (s_1^t, ..., s_n^t) : \exists i \in \{1, ..., n\} : p_s(s_i^t) = 0.$$

It seems suitable that in this later case, average mutation rates be 10 times larger than in **preservative selection** (a selection scheme is called preservative by Bäck, T. and Hoffmeister 91), i.e. proportional selection, ranking, if

$$\forall t \geq 0, \forall P(t) = (s_1^t, ..., s_n^t), \forall i \in \{1, ..., n\} : p_s(s_i^t) > 0.$$

Schaffer et all has proposed the following empirical expression :

$$p_m \approx \frac{1.75}{n\sqrt{l}}.$$

Fogarty (89) uses a deterministic decrease of mutation rates over time, such that

$$\lim_{t \to \infty} p_m(t) = 0.$$

Using GA the population slowly converges from the initial population, where the individuals are quite dissimilar, towards a state in which all individuals are almost similar, in this last state further improvement may only be possible by a favorable mutation.

An optimal strategy must maintain a balance between exploitation of the best regions found so far and continued exploration for potentially better payoff areas (Holand 75). When high mutation rates are used or local hill-climbing is added from a schema processing point of view these operators can disrupt schemas and disrupt the information contained in the population, but in practice these operators are often effective to improve optimization behavior.

GAs are not guaranteed to find an optimal solution, its effectiveness heavily depends on many factors, among them (and largely) of the population size (n). Goldberg 92 theoretically studies the optimal population size. Smith 93 proposes an algorithm of interest, too.

An important practical question is the optimal size of a population according the amount of real time available. Larger populations are processed in very much less time using Parallel GAs. A simple use of parallelism is the simultaneous production of candidates for the next generation. In order to probabilistically select parents for the crossover operation in case of partition the population of solutions and assignment of one subset of the population to the local memory of each processor is necessary, somehow, to use global information about the fitnesses.

Messy genetic algorithms (mGAs) have been proposed by Goldberg, Korb and Deb. 89, 91, opening new possibilities processing schemas: the chromosomes may be of variable length, redundant genes, free genes to complete, etc. A string for example in mGAs could be:

$$s_i = \{(3, 1), (4, 0), (4, 1), (3, 1)\}$$

where the first number gives a position and the other one the allele. mGAs use a "splice" and a "cut" operators replacing simple crossover. Both operators are used with probabilities defined and there are two phases in the process, a primordial phase where the population is initialized to contain all possible building blocks of a specified length and a juxtapositional phase where the population of good building blocks is enriched in several generations by means of reproduction like in standard GAs.

*12.2.1   Some general considerations commonly accepted by Genetic Community:
          Facts and Pitfalls*

- Generally GAs find nearly global optima in a variety of problems types, also complex spaces, for example, the search spaces multimodal.
- GAs can be inefficient when the cost is highly dominated by the evaluation of a large number of individual fitness functions, this inefficiency can be softened to a great extent when using sensible parrallel implementations.
- The success of GAs and their effectiveness in any application area can only be determined by experimentation.
- GAs using binary representation and single-point crossover and binary mutation are robust algorithms, but they are almost never the best algorithms to use for any problem.
- Since GAs are stochastic their performance is a more useful way to view the behavior of a genetic algorithm than a representation of the behavior of a genetic algorithm in a single run.
- A GA that satisfies the Schema Theorem does not necessarily a good optimizer. The effectiveness of GAs depending like we have seen of many factors, and mainly depending of the appropriated balance between explotation/exploration or selective pressure/diversity and optimal population size
- The coding of a problem holds an important key to a successful application.
- For a fixed GA with the same sets or multisets of values D and R for every domain function with the encoding makes easier to solve there is another domain function that it makes more difficult to solve.
- The GA-behaviour found in many experiments of Genetic community confirms the well-known contradiction between exploration and exploitation in global optimization. For unimodal functions exploitative search with high convergence rate is desired while for multimodal functions explorative search with a high convergence confidence towards a global optimum point is the objective.
- Never forget that GAs mimmic the behavior of Nature, so they try to get not one "supergood" individual but a population individuals of well adapted to the environment (fitness function).

12.3   Computational Implementation

Genetic algorithms has proven to be a valid and robust alternative in the optimization field. However, due to its intrinsically stochastic nature, there are several possible difficulties in its implementation.

In our case the test programs have been devised using ANSI-C which gives the best performance because if its capabilities of flexibility, power and transportability. On the other side, it seems that, almost all genetic programming is done in ANSI-C.

12.3.1   Program Description

The program used for the actual implementation can be viewed as a chain of several modules. Each one of them has a different task, such as initialization, memory allocation, calculation, evaluation, genetic operation, cleanup, etc.

Each module is composed of several submodules which take care of the different components of the process.

We could divide the program in the following main modules:

Parse Arguments.

Module which task is to read the configuration parameters from a user supplied file or command line.

Module Initialization and memory allocation.

In this part of the program we initialize several submodules such as: random number generator, initial main and auxiliary populations, transformations, fitness function and its interface, mutation, statistics, crossover and its masks, selection, etc.

We allocate memory for several structures, among them: statistical distribution, main and auxiliary population, crossover masks, fitness interface.

Initial Population.

In this module we build the initial population according to the configuration parameters. It can be build in a random way or being read from a user supplied file.

Genetic Operations.

In this module we perform the main loop of the Genetic Algorithm, it is dedicated to the specifical genetic operations: selection, crossover and mutation.

Statistics gathering and display.

This module gathers several statistics on performance of the algorithm for further reference and displays information about the partial results to show the progress of the optimization process.

Final Population.

This module saves the final population in a file for future use.

Clean Up.

Here all the variables, vectors and other memory structures are freed.

Interrupt Control.

This module controls the user interruption of the program, and saves its state for future restart.

12.4   Computational Details.

There are several interesting points in the details of the computational implementation which make the difference between a naïve algorithm and a "production quality" one.

Although GA coding can seem simple and straightforward, there are several sensitive points in which a more careful approach should be used.

12.4.1 Gray Coding.

In the correspondence into the search space done by the binary coding (genotype representation), it seems reasonable to keep good topological properties for this correspondence. However, when considering a simplistic translation of numbers to their binary representation, it appears that there are sharp jumps in the representation when the distance between numbers is small: i.e. the decimal number 256 in binary is (10000000), on the other side 255 is (01111111), thus we see that being small the distance between these two numbers, their binary representation is sharply different.

This effect is more noticeable when coupled with mutation or crossover operators, as a small variation of one bit "should give" a small distance between points, thus spoiling the inheritance of the good "building blocks".

This way, we check out this (possibly) deleterious effect of a simplistic approach: "Gray coding" can be an alternative to this problem.

If we consider the binary matrix:

$$G = \begin{pmatrix} 1 & 0 & 0 & 0 & 0 & 0 & 0 & 0 \\ 1 & 1 & 0 & 0 & 0 & 0 & 0 & 0 \\ 0 & 1 & 1 & 0 & 0 & 0 & 0 & 0 \\ 0 & 0 & 1 & 1 & 0 & 0 & 0 & 0 \\ 0 & 0 & 0 & 1 & 1 & 0 & 0 & 0 \\ 0 & 0 & 0 & 0 & 1 & 1 & 0 & 0 \\ 0 & 0 & 0 & 0 & 0 & 1 & 1 & 0 \\ 0 & 0 & 0 & 0 & 0 & 0 & 1 & 1 \end{pmatrix}$$

Then, we can easily realize that, given chromosomes $p_1 = (00001000)$ (value 8) and $p_2 = (00000111)$ (value 7), we have that, operating in 2's complement arithmetic, $Gp_1 = (00001100)$ and $Gp_2 = (00000100)$; which means only a change in the value of a bit in one position (a unit distance between the numbers means a unit distance in the strings).

Just to check out, for $p_3 = (10000000)$ (value 256) and $p_4 = (01111111)$ (value 255), we have $Gp_3 = (11000000)$ and $Gp_4 = (01000000)$; again a difference in one bit only.

Through this binary matrix multiplication we achieve a continuous mapping in the sense that the inverse image of the neighborhood of a point in the strings space is contained into a neighborhood of the inverse image of that point.

12.4.2 Crossover Alternatives.

When the search space is k-dimensional, we can view the chromosome string divided in 'k' sub-strings, each part corresponding to one variable of the search space.

It seemed suitable to implement a joined crossover strategy in which the "building blocks" for each variable should be maintained.

This way, we propose a crossover technique in which each k-part is crossed independently and simultaneously with the other parts.

For instance, being $k = 4$ and the length of a cromosome 16, if we have the following three strings:

$$p_1 = (1100111000011111)$$

$$p_2 = (0111010110110101)$$

$$cm = (1110100011001000)$$

where p_1 and p_2 are the 'parents' and 'cm' is the crossover mask, then the 'sons' would be:

$$p_1' = (1101110100111101)$$

$$p_2' = (0110011010010111)$$

The mask string indicates the crossover position in the separation point between 'ones' and 'zeroes'.

This way we can achieve an unconstrained all-directional 'movement' of the population all over the search space, keeping the 'building blocks' of the chromosome substrings.

12.4.3 Selection Strategy.

Sometimes it is said that a good Genetic Algorithm means a good selection strategy, we consider that there are two tendencies which are closely related to this strategy, one is exploration of the search space and the other one is the exploitation of the best in each generation.

In the first stages of the genetic algorithm there is the risk of premature convergence to a sub-optimal solution, so we should 'penalize' the best members of these initial populations to 'give a chance' to the not so good ones to 'explore' the search space.

To get this we propose a variable knob factor function; for instance, according to the evaluated fittness function 'f', the 'new' fittness value could be:

$$\tilde{f} = f - K \left(\frac{f - \overline{f}}{Range} \right)^{2l+1} \left[e^{\left(-\left(\frac{2n}{Nmax}\right)^{\frac{4Nmax+2n}{Nmax}}\right)} - \frac{1}{2} \right].$$

Where K is a positive constant, $Nmax$ is the maximum number of generations, $Range = max \mid f_i - f_j \mid$, "l" is a positive integer and "n" is the generation number.

This way we have several parameters to 'fine tune' the selection algorithm and avoid a premature (mis)convergence.

In this example the function 'favors' the not so good ones when '$Range$' is big combined with the algorithm being in the first stages of its evolution.

12.4.4 Random Numbers.

Being Genetic Algorithms stochastic, it is notorious the big influence of the underlying algorithm for the generation of random numbers.

There are methods for the generation of "real" random sequences based on specialized hardware, however for the general user the solution is to implement an algorithm which generates a pseudo-random sequence.

Seems odd to use the most deterministic device (computer) to build a random sequence: any computer program will produce a predetermined output, this way it will not be truly random, for this case we will use the term pseudo-random.

How could we define the vaporous concept of randomness?

A working approach could be that the deterministic program that builds a random series should be different and uncorrelated with the program which uses the sequence. This way two different generators should yield the "same" result, from a statistical point of view, when coupled with our GA, and that this result should coincide when the one obtained using a true random device.

From a pragmatic point of view an algorithm which is "good" for one application can show its weakness in another one.

There are several tests to prove the goodness of a pseudo random number generating algorithm, some of them are sustained on statistics and some other based on heuristics; a good pseudorandom number generator should pass all of them. However we should bear in mind that these are necessary conditions, never sufficient. One of the most important theoretical ones is the χ^2 criterion of non-correlation, among the heuristical ones we will mention the n-cube distribution for uniform deviates.

Classification

There are several methods for building pseudo-random sequences:

- Congruential
- Fibonacci
- Shuffle
- Cryptographic
- Other

Congruential:

These generators use a congruential relation of the type $p_{i+1} = kp_i + c$ (mod. m) "m" is the modulus, "k" and "c" are positive integers called multiplier and increment.

The series will repeat itself with a period which is not greater than m. The quality of the obtained sequence depends on the sensible election of "m", "k" and "c".

We can show a (not very good) example given in the ANSI-C standard which illustrates the use of these congruential generators:

```
unsigned long next=1;
int rand(void)
{
next= next*1103515245 + 12345;
return (unsigned int)(next/65536) % 32768;
}
void srand(unsigned int seed)
{
next=seed;
}
```

Most of the library implementations use this kind of generators. Their main disadvantages are short periods, poor fill in n-space and weak low order bit randomness.

In this sense it is "(in)famous" the subroutine RANDU with $k = 65539$ and $m = 2^{31}$

used in IBM mainframes for several years.

Some better alternatives are proposed by Park and Miller with $k = 7^5$, $m = 2^{31} - 1$ and $c = 0$. This algorithm modified by Schrage (to avoid overflow) can give a period approximately equal to 2.1×10^9.

Fibonacci

If we extend the recurrence to include more terms we obtain the Fibonacci generation scheme:

$$p_{n+1} = \sum_{i=n-l}^{n} (k_i p_i + c_i)$$

(mod. m); "l" is the lag.

These algorithms are more time consuming, Marsaglia proposes one with lag 17 with very good period properties. They seem to be more robust to n-space correlation problems.

Shuffle

They are based on the "shuffling" of the bytes of a sequence of numbers. We will mention Prof. Vose algorithm [Vose 92]

```
nrndm()
{
int i;
for (i = 0; i < 24; i++) rtab[i] -= rtab[i+31];
for (i = 24; i < 55; i++) rtab[i] -= rtab[i-24];
return 0;
}
initrand(s)
unsigned s;
{
int h,i;
unsigned j = s, k = 1;
rtab[54] = s;
for (i =1; i < 55; i++) {
h = (21*i)%55;
rtab[-h] = k;
k = j - k;
j = rtab[h];
}
nrndm(); nrndm(); nrndm(); nrndm(); rndx = 0;
}
```

Cryptographic

The engine here is cryptography. The sequence is based on cryptographic operations. Its properties are, somewhat, under discussion due to theoretical difficulties and gubernamental restrictions.

Here we present some comments about the implementation of a good cryptographic pseudo-random number generator due to W. Naylor:

> "As I understand it, a good encryption or random number generation algorithm must be based on a problem which is computationally difficult to solve. Actually, I want something stronger: it should be computationally difficult to even make

statistical statements about the solution other than obvious zero-knowledge statements.

Many NP-complete problems are easy to approximate or else make strong statistical statements about the solution (for example, 3-sat is easy to approximate by simulated-annealing and easy to guess most of the variables with high probability of correctness).

I choose the following problem as my foundation:

Solve a set of N simultaneous 2nd degree equations in N variables in MOD 2. Every variable is a bit which can take on only the values 0 and 1.

Here is an example problem for $N = 3$: $x*y+z = 1$; $x*z+y*z+z = 0$; $x + y + z = 0$.

This system has 3 solutions: $x = 0$, $y = 1$, $z = 1$; $x = 1$, $y = 0$, $z = 1$ and $x = 1$, $y = 1$, $z = 0$.

The general problem is easily shown to be NP-complete (reduction from SAT).

A NxN system can be regarded as an N-bit to N-bit mapping.

For example: $x * y + z = out1$; $x * z + y * z + z = out2$; $x + y + z = out3$; where x, y, z are inputs, $out[i]$ are outputs. The outputs behave erratically as a function of the inputs and the inputs solution behaves erratically as a function of the outputs. Approximating the general problem is also NP-complete (the proof is a bit more involved).

It seems that solving most systems requires 2^N time. The standard attacks for reducing the exponential growth factor – simulated annealing and branch-and-bound – are obviously useless here because of the erratic behavior. (Perhaps some algebraic attack might work?).

For a random number generator, we set up a machine with N bits of state. We produce $N+1$ 2nd order MOD 2 functions (non-sparse), perhaps using true physical randoms to generate the coefficients. We seed the state of the machine somehow. The machine iterates thus:

Fout(state1,state2,....,stateN) = output bit
F1(state1,state2,....,stateN) = next state1
F2(state1,state2,....,stateN) = next state2

...

FN(state1,state2,....,stateN) = next stateN

The argument that this machine is hard to break is obvious. Even if the adversary knows the state of the machine at time T, and the random number it generated, he cannot solve for the previous state of the machine without solving an $N \times N$ system, which seems to require $2N$ time.

The adversary doesn't know the state of the machine at any time, all he has is a long (say $O(N)$ string of random numbers it generated. This information is strictly less useful than knowing the state at the beginning of the string (say at time T) for solving for the state at time $T - 1$. He must solve for the state of the machine at some time in order to predict the other random numbers.

For this to be a good procedure, we need that the sequence of output bits is not only hard to break, but random in a fairly strong sense. I cannot see form this that even one bit is close to equally likely to be 0 or 1. Also, the functions $F1 - FN$ should come close to giving a 1-1 mapping. In addition, this does not necessarily make it cryptographically strong; it should be that way both forward and backward.

I only know of two methods in the literature claimed to have the desired properties. The simplest is the Shamir algorithm, which produces pseudo-random $[0, n - 1]$ numbers. The procedure is as follows:

Let p, q be primes, $n = pq$. Let d and e be numbers; one way is to have de congruent to 1 mod $p - 1$ and $q - 1$, which would make them an encoding- decoding pair for the RSA public key cryptosystem, but this is not necessary. It is necessary that d and e are relatively prime to $p - 1$ and $q - 1$. There are two seeds, x and y, which are between 2 and $n - 2$.

The procedure is: $x = x^d$ (mod n); $y = y^e$ (mod n): output $= x + y$ (mod n).

The Blum-Miccaeli procedure is simpler in principle, but slower. It has n as above, but p and q must be congurent to 3 mod 4. It would be best if p and q are Sophie Germain primes (this holds for the previous one also), which means that $(p - 1)/2$ and $(q - 1)/2$ are prime. The seed is one number between 2 and $n - 2$.

The procedure is: $x - x^2$ (mod n); output the least significant bit of x.

They produced this procedure because the Shamir procedure did not produce random bits. But let z be the output of the Shamir procedure, and consider $w = z$ XOR n. All bits of w (or z) after the leading one can be used. On the problem of complexity, I have looked into the complexity of generating non-uniform from uniform. It is surprisingly low, as I define it. Consider procedures of the following type, which I call infinite precision. After a procedure which terminates with probability one, some of the bits are output. All other bits are found by copying random bits.

A necessary and sufficient condition that one exists for a given density, even in k dimensions, is that there is a lower semi-continuous version of the density. For most common densities, the expected number of bits needed to carry out such a procedure (there are lots of procedures) is usually under 10".

Other

Other methods involve random bit generation through primitive polynomials modulo 2, intermixing of various procedures taking, for instance, a generator to decide between several other, etc.

Lastly there have been devised several other methods to build Random Number Generators. Here we give an example taken from Marsaglia, we include the full text of the private e-mail communication which illustrates in detail this new approach to the old problem:

"... Yet another RNG. Random number generators are frequently posted on the network; my colleagues and I posted ULTRA in 1992 and, from the number of requests for releases to use it in software packages, it seems to be widely used.

I have long been interested in RNG's and several of my early ones are used as system generators or in statistical packages. So why another one? And why here? Because I want to describe a generator, or rather, a class of generators, so promising I am inclined to call it "The Mother of All Random Number Generators" and because the generator seems promising enough to justify shortcutting the many months, even years, before new developments are widely known through publication in a journal.

This new class leads to simple, fast programs that produce sequences with very long periods. They use multiplication, which experience has shown does a better job of mixing bits than do +,- or exclusive-or, and they do it with easily- implemented arithmetic modulo a power of 2, unlike arithmetic modulo a prime. The latter, while satisfactory, is difficult to implement. But the arithmetic here modulo 2^{16} or 2^{32} does not suffer the flaws of ordinary congruential generators for those moduli: trailing bits too regular. On the contrary, all bits of the integers produced by this new method, whether leading or trailing, have passed extensive tests of randomness.

Here is an idea of how it works, using, say, integers of six decimal digits from which we return random 3-digit integers. Start with $n = 123456$, the seed. Then form a new $n = 672*456+123 = 306555$ and return 555. Then form a new $n = 672*555+306 = 373266$ and return 266. Then form a new $n = 672*266+373 = 179125$ and return 125, and so on. Got it? This is a multiply-with-carry sequence $x(n) = 672*x(n-1) + carry \ mod \ b = 1000$, where the *carry* is the number of b's dropped in the modular reduction. The resulting sequence of $3 - digit$ x's has period $335,999$. Try it.

No big deal, but that's just an example to give the idea. Now consider the sequence of 16-bit integers produced by the two C statements:

k=30903* (k& 65535)+(k>>16); return(k& 65535);

Notice that it is doing just what we did in the example: multiply the bottom half (by 30903, carefully chosen), add the top half and return the new bottom.

That will produce a sequence of 16-bit integers with period > 229, and if we concatenate two such:

```
k=30903 * (k& 65535)+(k>>16);
j=18000 * (j& 65535)+(j>>16);
return((k<<16)+j);
```

We get a sequence of more than 2^{59} 32-bit integers before cycling. The following segment in a (properly initialized) C procedure will generate more than 2^{118} 32-bit random integers from six random seed values

```
i,j,k,l,m,n:
k=30903* (k& 65535)+(k>>16);
j=18000* (j& 65535)+(j>>16);
i=29013 *(i& 65535)+(i>>16);
l=30345* (l& 65535)+(l>>16);
m=30903* (m& 65535)+(m>>16);
n=31083* (n& 65535)+(n>>16);
return((k+i+m)>>16)+j+l+n);
```

And it will do it much faster than any of several widely used generators designed to use 16-bit integer arithmetic, such as that of Wichman-Hill that combines congruential sequences for three 15-bit primes (Applied Statistics, v31, p188-190, 82), period about 2^{42}. I call these multiply-with-carry generators. Here is an extravagant 16-bit example that is easily implemented in C or Fortran. It does such a thorough job of mixing the bits of the previous eight values that it is difficult to imagine a test of randomness it could not pass:

$$x[n] = 12013 \; x[n-8] + 1066 \; x[n-7] + 1215 \; x[n-6] + 1492 \; x[n-5] + 1776 \; x[n-4]$$
$$+ \; 1812 \; x[n-3] + 1860 \; x[n-2] + 1941 \; x[n-1] + \text{carry mod } 216.$$

The linear combination occupies at most 31 bits of a 32-bit integer. The bottom 16 is the output, the top 15 the next carry. It is probably best to implement with 8 case segments. It takes 8 microseconds on my PC. Of course it just provides 16-bit random integers, but awfully good ones. For 32 bits you would have to combine it with another, such as

$$x[n] = 9272 \; x[n-8] + 7777 \; x[n-7] + 6666 \; x[n-6] + 5555 \; x[n-5] + 4444 \; x[n-4]$$
$$+ \; 3333 \; x[n-3] + 2222 \; x[n-2] + 1111 \; x[n-1] + \text{carry mod } 216.$$

Concatenating those two gives a sequence of 32-bit random integers (from 16 random 16-bit seeds), period about 2^{250}. It is so awesome it may merit the Mother of All RNG's title. The coefficients in those two linear combinations suggest that it is easy to get long-period sequences, and that is true. The result is due to Cemal Kac, who extended the theory we gave for add-with-carry sequences: Choose a base b and give r seed values $x[1], ..., x[r]$ and an initial 'carry' c. Then the multiply-with-carry sequence

$$x[n] = a_1 * x[n-1] + a_2 * x[n-2] + ... + a_r * x[n-r] + \text{carry mod } b,$$

Where the new carry is the number of b's dropped in the modular reduction, will have period the order of b in the group of residues relatively prime to $m = a_r * b_r + ... + a_1 * b_1 - 1$. Furthermore, the x's are, in reverse order, the digits in the expansion of k/m to the base b, for some $0 < k < m$.

In practice $b = 2^{16}$ or $b = 2^{32}$ allows the new integer and the new carry to be the bottom and top half of a 32- or 64-bit linear combination of 16- or 32-bit integers. And it is easy to find suitable m's if you have a primality test: just search through candidate coefficients until you get an m that is a safe prime (both m and (m-1)/2 are prime). Then the period of the multiply-with-carry sequence will be the prime (m-1)/2. (It can't be m-1 because $b = 2^{16}$ or 2^{32} is a square). Here is an interesting simple MWC generator with period $> 2^{92}$, for 32-bit arithmetic:

$$x[n] = 1111111464 * (x[n-1]+x[n-2]) + carry \bmod 232.$$

Suppose you have functions, say top() and bot(), that give the top and bottom halves of a 64-bit result. Then, with initial 32-bit x, y and carry c, simple statements such as

$$y = bot(1111111464 * (x+y)+c); \ x=y; \ c=top(y)$$

Will, repeated, give over 2^{92} random 32-bit y's. Not many machines have 64 bit integers yet. But most assemblers for modern CPU's permit access to the top and bottom halves of a 64-bit product. I don't know how to readily access the top half of a 64-bit product in C. Can anyone suggest how it might be done? (in integer arithmetic)".

12.5 Test Cases

General parameters

Population replacement: steady state keeping the best
Random number generation: shuffling
Initial population: duplicates not allowed
Transformation matrix: Gray code = I + SubDiag(1)
Number of runs: 10

Parameters for Test-cases 1

Problem description: function $\dfrac{sin(x)}{x}e^{sin(x)}$
Variable range: $[-300, 300]$
Accuracy: 16 bits per variable
Dimension: 1
Population size: 60
Crossover strategy: two point
Crossover rate: 0.85
Mutation rate: 0.73 %
Selection strategy: ranking

See Fig. 1 and Graph. 1

Parameters for Test-cases 2

Problem description: function $sin(50x) + \dfrac{x}{50}$
Variable range: $[0,5]$
Accuracy: 16 bit per variable
Dimension: 1
Population size: 60
Crossover strategy: two point
Crossover rate: 0.85
Mutation rate: 0.73 %
Selection strategy: modified window

See Fig. 2 and Graph. 2

Parameters for Test-cases 3

Problem description: function $Frac(50x) + \frac{x}{50}$
Variable range: $[0,1]$
Accuracy: 16 bits per variable
Dimension: 1
Population size: 60
Crossover strategy: two point
Crossover rate: 0.85
Mutation rate: 0.73 %
Selection strategy: ranking

See Fig. 3 and Graph. 3

Parameters for Test-cases 4

Problem description: function $\frac{sin(x)}{x}e^{sin(x)}$
Variable range: $[-5,5]$
Accuracy: 16 bits per variable
Dimension: 8
Population size: 150
Crossover strategy: two point
Crossover rate: 0.85
Mutation rate: 0.3 %
Selection strategy: proportional

See Graph 4-1

Parameters for Test-cases 4

Problem description: function $\frac{sin(x)}{x}e^{sin(x)}$
Variable range: [-5,5]
Accuracy: 16 bits per variable
Dimension: 8
Population size: 150
Crossover strategy: multidirectional
Crossover rate: 0.85
Mutation rate: 0.3 %
Selection strategy: modified window

See Graph 4-2

Parameters for Test-cases 4

Problem description: function $\frac{sin(x)}{x}e^{sin(x)}$
Variable range: [-5,5]
Accuracy: 16 bits per variable
Dimension: 8
Population size: 150
Crossover strategy: multidirectional
Crossover rate: 0.85
Mutation rate: 0.3 %
Selection strategy: ranking

See Graph 4-3

Optimum: The best of all runs.
Avg. Best: Arithmetic average of the best in all runs.
Range Best: The distance between the optimum and the 'worst' best.
Var. Best: Variance of the best for all runs.
Avg. Var.: Average (all runs) of the variances of the populations.

In these test cases we present, firstly, three unidimensional functions to optimize. The first one has a large interval of definition, the second one gives raise to a deceptive problem as it has many similar local maxima, the third one is deceptive and discontinous.

We show the asymptotic exponential behaviour of the algorithm in these instances.

The last three cases (4-1, 4-2, 4-3) illustrate a multidimensional optimization problem. It is noteworthy that, according to the area between the optimum line and the average-of-the-best line, the third case seems to achieve a better and faster convergence behavior.

As a last remark, we emphasize on the high importance of the choosing of a good pseudo-random number generator, multimodal-multidimensional problems treated with bad (ran() library) PRNG gave very poor results.

12.6   Some Applications

12.6.1   *Mesh Quality Improvement*

A high degree of automation of the mesh generation process is required in many

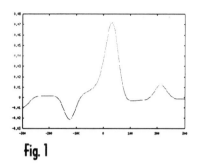

Fig. 1

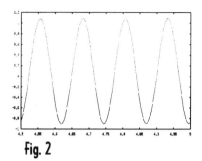

Fig. 2

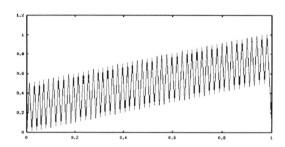

Fig. 3

Graph 1

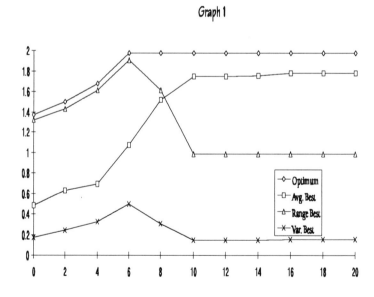

Figure 12.1

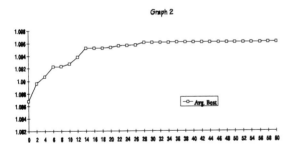

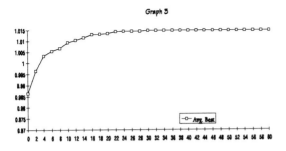

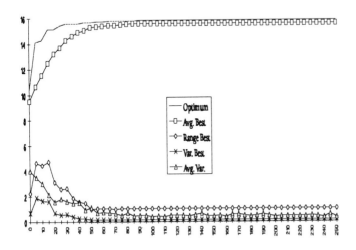

Figure 12.2

Graph 4-2

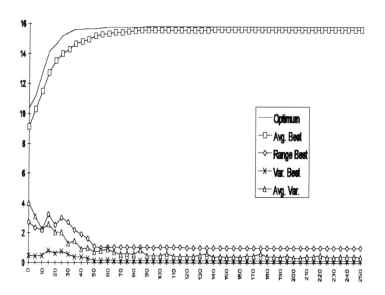

Graph 4-3

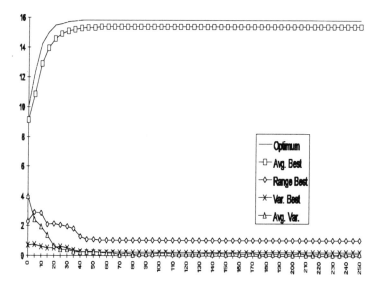

Figure 12.3

engineering applications using the general finite elements technique to solve boundary problems, specially interesting for non-convex dimensional domains.

A procedure for smoothing (regularization process making good geometrical characteristics for the triangles) has been developed with GAs in Cuesta 92. The methodology for a regularization process of unstructured meshes and adaptive remeshing in 2-D by simple GA on transonic flow simulation was proposed in Winter, Montero, Cuesta and Galan 94, which presents some possibilities of using GAs in different linked optimization problems.

We propose a regularization process of unstructures meshes. From a starting mesh, a new one is built employing GAs to minimized a fitness function which is based on geometrical conditions that allow to get better quality of the mesh. Several fitness functions are suggested depending on the proposed objetives to obtain a better mesh, including different geometrical considerations regasrding area, perimeter, angles, etc. of the triangles and error measurements based on error indicators or both of them.

The procedure may be generalized to three-dimensional unstructured meshes.

In non-convex two-dimensional domains, a method for smoothing and adapting meshes, initially built by a conventional mesher has been developed with GAs.

The smoothing technique based on the control by genetic algorithms of the nodes which form a triangle consists in the idea of finding a near location to the optimal location of a node inside a polygonal in such a way that the triangles formed by this node and two of the polygonal are as isosceles or equilateral as possible. In this sense, we have been designed the followings fitness functions to be minimized:

A function to get the central angles to be equal,

$$F_a = \sum_{i=1}^{NP} sin^2 \left(\frac{2\pi}{NP} - \alpha_i \right)$$

being NP the number of vertices of the polygonal containing the point to locate and α_i the angle corresponding to the inner vertex of the triangle "i".

A function to get that the triangles being as isosceles as possible,

$$F_b = \sum_{i=1}^{NT} (C_{1i} - C_{2i})^2$$

where NT is the number of triangles of the polygonal and C_{ji} the distance between the inner and the j-th point of the i-th triangle.

Also it is possible to use a combination of two fitness functions of different characteristics. Thus, in the context of adaptive remeshing we propose a fitness function that changes the density in those zones where available error indicators takes high values,

$$F_c = \sum_{i=1}^{NP} A_i \varepsilon_i$$

being A_i the area of the triangles and ε_i the error indicator of the i-th triangle.

And to minimize the functions

$$F_1 = \theta F_a + (1 - \theta) F_c \quad or \quad F_2 = \theta F_b + (1 - \theta) F_c \quad being \quad 0 \le \theta \le 1.$$

To minimize with Genetic Algorithms the fitness function, a population is represented by a set of chromosomes which are identified depending on theirs merit values, that are obtained evaluating a function that have been fixed in order to get the searched objetive. In this case, on the applications GAs to improve the quality of the mesh each cromosome means one location in the topological space and the geometrical representation of a point in two-dimensional space requires to fix two numbers to define its location with respect to coordinates axes, thus

$Location = < xy > = < 11010...10110 >$.

Both values (x, y) are written in binary system and a chain of 0 and 1 is obtained for each coordinate value.

The random number generation is very important in selection, crossover and mutation. A good selection of the random number generation algorithm is necessary.

12.6.2   Matching local solutions from domain decomposition

The problem to solve is to couple finite element solutions of incompressible Navier-Stokes and potential equations using domain decomposition and Genetic Algorithms, that is to say the coupling between the Navier-Stokes equations for unsteady incompressible viscous flows with the Laplace equation modeling inviscid incompressible potential flows.

The first step is to use finite element solvers for both models. The computational flow domain Ω is decomposed in two subdomains Ω_1 and Ω_2, where Ω_1 is the Navier-Stokes governed domain and Ω_2.

After appropiate time discretization let $\Omega_{12} = \Omega_1 \cap \Omega_2$ and γ_1, γ_2 denotes the overlapping region and the associated interfaces. The matching of physically local solutions in the overlapping domain Ω_{12} is based upon a minimization problem that takes the following optimal control formulation (in the sense of J.L.Lions [Lions, 68]):

Find ψ and ν such that $J(\overline{\psi}, \overline{\nu}) \leq J(\psi, \nu)$, $\forall \{\psi, \nu\}$, where

$$J(\psi, \nu) = \frac{1}{2} \int_{\Omega_{12}} | u - \nabla\Phi |^2 \, dx + \beta \left(\int_{\gamma_1} u_2 n + \int_{\gamma_2} u_1 n \right)^2 \qquad (12.1)$$

and (u, Φ) solutions of

$$\begin{cases} \alpha u - \nu\Delta u + u\nabla u + \nabla p = f & \\ \quad \nabla u = 0 & in \quad \Omega_2 \\ \quad u = \nu & on \quad \gamma_2 \end{cases} \qquad (12.2)$$

$$\begin{cases} \Delta\Phi = 0 & in \quad \Omega_1 \\ \Phi = \psi & on \quad \gamma_1 \end{cases} \qquad (12.3)$$

with appropiate boundary conditions and α, ν positive parameters.

Previous coupled solutions of problem (12.1)(12.2)(12.3) with derivative based conjugate gradient technique have been proposed [Dinh and al. 88]. We propose a method to solve the above matching problem by Genetic Algorithms (GAs).

For each individual, we can compute a fitness function which corresponds to its adaptation to the environment; in our case, the fitness function is the value of J and the individuals with low J values are very well fit to the problem.

Genetic Algorithms offer many advantages. There is no longer need to derivate J in order to find the minimum, so that enables us to deal with non linear operators. Furthermore, these algorithms are very robust because they do not focus on a single solution but compare at each generation all the indivuals of the current population, so they can avoid local minima.

12.7 Final Remarks

In this chapter, firstly we have described several general features, tools and comments on GAs. We have reviewed the main concepts with some comments and results from various chosen references and some important considerations related to computational aspects.

Following, we have presented some results from several simple test cases using several selection strategies and crossover alternatives in the context and possibilities that have been previously proposed.

We wanted this chapter to be, at least that was our aim, a helping short introduction guide on GAs, explaining several computational and practical aspects and addressed to a large and diverse audience including postgraduate students, practice engineers and people who would like to use and implement GAs for solving real-life applications.

Now a day, the research in the GAs field is very active, so this chapter should be viewed as an approximation and introduction to this wide, rich and complex world of the Genetic Algorithms.

Acknowledgements

We would like to express our most sincere thanks and consideration to our colleague **Prof. Juan Rocha** for his kind cooperation in the preparation and composition of this paper.

REFERENCES

[1] Bäck, T. - Hoffmeister,F. (1991) "Extented Selection Mechanisms in Genetic Algorithms", 92-99, Proceedings of the Fourth International Conference on Genetic Algorithms. Morgan Kaufmann Publishers, Los Altos, CA. Belew, R. and Booker, L. (Ed.)

[2] Bäck, T. (1992) "Self-Adaptation in Genetic Algorithms". Proceedings of the First European Conference on Artificial Life, 263-271. The MIT Press, Cambridge, MA. F.J. Varela and P.Bourgine (Eds).

[3] Baker J.E. (1985) "Adaptive Selection Methods for Genetic Algorithms". J.J. Grefenstette (Ed.). Proceedings of the First International Conference on Genetic Algorithms and Their Applications, 101-111. Hillsdale, New Jersey. Lawrence Erlbaum Associates.

[4] Baker, J.E. (1987) "Reducing Bias and Inefficiency in the Selection Algorithm", 14-21.Proceedings of the Second International Conference on Genetic Algorithms and Their Applications. Hillsdale, New Jersey. Lawrence Erlbaum Associates, Grefenstettte J.J.(Editor).

[5] Booker, L.B. (1982) "Intelligent Behaviour as an Adaptation to the Task Environment", Doctoral Dissertation, University of Michigan.

[6] Bratley P., Fox B.L., Schrage L. (1983) "A Guide to Simulation" Springer-Verlag.

[7] Brindle, A. (1981) "Genetic Algorithms for Function Optimization", Doctoral Dissertation, University of Alberta, Edmonton.

[8] Caruana, R.A., Schaffer,J.D. (1988) "Representation and hidden bias: Gray vs binary coding for genetic algorithms". Proceedings of the 5th International Conference on Machine Learning. 153-161. Morgan Kaufmann Publishers, San Mateo, California

[9] Cuesta, P. (1992) "Adaptive mesh generation in non-convex domain and applications". Doctoral Dissertation, University Las Palmas de Gran Canaria, Spain

[10] Davis L.(Editor), (1991) "Handbook of Genetic Algorithms". Van Nostrand Reinhold , New York.

[11] De Jong K.A., Sarma J. (1993) "Generation Gaps Revisited". Foundation of Genetic Algorithms, 19-27.

[12] De Jong K.A. (1975) "An Analysis of the Behaviour of a Class of Genetic Adaptive Systems". Ph. dissertation. University of Michigan, Ann Arbor, MI.

[13] De Jong K.A. (1993) "Genetic Algorithms are not Function Optimizers". Foundation of Genetic Algorithms, 5-17.

[14] Dihn, Q. V., Glowinski, R., Periaux, J. and Terrasson, G.(1988) "On The Coupling of Viscous and Inviscid Models for Incomporessible Fluid Flows Via Domain Decomposition". First International Sysmposium on Domain Decomposition Methods fot PDE, SIAM.

[15] Eshelman, L.J. and Schaffer,J.D. (1992) "Real-coded Genetic Algorithms and Interval-Schemata". 187-202, Foundations of Genetic Algorithms, 1, 69-93. Morgan Kaufmann Publishers.G.J.E. Rawlins (Ed.).

[16] Eshelman, L.J. and Schaffer,J.D. (1993) "Crossover's niche",9-14, Proceedings of the Fifth International Conference on Genetic Algorithms. Forrest,S. (ed.). Morgan Kaufmann Publishers, Los Altos, CA.

[17] Eshelman,L.J., Caruana, R.A. and Schaffer, J.D.(1989) "Biases in the Crossover Landscape", 10-19. Proceedings of the Third International Conference on Genetic Algorithms and their Applications Morgan Kaufmann Publishers, San Mateo, CA. Schaffer J.D. (Ed.).

[18] Fogarty , T.C. (1989) "Varying the probability of mutation in the genetic algorihtm". 104-109. Proceedings of the Third International Conference on Genetic Algorithms and their Applications Morgan Kaufmann Publishers, San Mateo, CA. Schaffer J.D. (Ed.).

[19] Forsythe G.E., Malcolm M.A., Moler C. (1977) "*Computer Methods for Mathematical Computations*" Prentice-Hall.

[20] G. Marsaglia and Tsang, (1983) "*A fast, easily implemented method for sampling from decreasing or symmetric unimodal density functions*", SIAM J.Sci.

[21] G. Marsaglia, (1994), private e-mail comm.

[22] Glover F. and Laguna M. (1993) "*Tabu search. Modern Heuristic Techniques for Combinatorial Problems*", C.R. Reeves C.R. (Ed.), Blackwell Scientific Publications, Oxford, 70-141.

[23] Golberg, D.E., Deb, K., Clark, J.H. (1992) "*Genetic Algorithms, noise, and the sizing of populations*". Complex Systems, 6, 332-362

[24] Goldberg D.E., Deb, K. (1991) "*A comparative Analysis of Selection Schemes used in Genetic Algorithms*". Foundations of Genetic Algorithms, 1, 69-93. Morgan Kaufmann Publishers.G.J.E. Rawlins (Ed.).

[25] Goldberg D.E., (1989) "*Genetic Algorihms In Search, optimization and Machine Learning*". Reading, Massachusets, Addison-Wesley.

[26] Goldberg, D.E. and Richardson, J. (1987) "*Genetic Algorithms with Sharing for Multimodal Function Optimization*", 41-49. Proceedings of the Second International Conference on Genetic Algorithms and Their Applications. Hillsdale, New Jersey. Lawrence Erlbaum Associates, Grefenstettte J.J. (Editor).

[27] Goldberg, D.E., Deb, K., Thierens, D. (1993) "*Toward a better understanding of mixing in genetic algorithms*". Journal of the Society of Instrument and Control Engineers, 32(1), 10-16.

[28] Goldberg, D.E., Deb, K., Kargupta, H., Harik, G. (1993) "*Rapid, accurate optimization of difficult problems using fast messy genetic algorithms*" Proceedings of the Fifth International Conference on Genetic Algorithms, 56-64.

[29] Goldberg, D.E., Kelsey M., Tidd, C. (1992) "*Genetic Algorithm:A Bibliography*", (IllGAL Report n. 92208), University of Illinois at Urbana-Champaign, Illionis Genetic Algorithm Laboratory, Urbana, USA

[30] Goldberg, D.E., Korb, B., Deb, K. (1991) "*Do not Worry, Be Messy*". Proceedings of the Fourth International Conference on Genetic Algorithms, 24-30. Morgan Kaufmann Publishers, Los Altos, CA. Belew, R., Booker, L. (Ed.).

[31] Goldberg, D.E., Korb, B., Deb, K. (1989) "*Messy Genetic Algorithms : Motivation analysis, and first results*". Complex Systems, 3(5), 493-530.

[32] Grefenstette, J.J., Baker, J.E.(1989) "*How genetic algorithms work : A critical look at implicit parallelism*", 20-27 .Proceedings of the Third International Conference on Genetic Algorithms and their Applications Morgan Kaufmann Publishers, San Mateo, CA. Schaffer J.D. (Ed.).

[33] Grefenstette, J.J. (1986) "*Optimization of Control Parameters for Genetic Algorithms*". IEEE Transactions on Systems, Man and Cybernetics SMC-16 (1), 122-128.

[34] Hesser, J., R. Männer (1991). "*Towards an optimal mutation probability in genetic algorithms*". Parallel Problem Solving from Nature, 23-32. Vol. 496 of Lecture Notes in Computer Science, Springer, Berlin. H.P. Schwefel, R. Männer (Eds).

[35] Holland H.J. (1975) "*Adaptation in Natural and Artificial Systems*" University of Michigan Press, Ann Arbor, MI

[36] Kahaner D., Moler C., Nash S. (1989) "*Numerical Methods and Software*" Prentice Hall.

[37] Kirkpatrick, S., E. Stoll, (1981) "*A Very Fast Shift-Register Sequence Random Number Generator*", Journal of Computational Physics, V.40.

[38] Knuth D.E. (1981) "*Seminumerical Algorithms: The art of Computer Programming*" Addison-Wesley.

[39] L'Ecuyer P., (1988) "*Communications of the ACM*" vol. 31, pp. 742-774.

[40] Lions, J.L. (1968) *"Controle Optimal des systemes gouvernes par des equations aux derivees partielles"*. Dunod, Paris.

[41] Marsaglia G., *"Comments on the perfect uniform random number generator"*, Unpublished Notes, Wash.. S.U.

[42] Meyer C., Matyas S. (1982) *"Cryptography: A New Dimension in Computer Data Security"* John Wiley and Sons.

[43] Michalewicz Z. (1994) *"Genetic Algorithms + Data Structures = Evolution Programs."* Springer Verlag. (Second, Extented Edition). New York.

[44] Miller,B.L. and Goldberg, D.E. (1995) *"Genetic Algorithms, Tournament Selection, and the Effects of Noise"*. IllGAL Report n. 95006, July 1995, University of Illinois at Urbana-Champaign, Illionis Genetic Algorithm Laboratory, Urbana, USA.

[45] Mühlenbein, H. , Schlierkamp-Voosen, D. (1993) *"Predictive Models for the Breeder Genetic Algorithm"* : I- Continuos Parameter Optimization. Evolutionary Computation., 1, 25-49.

[46] Muhlenbein, H., Gorges-Schleuter, Kramer O. (1988) *"Evolution Algorithms in Combinatorial Optimization, Parallel Computing"*, 7(65-88).

[47] Naylor, W. C. (1995), private e-mail comm.

[48] Park S.K., Miller K.W., (1988) *"Communications of the ACM"* vol. 31, pp. 1192-1201

[49] Press W., Teukolsky S., Vetterling W., Frannery B. (1992) *"Numerical Recipes in C: The Art of Scientific Computing"* Cambridge Univ. Press.

[50] Rawlins G.J.E. (1991) *"Introduction in Foundations of Genetic Algorithms"*. Morgan Kaufmann Publishers. G.J.E. Rawlins (Ed.).

[51] Schaffer, J.D., R.A. Caruna, L.J. Eshelman, R. Das (1989). *"A study of control parameters affecting online performance of genetic algorithms for function optimization."* Proceedings of the Third International Conference on Genetic Algorithms and their Applications, 56-64. Morgan Kaufmann Publishers, San Mateo, CA. Schaffer J.D. (Ed.)

[52] Schrage L. *"ACM Transactions on Mathematical Software"*, (1979) vol. 5, pp. 132-138

[53] Schwefel, H.P., (1981). *"Numerical Optimization for Computer Models"*, John Wiley, Chichester, U.K.

[54] Smith, S.F. (1993) *"Adaptively Resizing Populations : An algorithm and Analysis"*. Proceedings of the Fifth International Conference on Genetic Algorithms. Forrest,S. (Ed.). Morgan Kaufmann Publishers, Los Altos, CA.

[55] Syswerda G. (1991) *"A Study of Reproduction in Generational and Steady-State Genetic Algorithms"*,1, 100-103.Algorithms. Foundations of Genetic Algorithms. Morgan Kaufmann Publishers. G.J.E. Rawlins (Ed.)

[56] Syswerda G., (1989). *"Uniform Crossover in Genetic Algorithms"*, Proceedings of the Third International Conference on Genetic Algorithms and their Applications, 2-9. Morgan Kaufmann Publishers, San Mateo, Calif.. Schaffer J.D. (Ed.).

[57] Thierens, D. and Goldberg, D.E. (1994) *"Convergence models of genetic algorithm selection schems"*. Davidor, Y., Schwefel, H.P., and Männer, R. (Eds.), Parallel Problem Soving from Nature- PPSN III, 119-129, Berlin, Springer-Verlag.

[58] Vose, M. (1992). *"Modeling Simple Genetic Algorithms"*. Foundations of Genetic Algorithms 2., 63-74. Whitley (Ed.), Vail, CO : Morgan Kaufmann.

[59] Whitley, D. (1988) *"GENITOR : A Different Genetic Algorithms"* Proceedings of the Rocky Mountain Congerence on Artificial Intelligence, Denver.

[60] Whitley, D. (1989) *"The Genitor algorithm and selection pressure: Why rank-based allocation of reproductive trials is best"*, Proceedings of the Third International

Conference on Genetic Algorithms and their Applications, 116-121. Morgan Kaufmann Publishers, San Mateo, CA. Schaffer J.D. (Ed.)

[61] Whitley, D. (1991) *"Fundamental Principles of Deception in Genetic Search."* Foundations of Genetic Algorithms. Morgan Kaufmann Publishers. G.J.E. Rawlins (Ed.).

[62] Whitley, D., Mathias, K., Fitzhorn P., (1991) *"Delta Coding : An Iterative Search Strategy for Genetic Algoritms"*, 77-84, Proceedings of the Fourth International Conference on Genetic Algorithms, 24-30. Morgan Kaufmann Publishers, Los Altos, CA. Belew, R., Booker, L. (Ed.).

[63] Winter G., Montero G., Cuesta P., Galán M. (1994) *"Mesh generation and adaptive remeshing by genetic algorithms on transonic flow simulation"*, 281-287. Proceedings of Computational Fluid Dynamics'94. John Wiley and Sons Publisher.

[64] Wright, A.H. (1991) *"Genetic Algorithms for Real parameter Optimization"*, Rawlins, G. (Ed.). Foundations of Genetic Algorithms, 205-218. Morgan Kaufmann Publishers.

Part 2
Applications and Computational Implementation

13

Parallel Genetic Algorithms for Optimisation in CFD

DENIS DOORLY[1]

13.1 INTRODUCTION

This chapter is concerned with the use of genetic and parallel genetic algorithms for optimisation in computational fluid dynamics (CFD). It is intended to provide an introduction both to optimisation in this area, as well as to specific genetic algorithm (GA) applications. A brief sketch of the problem will thus be given before discussing issues of GA choice and implementation.

CFD attempts to model and predict fluid behaviour by numerically solving equations which approximate the physical laws governing a flow. We are all familiar with the complex patterns which can occur when a fluid is set in motion, from observations of the water in a bath as we splash about, or from looking at the flow of a river behind the pier of a bridge, for example. It is hardly surprising therefore that the computational effort required to solve many flow problems necessitates the use of a fast computer, to obtain a solution in reasonable time. Apart from obtaining a solution to predict some flow attribute, (such as the drag exerted on a given body), CFD may be used as a tool in designing a particular object. Computationally this is even more demanding, particularly where repetitive analysis, geometry modification, and optimisation are combined to automatically seek some 'optimal design'. Hence CFD has an enormous appetite for computational resources, and the use of parallel algorithms to meet these demands is now commonplace.

To maintain the efficiency of a parallel CFD code, there are strong reasons for ensuring an associated genetic algorithm is also rendered parallel. For a program

---

[1] Aeronautics Dept., Imperial College, London SW7 2BY U.K. e-mail d.doorly@ic.ac.uk

containing parallel and sequential portions, the maximum speedup S gained by using P processors instead of a single one is limited by Amdahl's law to:

$$S = \frac{1}{\frac{a}{P} + (1 - a)}, \quad 0 \le a \le 1$$

where a is the proportion of the program execution which is made parallel. Thus if a sequential genetic algorithm is coupled to a parallel CFD solver (to create an optimal design program for example), the maximum speedup will become limited by the time required for execution of the sequential portion. There is an additional motivation for using a parallel genetic algorithm, in that a particular type of parallelism (distributing the population) can have inherent beneficial effects, as discussed later.

Genetic algorithms can be used either in combination with CFD to optimise a design, or they can be more tightly coupled to CFD, for example to optimise some aspect of the solution process itself. In the former case, since genetic algorithms are general purpose optimisation methods they can be coupled to any flow solution method for optimal design. To be practical however, both how the GA is applied and adapted for CFD, and the type of flow analysis to which it is currently most suited need consideration.

The optimisation of flow solvers designed for parallel computers is an example of the second application. Generally this type of solver divides the computational domain into subdomains, with the flow in each subdomain being advanced concurrently on separate processors; 'domain decomposition' refers to the process. The assignment of computational elements to subregions, and the assignment of subregions to processors are equivalent to a form of graph partitioning; for example each processor can be represented as a vertex of a graph, and links between vertices represent interprocessor communication. The design and partitioning applications are used here as illustrative examples, with the material organised as follows.

After a brief outline of a representative CFD algorithm, genetic algorithms for optimising graph partitioning are discussed, and an example program outlined. Non-genetic algorithms for the partitioning problem are briefly considered as competitors or for hybridisation with the GA. The use of genetic algorithms for optimal design is then discussed. Both the above optimisation problems lead to consideration of parallel genetic algorithms; (preliminary) conclusions are finally summarised.

13.2   CFD ANALYSIS FOR AEROSPACE DESIGN

The choice of the most appropriate analysis method for a given flow problem depends on the requirements of accuracy, ability to represent complex geometry, and computational cost. Many methods which can satisfy the first and last of these are often incapable of dealing with the second requirement, which is essential for design studies on realistic configurations. Finite volume and finite element methods can handle geometrical complexity, and solve the full set of equations, (respectively the Navier Stokes equations for viscous flow, and the Euler equations for inviscid flow), but at a cost which may be prohibitive. Boundary integral or panel methods can likewise handle complex geometries and need at least an order of magnitude less computing time, but do not account for viscous effects or shock waves. Coupled panel and shear layer methods can represent viscous effects adequately in many applications, but the

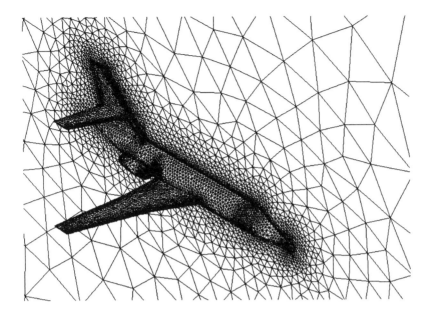

Figure 13.1   Unstructured mesh for aircraft configuration

full equations (considered for the most part here) are required where large separated regions exist, or in transonic flows for example.

13.2.1   Unstructured Finite-Volume and Finite-Element Methods

If conservation of mass, momentum and energy are applied to a control volume V bounded by a surface S, the Navier Stokes equations are obtained;

$$\frac{\partial}{\partial t}\int_V \vec{U}\,dV = -\int_S \vec{F}(U)\cdot d\vec{S}$$

Here $\vec{U}$ represents the vector of unknown flow variables (for a compressible fluid: density, momentum, and energy $[\rho, \rho u, \rho v, \rho w, \rho e]$), and the flux vector $\vec{F}(\vec{U})$ expresses transport of flow quantities across S and boundary acting terms. In the numerical solution procedure, the equations are discretised in space and the solution is advanced forward in time from an initial state, applying appropriate boundary conditions. In the finite volume formulation we consider, the spatial discretisation uses tetrahedral cells, within which values are assumed to vary linearly, and gradients are thus constant over a cell.

An example of an unstructured mesh for computation of the flow about an aircraft configuration is illustrated in fig. 13.1, showing the surface triangulation for a tetrahedral mesh (of approximately 80,000 nodes, and 350,000 elements) generated by the advancing front method, described in [PMP93] (The mesh was produced and kindly provided by Dr. J. Peiró). The discrete equations to update the value of the vector of flow variables at a node (such as "P" in the 2D mesh below) are then

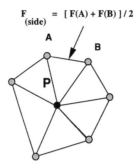

Figure 13.2

$$\frac{d}{dt}\,\vec{U}\Big|_{P} = \frac{1}{\Delta V} \sum_{face(i)} \vec{F}\Big|_{(i)} \cdot \vec{n}_{(i)} ds$$

where ΔV is a control volume about P, and the value of the flux vector on each face (i) with normal $n_{(i)}$, surrounding the node is obtained by interpolating the values at each extremity of the face. The gradient terms in the flux vector (either arising from viscous stresses, or artificially added to limit spurious oscillations near shock waves) are calculated from nodal values by applying the Green-Gauss relation to a suitable control volume. Details vary somewhat from method to method, but in any event the flux terms for the control volume surrounding each node (on the right of the above equation) may be assembled from the values of the fluxes at the mid points of the edges connecting the nodes. In performing the forward integration in time, explicit or implicit methods may be used, ie. the flux vector can be evaluated either wholly at the previous time level, or at least partly at the current level. For explicit integration, updating each node requires only the previous values of the flow variables at surrounding nodes, and can proceed independently of all the other nodal updates. With an implicit method, all the nodes are updated together, implying simultaneous solution. For this case however, we can assume that the implicitness is treated iteratively, (eg. via multigrid Jacobi iteration), or that the implicit treatment is broken into subdomains, with an iterative treatment of the equations linking subdomains (which are assumed to reside on different processors).

13.2.2 Parallel Computational Fluid Dynamics

Parallelisation of the CFD solution algorithm involves distributing the work (of updating nodal values) among the available processors; clearly for the algorithm to work efficiently the task is to minimise interprocessor communication whilst evenly distributing the computational load of updating the nodal values. In the case of a regular structured mesh, the decision as to how to distribute the mesh nodes among the processors may often be decided with ease in advance, eg. [HD90]. For an unstructured mesh, however, the problem of deciding how to apportion the elements is much more difficult. The number of nodes or elements assigned to a processor relates to the degree of parallelisation; in medium or coarse grain parallelisation, a given processor is responsible for updating a domain containing many if not thousands of nodes, whereas with fine grain parallelisation, a processor updates either a single node or edge, or possibly just the nodes associated with one cell.

Communication between processors is required to exchange flow variables along edges cut by the partition. For coarse grain parallelisation, this involves the exchange of 'halo data'. In the simplest case, the values at nodes which are excluded from a

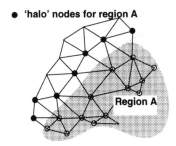

Figure 13.3   Halo data nodes (black) required to update solution in region A

given partition but which are required to update values inside the partition (near the boundary) are transferred from the neighbouring partitions, forming a 'halo' about the boundary. Since the flux associated with a given edge is used in updating the nodal values at each end of the edge, it need only be computed once. The graph connecting the nodes along the cut edges may therefore be treated as a directed graph, where an outward direction from one node implies that the processor which stores that nodal value is responsible for computing the shared flux. In fine grain parallelisation [HB92] each nodal vertex is assigned to a processor, and thus the elimination of redundant flux calculations by directing the edges of the graph is essential. Unfortunately however there can be a large load imbalance if one nodal vertex has many more outward pointing edges than other vertices. For a planar graph, there is a linear time algorithm [CE91] to assure no vertex has more than three outward directed edges, but there is as yet no bound for non planar graphs. For coarse grain parallelisation, the calculation of redundant fluxes is a very small proportion of the total workload, thus the graph is generally treatable as an undirected graph. (If need be, a form of directionality can be implemented by non-symmetric mapping of halo data, but we will not consider this).

13.2.3   Mesh Partitioning for Parallel CFD

The choice of cost function used to assess partitions, and the constraints imposed on all partitions exert a strong influence on the type of decomposition sought. In turn, the architecture of the target parallel computer and the type of problem guide the choices of cost function and constraints. It is assumed that the computer is of MIMD (Multiple Instruction Multiple Data) type, with equal amounts of distributed memory per processor. At one extreme, one might seek to utilise all the available memory. Exact sharing of the computational work on each processor (perfect load balance) is then enforced, and the problem is to minimise the communication cost. Usually however the amount of memory required is less than the sum total available, so there is a weaker constraint limiting the memory available for the largest partition. Thus a degree of load imbalance is tolerable provided the overall cost is lower. Suppose there are P processors, N nodes, and the decomposition leads to M domains. Assume M = P; also note that a domain may be the union of disjoint regions, though with a high communication penalty. The execution time is given by the 'worst performing' domain, i.e. the slowest processor and the cost can be written

$$C = max_i \{c_i\},$$

with

$$c_i = [n_i * w + \sum_{k=1,K} f(b_{i,k}, d_{i,k})].$$

Here n_i is the number of nodes mapped to region i, w is the average work required to update a node, $b_{i,k}$ is the number of boundary nodes requiring communication between regions i and k and $d_{i,k}$ is the interprocessor distance from processor i to processor k. Alternatively a cost function which expresses the communication cost, number of neighbours, and degree of load imbalance can be used, such as

$$C = \alpha_1 \cdot \sum_i (n_i - N/P)^r + \alpha_2 \cdot n_b{}^s + \alpha_3 \cdot n_K{}^t,$$

where r, s, t and $\alpha_1, \alpha_2, \alpha_3$ are exponents reflecting machine architecture and problem constraints; n_b, n_K are the number of boundary nodes and domains respectively. The optimisation thus seeks to achieve for each domain:

1. even load balance
2. minimum nodes on domain boundary
3. minimum number of neighbouring domains

It is clear that in the optimal decomposition, the effect of the second term alone is to minimise the surface area of the partition whereas the third term, in minimising the number of interprocessor connections, seeks to pipeline the regions into strips. Thus the communication criteria are in some sense in opposition. Assigning one of M colours to each node, the problem may be restated in terms of colouring the graph of nodal connections, so the complete decomposition is specified by the colour of each element.

13.3   GENETIC ALGORITHMS APPLIED TO CFD

13.3.1   Two Problems

The basis of genetic algorithms has been established by Holland [Hol75], and reference to this, Goldberg [Gol89], Davis [Dav91] and other chapters of this book should provide a comprehensive introduction and many examples. Here the use of genetic algorithms in CFD is introduced by outlining their application for two problems: mesh partitioning and optimal airfoil design. Copies of programs which implement the procedures described are available on e-mail request.

13.3.2   Mesh Partitioning

A simplistic approach to applying a GA to the partitioning problem would be to initialise a population of decompositions, encoded as chromosomes. Each gene comprising an individual chromosome (decomposition) could specify the colour of an associated node. However for 64 processors, 6 bits would be required to specify a unique colour, and a given decomposition (member of the population) would have a genotype more than 60,000 bits long for a 10,000 node mesh. This is simply too large for an affordable GA. Instead the GA may either be applied indirectly to optimise the parameters of an underlying nodal colouring procedure, or the GA may be used to improve an initial solution generated by an alternative procedure. In the former case, the reduced number of parameters of the colouring scheme provides a greatly compressed structure which is amenable to an affordable GA. As an example, the genetic algorithm has been applied to optimise the performance of a low level mesh partitioning scheme based on 'attracting points'; other low level schemes could easily be devised. The procedure is not particularly efficient (as presented here) for mesh decomposition, however it is useful as a simple way of introducing GA and parallel GA approaches.

   The basis of the attracting points method is to place a number of points (centres) equal to the number of processors at different locations in the mesh, nodes are then assigned to the nearest point (processor), to produce a given decomposition. (Any node exactly equidistant from two or more centres is arbitrarily assigned to one). Assignment conflicts are rare, so there is the advantage of a nearly unique relation between domains, and positions of the points, however the shape of the allowable regions is restricted, so only a subset of all possible solutions is reachable. The tasks required to assemble a genetic algorithm can be summarised as:

- *define chromosome encoding*
- *set up initial population*
- *evaluate each individual*
- *translate to fitness*
- *regenerate population by reproduction*
- *reproduction operator selection*
- *mutation*
- *crossover*

Defining the chromosome: the parameter set for the partitioning problem corresponds to the position coordinates of a set of attracting centres; the string comprising the successive x, y, and z coordinates of the set of centres makes up an individual or chromosome.

Initial population: an initial population of partitions is be set up by specifying different sets of attracting points. Real or bit string representations may be used; in the program a real position encoding is applied.

Evaluation: the cost of the decomposition specified by each member of the population is evaluated according to the second form of the cost function defined above.

Fitness: rather than simply use the inverse of the cost to define a fitness, the population members are first ranked. Based on rank, a transformed value of fitness is assigned to each member of the population.

Reproduction: the population is regenerated in each generation by reproduction. The scheme is elitist, without duplication. To select parents the Roulette wheel model is used with selection according to rank-transformed fitness.

Operator selection: the operations of mutation and crossover are separately applied and are selected with specified probabilities.

Mutation: if selected, the mutation operator applies (with low probability) a random displacement to each centre coordinate.

Crossover: two-point crossover is used.

Applications to Unstructured Meshes

 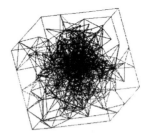

Figure 13.4   Partitioned unstructured box mesh, left shows surface, right all edge boundaries

The program is first applied to optimise the decomposition of an unstructured tetrahedral mesh of the interior of a cube, containing approximately 9,100 elements and 1700 nodes for a 12-processor computation. The above procedure is slightly modified by replacing the single point attractor with a cluster of three points, which allows greater flexibility in the shape of the decomposed regions though the computational cost increases. The resulting partition at the surface is shown in fig. 13.4. Although it appears to be quite sub-optimal, the problem is made more complex by clustering many of the elements about a point in the interior, (fig.13.4, showing edge boundaries). Comparison of the convergence of the generation best, and population average for the genetic algorithm using a population size of 100 is shown in fig. 13.5; for the same computing effort, a random search method typically produces hardly any improvement over the initial score. The processor loading and communications cost matrix are represented in fig. 13.5, where the diagonal elements represent the computational work per processor, and the off-diagonal terms represent the interprocessor communication load. Although not fully converged, a good decomposition is produced. The next example shows part of the surface decomposition for the aircraft configuration, for a 16 processor calculation after 400 generations. The clustering about the engine

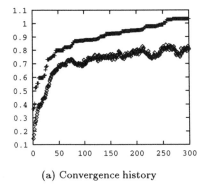

(a) Convergence history   (b) Processor loading (12 processors)

Figure 13.5

nacelle and near the wing leading edge again complicates the procedure, as shown. The decomposition is still far from optimal however, and more efficient GA approaches

Figure 13.6   Surface partitions of unstructured aircraft mesh

are considered later.

Non-Genetic Algorithm Approaches

The genetic algorithm described above, while simple to implement, will not produce a good decomposition in reasonable time if there are more than a few partitions. It is thus necessary to consider alternative approaches, either to initialise a genetic algorithm, or in combination with the genetic algorithm.

In *recursive coordinate bisection*, nodes are sorted into equal groups by position in one coordinate direction in one step; in succeeeding steps the coordinate direction is alternated and recursive subdivision applied. In *recursive graph bisection* distance is measured in the number of links required to connect nodes. Such techniques are fast, but the surface/volume ratio of some domains may be poor and the parallel efficiency suffers.

With the *greedy algorithm* [FL93], each node is weighted according to the number of connections; after randomly selecting a 'pivot node', the domain grows by adding boundary nodes (of lowest weight), promoting smoothness of the boundary. The procedure has been shown to be very fast [FL93], but it is most effective in minimising the communication cost associated with domain surface/volume ratio.

The *eigenvalue recursive bisection* procedure [BS92] is based on finding an eigenvector of the Laplacian matrix of the graph, and splitting the graph based on the magnitude of the components (corresponding to the nodes) of the eigenvector. At present it seems to be the most efficient sequential decomposition procedure.

Simulated annealing has also been applied to the partitioning problem. Comparative suggest that that given a sufficiently long run, this method produces the best decomposition.

Later in 13.5.3 the decomposition problem is reconsidered as an application of parallel genetic algorithms.

13.3.3   Optimal Airfoil Design

A few points on CFD design are given before considering the use of GAs in this area; further information can be found in [VKI94], [RJ95].

Optimal airfoil design can be categorised according to whether direct or inverse optimisation procedures are followed. In the direct approach, aerodynamic analysis is coupled to a numerical optimisation and geometry modification scheme, to search for an optimal design. In the inverse approach, a favorable pressure distribution is first sought, and the corresponding geometry is then determined by an inverse solution procedure. The former approach is more powerful, but far more demanding computationally; in the latter approach, the computation is greatly eased by restricting the scope of the design. There is a penalty however in that a specified pressure distribution may not be realisable, or may produce unforeseen disadvantageous flows. Furthermore, it usually involves modifying an initial design to approach the specified 'target' pressure distribution by an optimisation process, so that the distinction between methods can be somewhat blurred.

In both cases however, it is necessary to first define a cost function, which expresses the quantity it is required to minimise. Following Jameson [VKI94], write the cost function as $I = I(w, X)$ with w representing the flow, and X the geometry; examples are: $I = \frac{C_l^2}{C_d}$, and $I = \int (P_{target} - P_{design})^2 ds$.

Next consider the component parts of the optimisation:

1. *Flow analysis method:* the flow analysis method is a major impediment to the routine adoption of optimal design; the first decision relates to the type of analysis method to be used, i.e. Navier Stokes (expensive but more capable) or at the other extreme, a panel method (fast but potentially unreliable).

2. *Optimisation procedure:* traditional approaches use gradient search techniques, where in the simple steepest descent method, the geometry is modified in the direction of largest reduction in I. Since

$$\delta I = \frac{\partial I^T}{\partial w}\delta w + \frac{\partial I^T}{\partial X}\delta X$$

evaluation of δI requires δw, which is computed by recalculating the flow after making a small change in each design variable separately; the use of n design variables thus implies n flow evaluations. An adjoint equation approach has been developed [Jam94], in which the governing equations are introduced as a constraint, so a single solution of the adjoint equation yields the gradient. This approach considerably lessens the computational load of gradient-based optimisation procedures, and appears very suitable for combination with a GA.

3. *Geometry representation:* as it is necessary to minimise the number of design variables, the surface is usually parametrised in some way. Commonly used are B-spline representations, and localised shape functions, such as the Hicks-Henne [HH78] bump

$$b(x) = \left\{ \sin\left(\pi x^{\frac{log(.5)}{log(t_1)}}\right) \right\}^{t_2}, \quad 0 < x < 1$$

the width of the bump is controlled by t_2, and the maximum by t_1.

13.3.4   Application of Genetic Algorithms to Airfoil Optimisation

There have been a number of applications of GAs to airfoil optimisation, e.g. see [QD95],[OT95], [YI95]. The search ability and robustness of the GA appears suited to complex aerodynamic problems, where there may be highly nonlinear or discontinuous parameters. To begin, results using a simple GA optimisation program (requiring a few minutes on a fast PC) are outlined; both GA and CFD parts can readily be improved.

Simple GA airfoil optimisation

The surface of the airfoil is defined in the program by specifying the y ordinates of nodal points on upper and lower surfaces; an incompressible panel method solves for the flowfield. With leading and trailing edges fixed at $y = 0$, a string of 18 points defines a 20 panel airfoil, admittedly rather coarsely. A real number encoding is used for node height y, (scaled to an integer range). Mutation resets height to a random value in [-0.15,0.15]; population replacement avoids duplication, and is elitist with *(generation best)* and one *(mutated [generation best])* individuals automatically copied.

Two point crossover is used, with independent mutation and crossover operations, [Dav91]. Given an initial population of airfoils (of random y-ordinates), the GA may be used to search for a profile which matches a particular pressure distribution, as per inverse design. With the computed inviscid Cp distribution for a NACA 0012 airfoil input as a target pressure distribution, the results of applying the genetic algorithm (population 200) are shown in fig. 13.7. Fitness is defined as $\frac{1}{C}$, with

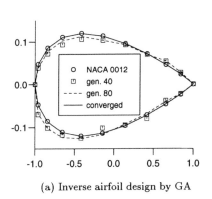

(a) Inverse airfoil design by GA

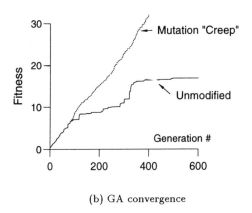

(b) GA convergence

Figure 13.7

$C = [\int (Cp_{ref} - \dot{C}p)^2 ds]^{\frac{1}{2}}$. (Geometry differences are hardly noticeable for $\frac{1}{C} \geq 15$). With the given, constant large amplitude mutation operator, the GA becomes less effective as the solution starts to converge, slowing the improvement rate. After 600 generations, the differences between target and achieved geometries are still noticeable.

If the mutation amplitude is scaled down (by 5,10,40 after generations 80, 160, 320), the convergence rate dramatically improves. Reducing mutation amplitude is a form of 'creep' [Dav91], and is useful near convergence, though in this case, a switch to gradient descent would be more efficient. It is readily demonstrated (by running the program with more panel locations) that the convergence rate drastically reduces, as the dimension of the search space increases. This and the requirement for smoothness (both to be aerodynamically acceptable, and to simplify the topography of the 'fitness landscape') make parametrisation of the geometry desirable.

Transonic airfoil optimisation, surface parametrisation

In [QD95], full potential CFD analysis and GA were combined to modify a baseline airfoil shape, and obtain shockless transonic flow. Shape modification was defined using 28 parameters each encoded in 7 bits, and single-point crossover was implemented. In airfoil optimisation often the optimal geometry may perform rather poorly outside a narrow range; multi-point optimisation (where the cost function combines objectives at several conditions) avoids selecting narrow optima. For example, two point GA optimisation was performed [QD95] to seek $max[1/(\frac{C_{d_1}}{C_{l_1}^2} + \frac{C_{d_2}}{C_{l_2}^2})]$ again for a transonic airfoil. Although effective, it needed nearly eight times the cost (in generated individuals) and more than 3 times the cost (in evaluations) of single-point optimisation.

An interesting way of defining geometry very suited to GA optimisation has been introduced by Yamamoto and Inoue [YI95], inspired by ideas of ontogeny. In their approach, a hierarchical tree structure is used to represent the entire geometry (fig. 13.8, from [YI95]). At the base or 'level 1' the position of the camber line (r_1, θ_1) and

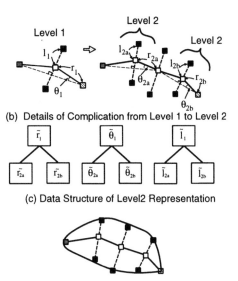

(b) Details of Complication from Level 1 to Level 2

(c) Data Structure of Level2 Representation

Figure 13.8   Encoding of airfoil geometry. Yamamoto and Inoue, (1995)
(Reproduced by permission).

thickness (l_1) are given; finer details, such as the camber and thickness at either end are specified in 'level 2' branches. The branching process continues to lower levels as required to specify the complexity of the geometry, which is thus recursively defined. Crossover involves local branch exchange, providing an effective way of combining schemata. In the work of [YI95], branching takes the form of local structure refinement via thickness and camber. Other of branch refinements could be employed (eg. separate branching for upper and lower surfaces, or final surface representation transferred to outermost branches). Although the same in principle it is worth identifying which form most efficiently provides building blocks for the GA. This work [YI95] also demonstrated the feasibility of full Navier Stokes GA optimisation of lift/drag for a supercritical airfoil, though the evaluations (70 generations, population of 64, and 160x50 grid) required substantial vector-parallel and parallel computer resources.

13.4   PARALLEL GENETIC ALGORITHMS

13.4.1   Classification of Parallel Genetic Algorithms

When implemented on a sequential machine, the GA normally operates on a single (global) population. On a parallel machine there is a choice between:
a) *preserving the use of a global population whilst parallelising the GA operations*,
b) *dividing the global population into several sub-populations and evolving a sub-population on each processor.*
Nang and Matsuo [NM94] provide a formal definition of the parallel GA with sub classifications of types a) and b). Essentially however, parallelisation may be used to:

• evaluate a single or several chromosomes in parallel

- reproduce population in parallel
- distribute population and run complete GA concurrently

Choice of single (panmictic) population or multiple demes

The first two of the above routes to parallelisation can be implemented using a single population (type a)), whilst the last uses sub-populations (type b)) is also referred to as the (parallel) distributed genetic algorithm (DGA). The issue of whether it is better to utilise a single, well-mixed ie. *panmictic* population or a number of smaller subpopulations, with some restrictions imposed on the mixing of these semi-isolated subpopulations (called *demes*) does not appear to be altogether settled, and may be problem dependent. Early DGA model studies [Tan89] demonstrated superior performance of the DGA using multiple small populations over that of a sequential GA with a single large population for a certain class of functions. Other studies have likewise shown improvements gained by use of demes; however [Bel95] found that the DGA does not always produce better results.

Amongst other factors the deme size migration rate and migration mode can have large effects, and there is as yet little to guide these choices. A recent study [GKHCP95] addresses the efficiency of multiple deme GAs for the extreme cases of ideal perfect mixing after deme convergence, and isolated non-mixing demes. The study supports the use of multiple demes, although it showed that for isolated demes, population sizes much below the single population optimum were inefficient on a serial computer. It also demonstrated GA success even with very small populations.

It has been suggested that the success of the DGA is alternately due to the 'shifting balance' theory of Wright [Wri32] or to 'species formation' and subsequent migration leading to 'speciation events', from the work of [EG72]. Premature convergence of the GA is at least partly avoided, in a sense either dynamically (by subpopulation 'drift' in the former case), and by preserving diversity in 'species' in the latter. Experience has shown that it can be difficult to avoid premature convergence of the GA, and 'nicheing' methods have been developed to preserve diversity, eg. the sequential niche technique [BBM93], restricted tournament selection [Har95]. The former is unsuitable for parallelisation, whereas the latter could be implemented in a parallel panmictic GA. Although the DGA may promote species formation it may also be useful to consider some form of niche protection, but such issues will not be considered here.

13.4.2 Constructing a Parallel Genetic Algorithm

The DGA is in a sense naturally parallel; it has an inherent advantage over the parallel single panmictic GA in that the restricting interactions between individuals in different subpopulations limits the interprocessor communication cost.

Parallel single panmictic population GA

The major issue relates to efficiency of parallel implementation, since the basic GA properties are unchanged. Assuming a total population of N, and P processors, several methods [NM94] make each processor responsible for replacing N/P individuals. However, rather than broadcasting complete information on each partial population at every generation, the parallel implementation of Tajima [Taj95], requires only fitness values to be broadcast. Local selection from the entire population fitness can be performed and then only requested chromosomes are broadcast.

Parallel DGA

There are alternative parallel DGA implementations, according to migration strategy. In the *island parallel* approach, each processor exchanges individuals with randomly chosen processors, (usually at random time intervals). It is thus suited to asynchronous MIMD implementation. For the *stepping stone* model, migration exchange only occurs between geographical neighbours (eg. North, South, East, West in a 2D processor array). Migration frequency may vary from

- never - effectively a partitioned GA
- only after local convergence [CMR91]
- randomly or at regular intervals

13.5   PARALLEL GENETIC ALGORITHMS IN CFD

13.5.1   Overview

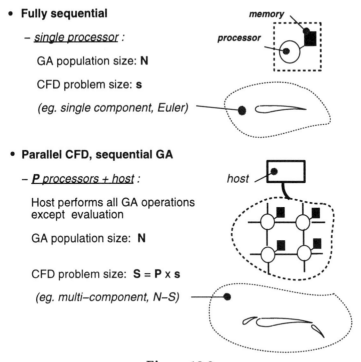

- **Fully sequential**

  - *single processor :*

    GA population size: **N**

    CFD problem size: **s**

    (eg. single component, Euler)

- **Parallel CFD, sequential GA**

  - *P processors + host :*

    Host performs all GA operations
    except evaluation

    GA population size: **N**

    CFD problem size: **S = P x s**

    (eg. multi–component, N–S)

Figure 13.9

A program combining a GA and CFD analysis can be structured so execution is purely sequential, either one operates in parallel, or both do so. Figs. 13.9 and 13.10

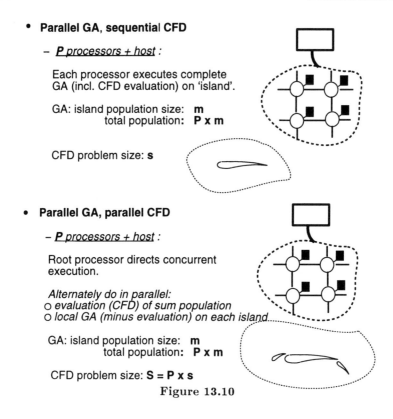

- **Parallel GA, sequential CFD**

    - *P processors + host* :

        Each processor executes complete
        GA (incl. CFD evaluation) on 'island'.

        GA: island population size: **m**
                       total population: **P x m**

        CFD problem size: **s**

- **Parallel GA, parallel CFD**

    - *P processors + host* :

        Root processor directs concurrent
        execution.

        Alternately do in parallel:
        o *evaluation (CFD) of sum population*
        o *local GA (minus evaluation) on each island.*

        GA: island population size: **m**
                       total population: **P x m**

        CFD problem size: **S = P x s**

Figure 13.10

illustrate the possibilities, for the example of GA-based airfoil or wing optimisation. In the simplest case (fig. 13.9), parallelism is used merely to accelerate the CFD and/or allow larger problems to be attempted, for example the analysis may change from an Euler computation of a single airfoil, to a Navier Stokes analysis of a multi-component airfoil. Whilst the efficiency will be limited by Amdahl's law, this may not be much of a problem if the CFD analysis greatly dominates the computing time. In the third case, (top of fig. 13.10) the execution is fully parallel, although in effect a distributed sequential CFD analysis is performed on each processor. This form of implementation is ideally suited to a parallel DGA; results using the program of section 13.3.3 adapted to a stepping-stone population exchange DGA will be discussed below. Finally, both CFD and GA can be implemented in parallel. There are several possibilities: e.g. evaluation (CFD), and GA operations may alternately execute with either panmictic or distributed populations, or processors may be grouped into clusters, with each cluster running a DGA.

13.5.2 Parallel Genetic Algorithms for Design Optimisation

Most of the applications of GAs to airfoil design have used a single population, with parallelism restricted to the evaluation by CFD, as in the lower part of fig. 13.9. It is probably simplest to implement a parallel GA-CFD optimisation using the DGA model with multiple sequential CFD evaluations as for the top example in fig. 13.10.

To do so, requires sufficient memory to allow each processor to execute independently. For panel method analyses it is unlikely that there will be insufficient memory on each MIMD processor, but there may not be sufficient to permit independent Navier-Stokes or even Euler (in 3D flow) execution. In that case, a more complex fully parallel GA-CFD implementation is necessary, at least for direct optimisation.

If however, an inverse design method of the form used by [OT95] is employed, the GA is solely used to design the optimal target pressure distribution. The method finds the shape of the pressure distribution curve which best provides a specified integrated lift, and meets a set of semi-empirical criteria; the CFD solution is merely used to recover the designed shape. For this method, it is easy to adapt the GA to be parallel, and there is no restriction on the type of CFD solver, as it is only used after completing the GA optimisation.

Optimisation using parallel DGA panel program

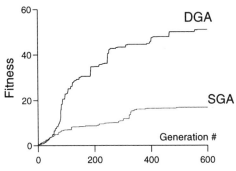

Figure 13.11   Performance of DGA

The program of 13.3.3 has been adapted to a parallel DGA with stepping-stone migrant exchange. To compare sequential panmictic and DGA results, the NACA 0012 inverse design problem is again considered. The DGA uses the same total population size of 200, which is split into 20 demes of 10 individuals, each mapped to a separate processor. After every five generations the fittest individuals from each island migrate to north, south, east and west neighbours, replacing the least fit locals. Fig. 13.11 compares convergence rates for the DGA and single population (labelled SGA) applied to the problem (both without using creep). In a series of such runs, the DGA consistently outperformed the SGA. Given that the parallel DGA also requires a fraction of the execution time, to process the same number of generations, the improvement is very large.

13.5.3   Parallel Genetic Algorithms for Mesh Partitioning

Post-generation mesh partitioning

The mesh partitioning methods described earlier are intended for an existing generated mesh. For optimisation of any basic method by a parallel DGA which uses independent evaluation on each processor, (analagous to the top paradigm in fig. 13.10) there is the disadvantage that each processor needs sufficient memory to hold the entire mesh, which is often impractical. An advantage of the attracting centres method of 13.3.2 however, is that the evaluation can easily be performed in parallel. Since the assignment is purely based on a distance metric, the elements may be distributed in any fashion among the processors; each processor being responsible for 'colouring' a portion of the elements. Either form of parallel GA may be implemented, but the basic procedure is still inefficient. It may be improved by recursively dividing the domain among two or three centres at a time; the GA can be applied to optimise each stage of the process.

Other methods are based on improving fast approximate partitions by exchanging interface nodes or elements, until the partition is optimal; simulated annealing (SA) has for example been used [FMB95] for the element exchange optimisation. Combining the parallel GA with SA, as in [MG95] appears to be an attractive procedure for accelerating such a process, but remains to be tried.

Parallel generation and partitioning

The above approach to mesh partitioning for parallel CFD computations suffers two serious deficiencies. In the first place, it is assumed that there is sufficient memory available to perform the decomposition on a single processor, the host or root processor in fig. 13.9, 13.10, whereas for very large computations on a MIMD machine, the complete mesh may not fit on any single processor. Secondly, the time required to produce the decomposition must be greatly reduced, particularly if adaptivity is employed, since remeshing and thus re-balancing may be required frequently. Both issues were addressed by Khan and Topping [KT93], who proposed a combined GA-optimised recursive bisection strategy and neural net mesh density predictor, termed the subdomain generation method or SGM [TK93], [BTK94].

The basic idea is to partition a coarse background mesh into subdomains containing relatively small numbers of elements, and subsequently generate a fine mesh in each subdomain. The partitioning of the background mesh must be such that each subdomain will contain approximately equal numbers of elements and interfacing boundary edges will be minimised. In their approach, a trained neural network was used to estimate the number of elements of the fine mesh generated per coarse mesh element by an adaptive mesh generator, given the nodal coordinates and mesh parameter values for each background mesh element. The coarse mesh decomposition is by recursive bisection, accomplished by a GA-optimized greedy algorithm. They report difficulties in training the neural net for a wide range of mesh densities, but achieved encouraging results. Clearly there is considerable scope for the application of parallel GAs to the partitioning problem.

13.6   CONCLUSION

In summary:

- for parallel CFD, a parallel GA implementation maximises combined efficiency,

- parallel computing makes the GA more affordable,
- the parallel distributed GA (parallel DGA) is simple to implement,
- the parallel DGA may outperform the GA in optimal airfoil design,
- to maximise parallel CFD efficiency, the GA may be used in optimising the parallel partitioning and generation of subdomain meshes.

REFERENCES

[BBM93] Beasley D., Bull D. R., and Martin R. R. (1993) A sequential niche technique for multimodal function optimization. *Evolutionary Computation* 1(2): 101–125.

[Bel95] Belding T. C. (1995) The distributed genetic algorithm revisited. To appear in: *Proceedings of Sixth Int'l. Conference on Genetic Algorithms.* Morgan Kaufmann.

[BS92] Barnard S. T. and Simon H. D. (1992) A fast multilevel implementation of recursive spectral bisefction for partitioning unstructured problems. NAS Systems Division. Moffett Field, CA.

[BTK94] Bahreininejad A., Topping B. H. V., and Kahn A. I. (1994) Subdomain generation using multiple neural networks models. Heriot-Watt University, Edinburgh, UK.

[CE91] Chrobak M. and Eppstein D. (1991) Planar orientations with low out-degree and compaction of adjacency matrices. *Theoretical Computer Science* 84: 243–266.

[CMR91] Cohoon J. P., Martin W. N., and Richards D. S. (1991) Genetic algorithms and punctuated equilibria in vlsi. LNCS-496, pages 134–144. Springer-Verlag.

[Dav91] Davis L. (1991) *Handbook of Genetic Algorithms.* Van Nostrand Reinhold.

[EG72] Eldredge N. and Gould S. J. (1972) Punctuated equilibria: an alternative to phyletic gradualism. in Schopf, T. J. (Ed.) Models of paleobiology, pages 82–115.

[FL93] Farhat C. and Lesoinne M. (1993) Automatic partitioning of unstructured meshes for the parallel solution of problems in computational mechanics. *International Journal for Numerical Methods in Engineering* 36: 745–764.

[FMB95] Farhat C., Maman N., and Brown G. W. (1995) Mesh partitioning for implicit computations via iterative domain decomposition: impact and optimisation of the subdomain aspect ratio. *Int. J. Num. Methd. Engng.* 38: 989–1000.

[GKHCP95] Goldberg D. E., Kargupta H., Horn J., and Cantu-Paz E. (1995) Critical deme size for serial and parallel genetic algorithms. IlliGAL Report No. 95002. Illinois Genetic Algorithms Laboratory, University of Illinois.

[Gol89] Goldberg D. (1989) *Genetic Algorithms in Search Optimization and Machine Learning.* Addison-Wesley.

[Har95] Harik G. (1995) Finding multimodal solutions using restricted tournament selection. Proceedings of the International Conference of Genetic Algorithms. ICGA95.

[HB92] Hammond S. W. and Barth T. J. (1992) Efficient massively parallel euler solver for two-dimensional unstructured grids. *AIAA Journal* 30(4): 947–952.

[HD90] Hall R. C. and Doorly D. J. (April 1990) Parallel solution techniques for compressible transonic flow. Proc. 12th Int. Conf. on Numerical Methods in Fluid Dynamics. Oxford.

[HH78] Hicks R. M. and Henne P. A. (1978) Wing design by numerical optimization. *J. Aircraft* 15(7): 407–412.

[Hol75] Holland J. H. (1975) *Adaptation in Natural and Artificial Systems.* Ann Arbor: Univ. Michigan Press.

[Jam94] Jameson A. (April 1994) Optimum aerodynamic design via boundary control. AGARD-VKI Lecture Series. Von Karman Institute, Belgium.

[KT93] Kahn A. I. and Topping B. H. V. (1993) Subdomain generation for parallel finite element analysis. *Computing Systems in Engineering Vol. 4* 4: 473–488.

[MG95] Mahfoud S. W. and Goldberg D. E. (1995) Parallel recombinative simulated annealing: a genetic algorithm. *Parallel Computing* 21: 1–28.

[NM94] Nang J. and Matsuo K. (1994) A survey on the parallel genetic algorithm. *J. SICE* 33(6): 500–509.

[OT95] Obayashi S. and Takanashi S. (1995) Genetic optimization of target pressure distributions for inverse design methods. AIAA-95-1649-CP.

[PMP93] Peraire J., Morgan K., and Peiró J. (1993) Multigrid solution of the 3d compressible euler equations on unstructured tetrahedral grids. *Int. J. Num. Meth. Engng.* 36: 1029–1044.

[QD95] Quagliarella D. and DellaCioppa A. (1995) Genetic algorithms applied to the aerodynamic design of transonic airfoils. *J. Aircraft* 32: 889–891.

[RJ95] Reuther J. and Jameson A. (1995) A comparison of design variables for control theory based airfoil optimization. Proc. 6th ISCFD Conf., pages 101–107. Lake Tahoe NV USA.

[Taj95] Tajima K. (September 1995) Improvement of interprocessor communication in a parallel genetic algorithm retaining sequential behaviour. Fourth Parallel Computing Workshop. Imperial College London.

[Tan89] Tanese R. (1989) *Distributed Genetic Algorithms for Function Optimization*. PhD thesis, University of Michigan.

[TK93] Topping B. H. V. and Kahn A. I. (1993) Subdomain generation method for non-convex domains. Information Technology for Civil and Structural Engineers. Heriot-Watt University, Edinburgh, UK.

[VKI94] (April 1994) Optimum design methods in aerodynamics. AGARD-VKI Lecture Series. Von Karman Institute, Belgium.

[Wri32] Wright S. (1932) The roles of mutation,inbreeding,crossbreeding and selection in evolution. volume 1 of *Proceedings of the Sixth International Congress of genetics*.

[YI95] Yamamoto K. and Inoue O. (1995) Applications of genetic algorithm to aerodynamic shape optimization. Paper AIAA-95-1650-CP.

14

Genetic Algorithm for Load Balancing in the Process Industry

TERENCE C. FOGARTY AND FRANK VAVAK[1]

14.1 INTRODUCTION

In this chapter we report work on two industrial applications of the genetic algorithm that involve optimising the load between a number of parallel processing elements of a production system. Because both the applications are time varying systems, possibilities of tracking changing environment has to be considered and an adaptive strategy is needed. The first application project presented describes use of the genetic algorithm to balance the load of sugar presses to optimise energy consumption. The other application of the genetic algorithm deals with a problem of balancing combustion between multiple burners in furnaces of a boiler plant which is essential for the energy efficient operation of the boilers.

14.2 CONTROL APPLICATIONS OF THE GENETIC ALGORITHM

The genetic algorithm is a proven search/optimisation technique (Holland 1975) based on an adaptive mechanism of biological systems, i.e. on the process of evolution and is often used in conjunction with other methods. In contrast to other search methods, the genetic algorithm is a robust scheme because it simultaneously evaluates many points in the search space (parallelism inherent to the genetic algorithm) and is more likely to converge toward the global solution. Moreover, it needs no assumptions that the search space is differentiable or continuous and linear in parameters and it can tolerate noisy function evaluations. The genetic algorithm can be seen as a randomised search technique which uses a random choice as a tool to guide a highly exploitative search through a coding

---

1. University of the West of England, Bristol, BS16 1QY, UK
   {tc-fogar, f-vavak}@uwe.ac.uk

of a parameter space. Given a set of points in some search space and a value or cost for each point it selects from that set, with replacement, points with a probability proportional to their value or cost. It then uses the genetically inspired operators of crossover and mutation on that set to generate a new set of points to test. Each point in the load balancing case is a list of parameter values which define a model of the system or a set of load ratios as described in the following paragraphs.

In control engineering systems, the genetic algorithms should be harnessed to address those problems which are not susceptible to efficient solution by recognised conventional approaches. The genetic algorithm can be used as an optimisation tool or as a basis of more general hybrid adaptive systems.

Broadly speaking, the application of the genetic algorithms in control engineering problems may be classified in two main areas: off-line design and on-line adaptation, learning and optimisation. In on-line application, however, the direct evaluation (through experimentation) of weak individuals may have dangerous consequences on the system controlled. Therefore a genetic algorithm is often used to operate on a model of the system, influencing the control operations indirectly. An area of the control engineering where the genetic algorithm can be successfully applied on-line, as well as off-line, is system identification (Dumont and Knistinsson 1992). Parameters of the objective function (system model) can be the system parameters in their natural form, poles and zeros or any convenient set of transformed variables.

14.3 LOAD BALANCING PROBLEMS IN INDUSTRY

There is a wide range of industrial activities that involve load balancing problems and there is currently no general approach to such problems.

Multiple element propulsion systems are frequently used in rail and sea transportation, both are areas where energy efficiency is important. Such systems require that a load be distributed across the propulsion elements which have different non-linear characteristics, and may be of differing types altogether.

Chemical production and oil refining plant frequently involve a mixture of serial and parallel processing elements. At each parallel stage there exists a situation similar to that of the beet processing described in this chapter where there is a variable load which must be distributed between multiple reaction vessels in the most efficient way. These activities are large energy users and so may provide significant cost savings for relatively small improvements in efficiency.

Power production in general has potential for optimising the load being met by a set of generators, but combined cycle power plant may be of particular interest. For example, a generator incorporating a gas turbine and a steam turbine fuelled by hot waste gases from the gas turbine must distribute the load between two turbines in an efficient way. The increasing competition in this industry in general, and between small combined cycle generator suppliers makes this a particularly interesting prospect.

There are also many food processing type applications and although there is a general tendency in this industry to rely on trusted techniques this may in fact be turned to advantage. The approach when the genetic algorithm is used to generate a model of the system allows the user to understand the internal operation of the optimiser and if necessary to countermand its "decisions".

14.4   LOAD BALANCING IN THE PRESSING OF SUGAR PULP

The objective of load balancing of the sugar beet press station is to maximize the percentage of dry substances in the pressed pulp produced by the presses. Reduced moisture of the pressed pulp improves energy efficiency of the driers which dry the pulp to be used as animal feed.

Analysis at two British Sugar plants has revealed significant variability in the effectiveness of the pressing process. It is believed that automatic optimisation of the load balancing operation with the aid of a moisture meter will improve the effectiveness of a press station from the present average weekly figures to half way towards the present best weekly figures.

The press station (Figure 14.1) consist of 8 continuous screw presses working in parallel. Performance of the presses (Figure 14.2) is non-linear and will drift as a result of the variation of upstream process as well as drain blockage and wear and tear. The faster is the press running, the less efficient it is. Whole sugar beet throughput of the plant is processed by the pressing station and load balancing in the press station is, at present, a largely manual operation. The dryness of the pressed pulp is not measured (the moisture meter in the Figure 14.1 is not used for manual control of the station) and the main requirement of the station operator is to maintain throughput. Presses 1, 2, 3 and 4 are running at a constant speed and the operator sets the speeds of the variable speed presses 5, 6 and 7 depending on the mass of sugar beet which is currently processed by the plant. Smaller changes of throughput are controlled by a traditional PID controller (proportional, integral and derivative control action). The input of the PID controller is the level of pressed pulp in press number 8 input (container/chute).

14.5   TWO APPROACHES

Two methods which use the genetic algorithm to balance the load of the presses in the sugar beet pressing station has been tested. The first approach uses the genetic algorithm to identify a mathematical model of the station which is subsequently used to maximize the percentage of dry substances in the pressed pulp produced. The second approach implements the genetic algorithm in direct control of the station with the identical goal of reducing the moisture content of the pressed pulp while meeting the constraint of factory throughput.

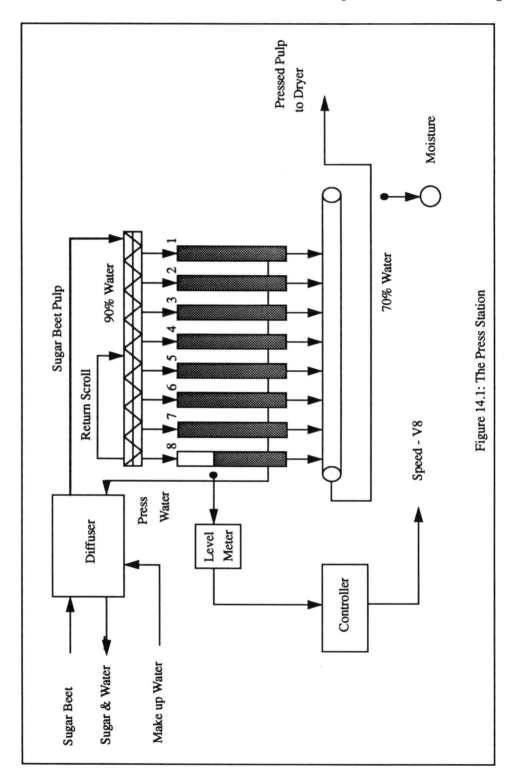

Figure 14.1: The Press Station

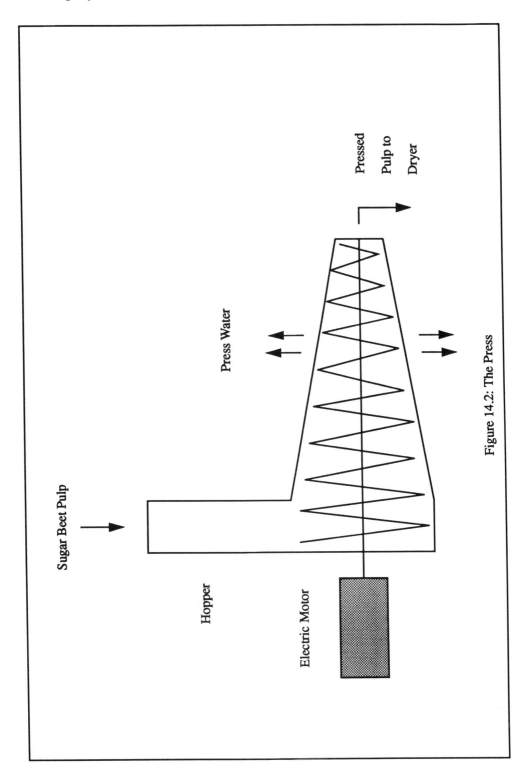

Figure 14.2: The Press

14.6   SYSTEM IDENTIFICATION FOR LOAD BALANCING IN THE PRESSING STATION

In order to optimise the load balancing of the press station using this approach, knowledge of a system model is required. It will enable a determination of the optimal load balance, i.e. optimal speed ratios R5, R6 and R7 (R5 = V5/V8, R6 = V6/V8, R7 = V7/V8; where V5, V6, V7, V8 are speeds of the presses) by using feedback signals from the system.

Analytical optimisation can cope with a simplified time invariant model of a press station. Assuming that the characteristic of each press (in terms of percentage dry substances in pressed pulp versus speed, and capacity versus speed) is known (linearly approximated) and remains unchanged, the optimum operation profile can be solved. As it is known that the performance of the presses is non-linear and time varying, it is necessary to implement an adaptive control strategy for the problem of optimisation of load balancing to be solved.

When using the system identification approach each point of the search space of the genetic algorithm is a list of parameter values which define a model of the presses. Each parameter is represented by a string of twelve bits. The best model will be used by the analytical optimisation to balance the load of the presses around the current operation condition. It will be evaluated in terms of how well it does this for comparison with other possible models.

The evaluation function used to determine the resulting cost of a chromosome (encoded list of the system parameters) has the following form:

$$\text{cost} = \frac{\displaystyle\sum_{i=1}^{\text{MINUTES}} (W - W_{\text{measured}})^2}{\text{MINUTES}}$$

- W= value of pressed pulp moisture computed from the objective function
- W measured = measured value of pressed pulp moisture
- MINUTES = number of input data records the genetic algorithm is working with

To adapt the parameters of a linearised system model to the changing environment, the genetic algorithm uses a windowed system input and output data ("window" here means data acquired during a certain time interval) as its input. The input of the genetic algorithm are values V5,V6,V7,V8 and W measured. W is calculated using the model of the pressing station as follows:

$$W = \frac{M_1 W_1 + M_2 W_2 + M_3 W_3 + M_4 W_4}{M}$$

- $M_i = C_i + D_i V_i$ is throughput of the i th press
- $W_i = A_i + B_i V_i$ is moisture of the press pulp for the i-th press.

- A_i, B_i, C_i and D_i are parameters of the system model.

The search space of the genetic algorithm for the system identification is large and processing time has been a limiting factor for on-line use of this method so far. Nevertheless, preprocessing data and increasing of its effective size (data compression while selecting a representative data set) used by the genetic algorithm can improve performance of the genetic algorithm based control system. Although the off-line testing of the genetic algorithm for system identification still continues on simulation and real data a general specification of an on-line model building and control system has been considered. The "current" model will be used by a controller until an error value (difference of measured and predicted moisture) exceeds an allowed limit. In the meantime, an up-dated model is being built by the genetic algorithm using input values V5, 6, 7, 8 and moisture measured on line.

14.7   DIRECT OPTIMAL CONTROL OF THE PRESSES

This approach to the load balancing problem uses the genetic algorithm to generate the required speed ratios R5, R6 and R7 directly and evaluation of the members of the population, i.e. chromosomes representing the three ratios, is done by means of measurement of pressed pulp moisture after a control action corresponding to the values of the ratios has been performed. Direct evaluation (through experimentation) of the weak individuals is not of critical importance for this application.

A chosen range of the generated ratios (0.85 to 1.15) ensures that no press can increase its speed over maximum value acceptable or stop working Each of the three ratios is encoded into a chromosome as four elements of a character array which have a value of either 0 or 1.

One problem, when compared with the system identification approach, is that this method cannot use the measured value of pressed pulp moisture alone as the cost function. The mass of sugar beet pulp processed by the press station is not constant over time and because the readings of speeds of the presses are not used by the genetic algorithm when generating sets of ratios directly, these changes have to be considered as disturbances acting on the input level of the system. Another disturbance acting on the system which has to be taken into account is a Ph factor of press water. Increase in the Ph value has a negative effect on pressability of sugar beet pulp and results in increased speed of the presses and consequently in increased value of pressed pulp moisture.

A modified evaluation function has to be used to eliminate/minimize the above described disturbances. Several forms of the evaluation function have been tested in an on-line control system. The best results so far in terms of convergence to an optimal set of ratios has been achieved by minimising the following cost function:

$$cost = \frac{W}{SW * Ph}$$

- W = measured value of pressed pulp moisture (wetness
- SW = value of sliced weight, i.e. mass of sugar beet pulp processed by the press station; it is measured at the front end of the diffuser (see Fig. 14.1)
- Ph = Ph factor of press water; it is measured behind an outlet of press water from the press station

An incremental genetic algorithm used for off and on-line tests is working with a population of 30 chromosomes, a crossover probability of 1.0 and a string mutation probability of 0.01. It uses tournament selection with a probability of accepting the winner of 70%. During simulation (off-line) tests with constant mass the genetic algorithm converged for different random seed values to the optimal solution after an average value of 140 generations.

The first on-line tests were run at the Bardney Sugar Factory - British Sugar plc during the last sugar campaign. The values of pressed pulp moisture, sliced weight and those of the Ph factor of press water are averaged over 3 minutes and the genetic algorithm generates a new set of ratios every 15 minutes. On the real system it is also necessary to take a time delay ("transport" delay between the time of reading values of sliced weight and an instant when sugar beet pulp is conveyed to the presses to be processed) into account. Similarly, there is a time shift between alteration of the ratios and a corresponding change in measured value of pressed pulp moisture. Values of 60 and 8 minutes are used as the two transport delays respectively. Figure 14.3 shows an example of results for on-line application of direct optimal control with the genetic algorithm using the modified evaluation function described in the previous section (moisture of pressed pulp cannot be used as an indicator of the system performance directly in this application because its values depend on time varying values of sliced weight and Ph factor). The graphs of ratios R5 and R6 are shifted by +0.5V and -0.5 V respectively to make the figure clearer. It can be seen that the genetic algorithm provides nearly converged set of ratios after approximately 160 - 300 evaluations. Further tests of the on-line control system with a modified evaluation function will be undertaken during the next sugar campaign. The new evaluation function is believed to minimize oscillations of the ratio values as a result of assumption about linear relation between changes of pressed pulp moisture and changes of sliced weight and Ph factor.

The suggested on-line optimal control system layout for the direct control is shown in figure 14.4. The block validating the currently used set of ratios has not been implemented in on-line testing yet. It will evaluate performance of the system based on the actual values of pressed pulp moisture, sliced weight and Ph factor of press water and will restart the genetic algorithm when it exceeds a limit value.

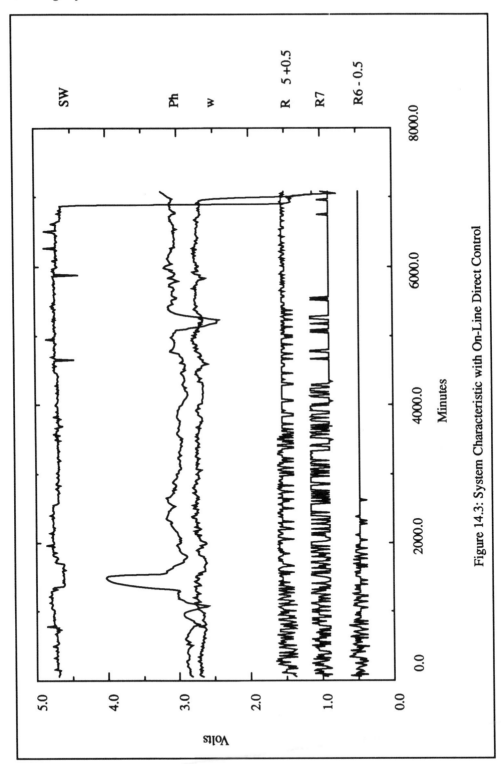

Figure 14.3: System Characteristic with On-Line Direct Control

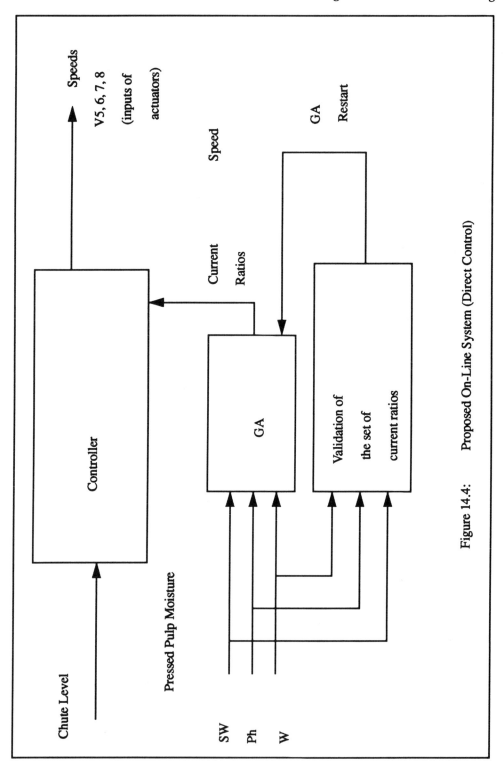

Figure 14.4:   Proposed On-Line System (Direct Control)

It is apparent that necessary restarting of the genetic algorithm from a random initial point to accommodate time changes of the working characteristics of the press station is inefficient and is a rather limiting for the use of the GA for "tracking" optima in changing environments. Studies which has been carried out in the discussed area (Dasgupta, McGregor 1992, Harvey 1992, Cobb, Grefenstette 1993) looked into two modification strategies - increasing effective diversity in the population and expanding the memory of the GA to build up a repertoire of ready responses for environmental conditions. The aim of our current research is to design a general technique for the genetic algorithm which will enable tracking optima in changing environments while this feature will not be to the detriment of the ability to provide sound results for stationary environments.

Selection of an incremental version of the genetic algorithm and a choice of the replacement strategy where the worst member of the population is replaced by an offspring was also adopted to obtain an optimal performance of the genetic algorithm in a changing environment.

14.8   LOAD BALANCING OF MULTIPLE BURNER BOILER

The automatic and continuous optimisation of combustion in large steam rising boilers using oxygen trim techniques is essential for energy efficient operation of boilers, because any pre-set balance (i.e. an optimal fuel-to-ratio balance when energy losses are minimal) is quickly lost in the harsh conditions found close to the combustion zone. In single burner system it is relatively easy to optimise air flow to the burner by measuring the levels of oxygen found in the boiler flue. However in multiple burner systems this method is less effective, since the waste gas in the flue reflects only the overall air-to-fuel ratio and not that of individual burners.

The GA-based control system for this application enables the oxygen flow to each burner to be continuously trimmed to achieve an optimal setting of the burners relative to each other using a single waste gas analyser - Figure 14.5.

The control system similar to the one used for the direct control of the sugar beet press station has been successfully tested on simulation of a multiple burner boiler with constant load. Value of stackloss (based on values of Oxygen, CO and temperature in the stack/common flue) was used as an evaluation function in this case. During initial on-line tests were used only two position of air registers: 100% of incoming air and 92% of incoming air. The tests proved that the genetic algorithm based control system can cope well with noisy environment. Despite the fuel load during the tests was changing significantly (i.e. noisy evaluation of the fitness function), the genetic algorithm found an optimal setting of air registers which was confirmed during experiments when register positions were set manually. Figure14.6 shows an example of on-line run of the genetic algorithm based optimiser where ar1, ar2, ar3 and ar4 stand for air register position (100% means fully open), T stands for temperature in the stack in degrees of Celsius, CO stands for number of particles of carbon monoxide per

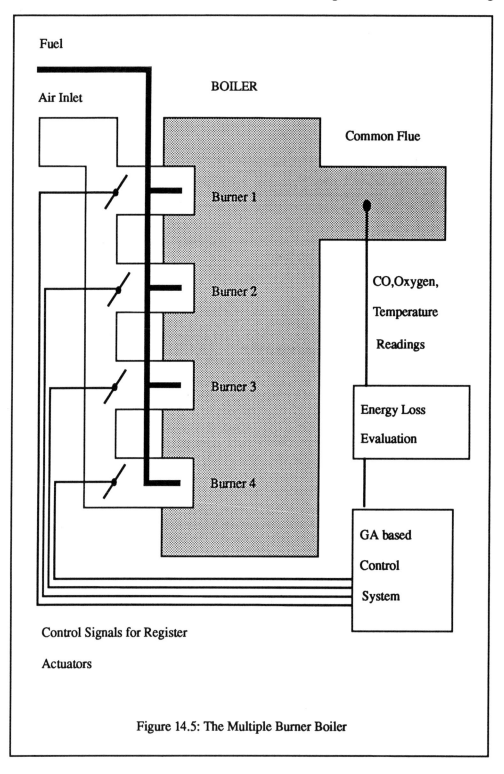

Figure 14.5: The Multiple Burner Boiler

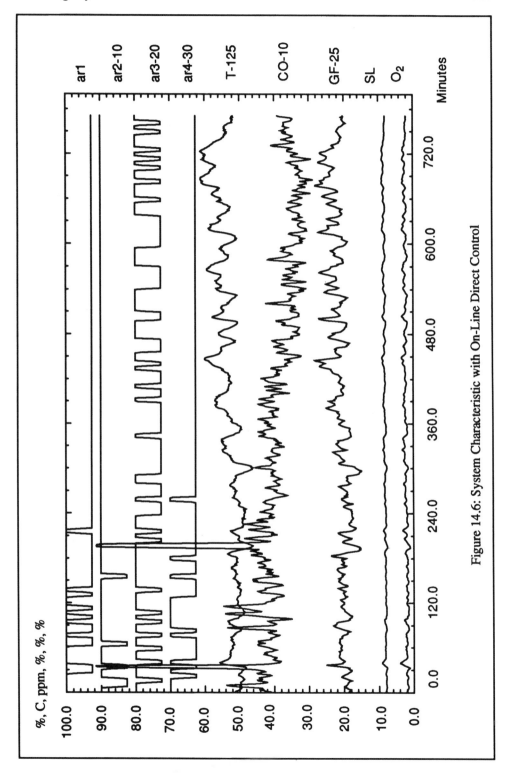

Figure 14.6: System Characteristic with On-Line Direct Control

million (ppm), GF means percentage of maximum gas flow/load, SL stands for percentage of energy lost in the stack (stack loss) and O_2 indicates percentage of oxygen in the stack.

Period of 420 minutes approximately corresponds to 50 generations of the genetic algorithm used. At this time position of air registers 1,2 and 4 are converged to settings 92%, 100% and 92% respectively. Setting of the register no. 3 was stabilized at the value 100% after 80 more generations.

14.9 CONCLUSIONS

Results which have been obtained so far both on simulations and on-line tests confirm that the genetic algorithm is a suitable method for solution of load balancing problems. Both approaches to optimal load balancing have their advantages and disadvantages described in the previous text and neither of the methods can be said to be superior over the other.

The main advantage of the system identification approach is the knowledge of mathematical description of the model of the controlled system which can be used to provide feedback to the plant managers. The values of the parameters can provide clues for fault diagnosis of the press station and early warning system to detect any degradation of the upstream process can be implemented. In contrast to this approach, the search space of the genetic algorithm used for the direct controlapproach is generally smaller, the genetic algorithm can use a small population and convergence to the optimal solution of the problem is faster without a relatively long period of prior data collection with random variation of speeds of the presses. On the other hand there is a need to monitor sliced weight and Ph factor of press water and use these values for modification of the cost function.

The preliminary evaluation of on-line tests of direct optimal control of the sugar beet press station based on data obtained so far indicates that the direct control improves performance of the press station by increasing the percentage moisture removed from the sugar beet residue during pressing by 0.623% - 0.749%. This result corresponds with British Sugar plc initial estimates of 0.635% and 0.835% improvement based on figures obtained from two different sugar plants.

It is believed that current period of on-line evaluation of the genetic algorithm-based optimiser for load balancing of the multiple burner boiler will prove a significant improvement of energy efficiency of the boilers.

The main aim of the rest of the project is to continually evolve an optimal solution to the load balancing problems which will limit the need for repeated initialization of the genetic algorithm from a random starting point in the search space when disturbances acting on the system cause variation of the parameters of the controlled systems.

REFERENCES

Holland J H (1975) "Adaptation in Natural and Artificial Systems".

Dumont and Knistinsson (1992) "System Identification and Control Using Genetic Algorithms" IEEE Transactions on Systems, Man and Cybernetics, 22 (5): 1033 - 1046, September.

Fleming P J, C.M.Fonseca (1993) "GAs in Control Systems Engineering: a Brief Introduction" - GA for Control Systems Engineering - Colloquium organised by Professional Group G7

Goldberg D E (1989), "Genetic Algorithms in Search, Optimisation and Machine Learning"- Addison Wesley

Dasgputa D, McGregor D(1992) "A Structured GA" - Technical report IKBS-8-92.

Harvey I(1993) "Species Adaptation GA" - 1st European Conference on Artificial life.

Cobb H, Grefenstette J(1993) "GA for Tracking Changing Environments" - 5th International Conference on GA.

Fogarty T C (1989) "An Incremental Genetic Algorithm for Real-time Learning" in 'Proceedings of the 6th International Workshop on Machine Learning', pp 416- 419.

Fogarty T C (1988) "Rule-based optimisation of combustion in multiple burner furnaces and boiler plants" in Engineering Applications of Artificial Intelligence, vol:1, issue:3, p.203-9.

Fogarty T C, Vavak F, Cheng P (1995) "Use of the Genetic Algorithm for Load Balancing of Sugar Beet Presses" - 6th International Conference on GA.

Fogarty T C (1989) "Learning new rules and adapting old ones with the genetic algorithm" in 'Artificial Intelligence in Manufacturing: Proceedings of the Fourth International Conference on Applications of Artificial Intelligence in Engineering', edited by Rzevski,G., Comput. Mech. Publications, Springer-Verlag, p.275-290.

Rechenberg (1973) "Evolutionstrategie: Optimierung technischer Systeme nach Principien der Biologischen Evolution".

Sutton R S (1991) "Reinforcement Learning for Animats" - 1st International Conference on Simulation of Adaptive Behaviour.

15

Solving Large Parameter Optimization Problems Using A Genetic Algorithm with Stochastic Coding

K.KrishnaKumar[1], S.Narayanaswamy [2], and S.Garg[3]

15.1   INTRODUCTION

There are many problems in engineering that require optimizing a function that depends on a large number of parameters. Problems such as these pose a great challenge to the engineer due to the large parametric space to choose from, the possibility of large infeasible and non-uniform areas, and the presence of multiple peaks. To solve these problems, one resorts to either analytical gradients or numerical searches. Some of these techniques could be time consuming either in developing the optimization approach or in their execution, and may lead to local optimum. In some instances, the problem is simplified to satisfy the optimization technique, thus sacrificing performance criteria that was desired to begin with. In the recent past, genetic algorithms have gained popularity in solving optimization problems while treating the function to be optimized as a black box [Gol89, KG92, KSM94, ZCJ92, KD92] . Genetic Algorithms (GA) have been shown to do well on problems

[1] The University of Alabama, Box 870280, Tuscaloosa, AL 35487-0280, USA;
   e-mail: kkumar@coe.eng.ua.edu
[2] The University of Alabama, Box 870280, Tuscaloosa, AL 35487-0280, USA
[3] NASA Lewis Research Center, 21000 Brookpark Road, MS 77-1, Cleveland OH 44135,
   USA; e-mail:garg@ariel.lerc.nasa.gov
   This research is being supported by a NASA Lewis Research Center Grant (NAG3-
   1564)

that contain small number of parameters and are difficult to solve using traditional techniques [KG92, KSM94, ZCJ92, KD92] . In this chapter, we will examine what causes GA to fail for large problems and we outline, implement, and evaluate a GA technique that retains the benefits of GA in solving large parameter optimization problems.

Our interest in using Genetic Algorithms for large parameter optimization problems was in part motivated by the integrated flight propulsion controller (IFPC) optimization problem as researched at NASA Lewis research center [Gar93b, Gar93a, GS93]. The performance index (PI) for the IFPC optimization problem is defined in terms of '320' state-space parameters of the partitioned airframe and propulsion controllers. The success of GA application to a problem relies heavily on the extent to which the solution space could be sampled. Large string lengths require a large population size leading to excessive function evaluations. To effectively apply GA to this problem, a procedure utilizing the non-optimal partitioned airframe and propulsion controllers was devised. In this procedure, the GA would yield the percent variation from the initial partitioning parameter values. Since the likely ranges of controller parameters were not available, this approach would reduce the string length while providing a reasonable region of search for each parameter. Still, the string lengths for 320 parameters dictate large population sizes making GA approach computationally prohibitive. The main motivation was to devise modified GA techniques that retain the random yet structured approach of Genetic Algorithms but at the same time alleviate the problems associated with large number of parameters.

This chapter is organized as follows: first we present the difficulties associated with GA for large problems. Next, the modified GA-based approach for solving large parameter optimization problems is presented and tested using a difficult test function. We also present two other techniques that were presented by the authors [KSGN95] in an earlier paper. Finally, we present the IFPC optimization problem and conclude by presenting the results of the modified GA technique to this problem.

15.2 BACKGROUND

15.2.1 Building Blocks and Schema Theorem

The processing power of GA is understood in more rigorous terms by examining the growth rates of the various schemata or similarity templates contained in a population [Gol89]. It is important to understand the concepts of schemata to better understand the GA mechanism. Some of the ideas of schemata are summarized below

1. A schema matches a string if at every location in the schema a '1' matches a '1', a '0' matches a '0', or a '*' matches either.
   Ex: Consider a string A = 101; The following schemata convey the same information as string A: A1=10*; A2=*01; A3=***; and A4=1**.
2. For alphabets of cardinality k, there are $(k + 1)^L$ schemata (L=length of the string).
3. A population of size N contains somewhere between 2^L and $N * 2^L$ schemata.

Table 15.1   Schemata for a 3–bit String

String	Order 0 Schema	Order 1 Schema	Order 1 Schema	Order 1 Schema	Order 2 Schema	Order 2 Schema	Order 2 Schema	Order 3 Schema
101	***	1**	**1	*0*	10*	1*1	*01	101
110	***	1**	**0	*1*	11*	1*0	*10	110
001	***	0**	**1	*0*	00*	0*1	*01	001

Total unique Schemata per string = 8; Total unique Schemata for all three strings = 18

4. Order of a schema, $O(H)$, is defined as the number of 0's and 1's in a schema. The defining length, $d(H)$, of a schema is the distance between the first and the last specific digit.
   Ex: Consider a schema $A1 = 1**1$; $O(A1) = 2$ and $d(A1) = 3$.
5. Holland's schema theorem: Low order, short defining length schema that are also fit (known as *building blocks*) increase exponentially in the population. His estimate of the number of schemata processed in a population of n strings is $O(n^3)$. This processing power gives the GA the computational advantage over the other traditional techniques.

15.2.2   Difficulty With Large Problems

GAs do not require gradient information and have a built-in global mechanism of search. Hence, GAs are well suited for many optimization problems. However, when a problem has several parameters, large string lengths are required for binary coding of all the parameters. For the GA search to be effective, at the very least, every possible point in the search space must be reachable from the initial population through crossover only. Mutation in general is a secondary operator and one cannot rely on it for reaching the whole search space.

As an example, let us consider reaching a point given by the three bit code 111. If the initial population consisted of 110, 000, and 010, there is no way crossover alone can reach the optimal solution. This is due to the fact that none of the three strings have a 1 in the third bit location. In terms of schema, the schema **1 is not present in the initial population. One has to rely on mutation to obtain the correct bit. When this is examined for larger strings, one can approach the analysis using probability measures. The question here is what is the probability of arriving at the right string for a given initial population that is randomly distributed? The task is more complicated if mutation is included. We will consider examining the probability of finding q^{th} order schemata in a given initial random population. By examining this, we are assuming that we have only crossover and crossover can efficiently manipulate schemata to arrive at the necessary string.

Consider a GA population of size "M". The probability that a schema with "q" fixed positions (q^{th} order schema) does not belong to the schema set "s" of a certain string of length "L" is given by:

$$p(q \in s) \;=\; 1 - [\tfrac{1}{2}]^q \tag{15.1}$$

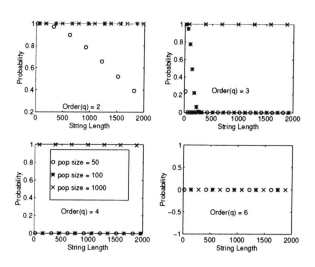

Figure 15.1   Probabilities of finding q^{th} order schemata in a given population

The probability that a schema with "q" fixed positions does not belong to the schema set "S" of all strings of length "L" in the GA population is given by:

$$p(q \in S) \;=\; [1 - (\tfrac{1}{2})^q]^M \tag{15.2}$$

Therefore, the probability that all schemata with "q" fixed positions that belong to the optimal string of length "L" will be available in the GA population of size "M" is given by:

$$p(q \in S^*) \;=\; \{1 - [1 - (\tfrac{1}{2})^q]^M\}^{\binom{L}{q}} \tag{15.3}$$

where $\binom{L}{q} = \frac{L!}{q!\,(L-q)!}$ is the number of schemata with "q" fixed positions in a string of length "L".

The probability function is plotted in Figure 15.1 for varying q, L, and M. Consider a problem in which '200' parameters are to be optimized with each parameter represented using '10' binary bits. This would lead to a string length of '2000'. In an initial population of 100, the probability that all optimal schemata of orders '2', and '4' are present are '0.999', and '0.0' respectively. This implies that the problem should be solvable by manipulating mostly 2nd order schemata. Since such optimization problems are very rare, this indicates the need for enormously large population size and hence large number of function evaluations for effective GA search. To conclude, for successful application of GA to such large problems with a reasonable number of function evaluations, the simple GA process must be modified. In the next section we propose and evaluate a modification of the simple GA process that would increase the effectiveness of GA when applied to large problems.

15.2.3 Previous Approaches For Overcoming This Difficulty

In this section, we present two other approaches that were found to provide reasonable success on large optimization problems [KSGN95].

Sensitivity Based Methods

For problems with a large number of parameters, the string length can be reduced by optimizing successively for few parameters that are relatively more sensitive with respect to the fitness function. Therefore, the GA optimization could be carried out with a reasonable population size and function evaluations.

Usually, in most problems, an approximate or a starting solution would be available. Problems for which this approximate solution is unavailable, a starting guess is sufficient. The sensitivities of parameters are computed for certain perturbations of the parameter values from the available reference solution. If the reference solution has been guessed, then investigating sensitivities of the function for more than one perturbation of parameter values would improve the resulting estimate. The multi-step GA search process is described below.

Consider optimizing a certain function of "N" parameters. This optimization could be broken into certain number of stages "s", optimizing for N_i parameters (i = 1,2,...,s) at any one time, with $N_i \subset N$.

1. We conduct a sensitivity study to investigate sensitivities of the function to be optimized (Performance Index, PI) to a predefined variation in each of the parameter (p_i) from their reference values.

$$S_i = \frac{\{|PI(p_i + 0.1p_i) - PI(p_i)| + |PI(p_i - 0.1p_i) - PI(p_i)|\}}{2PI(p_i)}$$

   Based on this sensitivity study, we choose most sensitive parameters for optimization using GA.
2. On convergence (no further improvement in best PI over a few generations) of the GA search, we repeat the sensitivity study with the best solution from previous step as the reference solution. On the basis of this sensitivity study, we choose next N_i (i = 1,2,...,s) parameters that are to be optimized and restart the GA routine.

We repeat Step 2 until the fitness reaches a satisfactory level.

In this method, since we optimize for only a few most sensitive parameters at any one time, large string lengths are no longer required. Also, since we select parameters with relatively higher sensitivities for optimization, a good improvement of the fitness could be realized within fewer function evaluations. The effectiveness of this approach generally depends on the availability of a good reference solution. If all or many of the parameters in the optimization problem are highly correlated this approach would fail.

GA-Local Search Hybrid Optimization

Previous studies have shown that GA/local search hybrid techniques often outperform either methods operating alone [Sco93]. In this approach, GA would search for optimal solution at a global level, while a direction or gradient search algorithm could be employed to search locally around each solution with good fitness provided by GA. This would allow us to have a coarser encoding structure, thereby reducing the string length for large problems.

In this method, after a population goes through the GA operators of selection, crossover, and mutation, "k" strings with best PI are chosen from the GA population and decoded into parameters. Each of these "k" solutions are used as starting guesses with a local search algorithm such as a direction or a gradient search method. The resulting optimized solutions are recoded into binary strings, and "k" strings with low fitness in the GA population are replaced by these newly recoded strings and the GA process is continued.

Since the direction search around solutions with better PI provides building blocks for the problem which would have taken a few more generations for GA to find, one would expect the technique to perform well. Unfortunately, if the solution obtained through the local search is not better than the GA provided solution, this method would result in a waste of computational effort.

15.3 GA WITH STOCHASTIC CODING

Studies conducted using dynamic coding, in which a search region is narrowed down as the search evolves, have usually failed. The attraction in dynamic coding is the shorter strings and hence fewer function evaluations. The failure of these techniques is mainly caused by discarding the regions that are not promising as evaluated based on GA performance. This leads to a possibility of losing the region that might actually have the global optimum. It will be desirable to have a dynamic coding scheme that does not discard any portion of the region but at the same time shifts emphasis to different regions of the search space. We propose a GA technique with stochastic coding. Some of the features of this approach are presented below.

1. Each discrete possibility, as decoded from the binary string, represents a region and not a single value.
2. The definitions of the regions are dynamic.
3. Region definitions are altered based on the GA evolution.
4. No region is absolutely discarded.
5. Search Region is not explicitly constrained.
6. GA never absolutely converges.

15.3.1 The Basic Algorithm

The basic algorithm is presented similar to the steps presented by Schwefel [Sch94].

STEP 0 : (Initialization) A given population consists of n individuals. Each is characterized by its genotype consisting of p genes, which determine the vitality, or fitness for survival. Each individual's genotype is represented by a binary bit string representing the object parameters value regions by means of an encoding scheme consisting of multivariate Gaussian distributions with mean vector μ and variance matrix Σ (see Figure 15.2).

STEP 1 : (Phenotype Variation) Each individual parent produces m offspring's on average, so that a total of mn new individuals are available. The phenotype of a descendant differs only slightly from that of its parents. The genotype, however, remains the same.

STEP 2 : (Filtering) Only one out of every m offspring become parents giving a total of n new individuals. The chosen phenotypes are used to redo the coding. The parameter values become the mean vector and the variance matrix is altered based on the fitness evolution.

STEP 3 : (GA Selection) Two parents are chosen with probabilities proportional to their relative position in the current population, either measured by their contribution to the mean objective function value of the generation (proportional selection) or by their rank (e.g., linear ranking selection).

STEP 4 : (Recombination) Two different offspring are produced by recombination of two parental genotypes by means of crossover at a given recombination probability p_c. Both of those offspring are taken into further consideration. Steps 3 and 4 are repeated until n individuals represent the next generation.

STEP 5 : (Mutation) The offspring eventually (with a given fixed and small probability p_m) undergo further modification by means of point mutations working on individual bits, either by reversing a one to a zero, or vice versa; or by throwing a dice for choosing a zero or a one, independent of the original value.

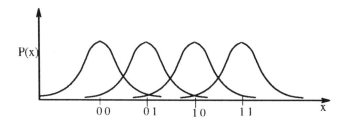

Figure 15.2   Stochastic GA Coding.

In the simplest approach, a given binary string represents a region of the search space as shown in Figure 15.2. To obtain a single point from this region, we draw a random sample from these regions using a multivariate Gaussian distribution. The statistics are initially defined with the mean equal to the midpoint of the subdivided region and standard deviation equal to half the interval of the region. As the search evolves, the strings will slowly converge to the most promising regions. At this point we have two choices: (a) Sample the most promising region with no modifications to the normal distribution; or (b) Adapt the statistics of the normal distributions and restart the GA search. The first approach is the simplest one and we have adopted this in this study. One approach for adapting the statistics will be to adopt the adaptation rules used in Evolutionary Strategies [Sch94]. This approach is more promising and research is needed to formalize a process for achieving this. Also, ways to correlate the statistics of the search regions needs to be investigated.

15.3.2 GA with Stochastic Coding, GA, and Evolutionary Strategies

Evolutionary Strategies (ES), as outlined by Schwefel, [Sch94], do not rely on manipulating building blocks and rely mostly on guided mutations. On the other hand, genetic algorithms rely mostly on building blocks manipulation through genetic operators such as crossover and inversion. GA with stochastic coding bridges the gap between ES and GA by providing a means to do both. As the number of bits representing each of the parameters increase (smaller Σ), the technique approaches simple genetic algorithms. When the number of bits equals zero (i.e., no genotype, only phenotype representation) the technique is equivalent to evolutionary strategies. This merge provides the flexibility needed to handle large problems, as well as problems with mixed types. In the results presented below, we see that GA with stochastic coding does better than GA or ES operating alone. All the tests were conducted without adapting the statistics of the multivariate Gaussian distribution.

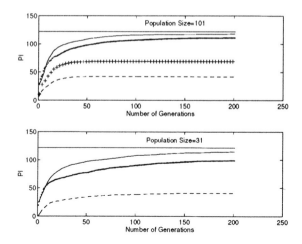

Figure 15.3 GA Search Comparison for the Test Function. In the above
figures, solid lines denote cases with Number of bits per parameter(P_B)=5;
thick solid lines denote cases with P_B=3, dashed lines denote cases with P_B=1;
and finally '+' symbol denotes regular GA.

15.3.3 A Test Function

We use a difficult test function to compare the above technique with earlier studies. This test function was proposed by [Tan94].

Table 15.2   GA Parameters used

	Regular GA	Pure Local[#]	Stochastic [##] GA	Sensitivity GA	GA–Local[$]
Generations	750	Not Applicable	200		400
# of parameters	100	100	100	100	100
Parameter's range	3 to 13	N/A	3 to 13	3 to 13	3 to 13
Chromosome length	1000	N/A	500	300	300
Population size	51	N/A	31	51	51
# of function evaluations	38,250	90,013	12,000	11,259	37,493
Best PI; Global Optimum=121.59	69	65	115	114	105

[#] Solution converged; [##] Case 3 ($N=31$; $P_B=5$) of Stochastic GA was used. Note that Case 6 gave a better result but had a larger N.
[$] Local search was carried out once in every 8 generations.

$$\underset{x_j}{max} \; PI = \sum_{i=1}^{d} [\sin x_j + \sin \frac{2x_j}{3}] ; \quad 3 \le x_j \le 13, \; j = 1, ...d.$$

For d=1, the function has two local maxima. It can be shown that for a given dimension d, the total number of local maxima is equal to 2^d. The global maximum is at $x_j = 5.3714$, j=1,...,d and the global maximum is equal to 1.2159d.

For our comparison we have used d=100, implying a total of 2^{100} local maxima and the absolute maximum is 121.59. In table 15.2, we present the details of the genetic algorithm implementation and the maximum obtained along with the number of function evaluations used. It is seen that the Regular GA did not do very well and neither did the pure local search. The modified GA approaches did well, although none of them found the true maximum. This is entirely attributed to the complexity of the search space (number of local maxima=2^{100}). While using GA with stochastic coding, we have tried six different cases. These are: (1) Population Size (N)=31; Number of bits per parameter(P_B)=1; (2) $N=31$; $P_B=3$; (3) $N=31$; $P_B=5$; (4) $N=101$; $P_B=1$; (5) $N=101$; $P_B=3$; (6) $N=101$; $P_B=5$. For phenotype variation (step 2), we have chosen $m=2$. This implies that we select two slightly different parameter sets and pick the best one as the phenotype. Figure 15.3 presents the evolution curves for all of the GA with stochastic coding runs. It is noted here that all the cases were run 5 times using different random seeds and the results presented show the average values for these 5 runs. It is clear from these figures that the combination of GA coding with GA operators and ES-like phenotype variations perform much better than the individual approaches.

15.4 APPLICATION TO AN AIRCRAFT CONTROLLER OPTIMIZATION PROBLEM

The fighter/military aircraft of future will be designed for highly enhanced maneuvering and handling capabilities. Such a need has led to the use of forces and moments produced by the propulsion system to augment the flight control functions. The Advanced Controls Technology Branch of NASA Lewis research center is actively involved in the development of technologies that will lead towards aircraft with enhanced maneuvering capabilities. An example of such an aircraft is the Short Take-off and Vertical Landing (STOVL) aircraft. There is a high degree of coupling between airframe and propulsion system in a STOVL aircraft because propulsion system forces and moments are used in flight control. Separately designed controllers for airframe and propulsion subsystems are not adequate as such a design would not account for interactions between propulsion and airframe systems of the aircraft. An Integrated Flight Propulsion Controller (IFPC), on the other hand, would account for all subsystem interactions. However, such an IFPC design results in a high-order controller that poses implementation problems. Usually, the airframe and propulsion subsystems are manufactured by different manufacturers. The engine manufacturer must be able to conduct extensive testing on the propulsion system even before it is installed in the airframe. This calls for a decentralized implementation of the IFPC system. The approach adopted at NASA Lewis consists of first designing a centralized controller, so that all subsystem interactions are accounted for in the initial design stage, and then partitioning the centralized controller into separately implementable airframe and propulsion subcontrollers for individual subsystems [Gar93a]. Then a parameter optimization procedure is used to match the closed-loop performance and robustness characteristics of the partitioned subcontrollers with those of the original centralized controller.

15.4.1 Benefits of a GA approach for these classes of problems

A gradient-based analytical optimization algorithm was proposed and used by Garg and Schmidt [GS93] . These analytical gradients are difficult to obtain due to ill conditioning of the system matrices and the matrices must be normalized before the optimization is carried out. Appropriate selection of scaling factors for normalization would call for certain prior experimentation with the system matrices. Also, such an approach provides just a single solution at the end of optimization process whereas availability of alternate solutions would provide the designer with additional implementation choices. Since GA-based optimization does not require such gradients, it can be applied with no modifications to the problem when either the PI has changed or the aircraft system has changed. In our investigation, we solve this optimization problem using the genetic algorithm with stochastic coding search procedure.

15.4.2 The Aircraft Model

The vehicle considered in this study is a representative of the delta winged E-7D supersonic STOVL airframe powered by an enhanced version of high performance turbofan engine [Gar93b] . The aircraft is equipped with the following controls:

- ejectors to provide propulsive lift at low speeds and hover; a two-dimensional Convergent-Divergent (2D-CD) vectoring aft nozzle with after burner for supersonic flight
- a vectoring ventral nozzle for pitch control and lift augmentation during transition.
- jet Reaction Control Systems (RCS) for pitch, roll, and yaw control during transition and hover.

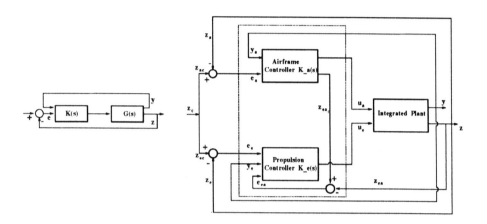

Figure 15.4 Centralized and De-centralized Hierarchical Controllers.

The flight phase considered in this study is the decelerating transition during approach to hover landing. Since in this phase the control of aircraft is transitioning from aerodynamically generated forces and moments to those generated by propulsion system, the control design is very challenging.

The linear integrated model was obtained corresponding to steady-state level flight at a trim speed of $V_0 = 80$ knots and a flight path angle of $\gamma_0 = -3^0$, at NASA Lewis [Gar93b] . The aircraft longitudinal dynamics and engine dynamics small perturbation model is of the form

$$\dot{x} = Ax + Bu; \; y = Cx + Du,$$

where the state vector is

$x = [N2, N25, Tmhpc, Tmpc, Tmhpt, Tmlpt, u, w, q, \theta, h]^T$

with

$N2 = $ Engine Fan Speed, rpm

$N25 = $ High Pressure Compressor Speed, rpm

$Tmhpc = $ High Pressure Compressor Metal Temperature,

$Tmpc = $ Burner Metal Temperature,

$Tmhpt = $ High Pressure Turbine Metal Temperature,

$Tmlpt = $ Low Pressure Turbine Metal Temperature,

$u = $ Forward velocity, ft/s

$w = $ Vertical velocity, ft/s

$q = $ Pitch Rate, rad/s

θ = Pitch Attitude, rad
γ = Flight Path Angle, rad
h = Altitude, ft

The control inputs partitioned into airframe and engine control inputs are
$u_a = [\delta_e, AQR, ANG79, ANG8]^T$
$u_e = [WF, A8, ETA, A78]^T$
with
δ_e = Elevator Deflection, degree
AQR = Pitch RCS Area, sq.in
$ANG79$ = Ventral Nozzle Vectoring Angle, degree
$ANG8$ = Aft Nozzle Vectoring Angle, degree
WF = Fuel Flow Rate, lbm/hr
$A8$ = Aft Nozzle Area, sq.in
ETA = Ejector Butterfly Angle, degree
$A78$ = Ventral Nozzle Area, sq. in

The controlled outputs and the measurement feedback for the airframe and engine
systems are
$z_a = [V_v, Q_v, \gamma]^T$; $z_e = N2$
(where $V_v = V + 0.1\dot{V}$ and $Q_v = q + 0.1\dot{q}$)
and
$y_a = [V, \dot{V}, \theta, q]^T$; $y_e = N2$
respectively. The interface from the propulsion system model to the airframe model is
defined by the gross thrust from the three engine nozzle systems, i.e.
$z_{ea} = [FG9, FGE, FGV]^T$
where
$FG9$ = Aft Nozzle Gross Thrust, lbf
FGE = Ejector Gross Thrust, lbf
FGV = Ventral Nozzle Gross Thrust, lbf

The actual system matrices used are presented in [Swa94] .

15.4.3 Partitioning of the Centralized Controller

A centralized controller design yields a high-order IFPC system that poses
implementation problems. Therefore a decentralized implementation of the IFPC
system is warranted. The control structure for the IFPC system is hierarchical with the
airframe controller generating commands for both the aerodynamic control surfaces
and propulsion system. The centralized and the decentralized hierarchical control loops
are presented in Figure 15.4. In this figure, subscripts and superscripts "a" and "e"
refer to airframe and propulsion system quantities respectively, subscript "c" refers
to commands, and the variables "z" are the controlled outputs with subscripts "e"
denoting the corresponding errors. The intermediate variables z_{ea} refers to propulsion
system quantities that affect the airframe, such as propulsion system generated forces
and moments. The controlled plant $\tilde{G}(s)$ could be put in the form

$$
\left[\begin{array}{c} \left\{ \begin{array}{c} z_a \\ y_a \end{array} \right\} \\ z_{ea} \\ \left\{ \begin{array}{c} z_e \\ y_e \end{array} \right\} \end{array} \right] = \left[\begin{array}{cc} G_{aa}(s) & G_{ea}(s) \\ G_{ea}^a(s) & G_{ea}^e(s) \\ G_{ea}(s) & G_{ee}(s) \end{array} \right] \left[\begin{array}{c} u_a(s) \\ u_e(s) \end{array} \right] = \hat{G}(s) u(s)
$$

and the centralized controller $K(s)$ in the form

$$
\left[\begin{array}{c} u_a \\ u_e \end{array} \right] = K(s) \left[\begin{array}{c} \left\{ \begin{array}{c} e_a \\ y_a \end{array} \right\} \\ \left\{ \begin{array}{c} e_e \\ y_e \end{array} \right\} \end{array} \right] ; \; K(s) = \left[\begin{array}{cc} K_{aa}(s) & K_{ae}(s) \\ K_{ea}(s) & K_{ee}(s) \end{array} \right]
$$

with the partitioning of "u" and the columns of K(s) have been rearranged to represent the grouping of the airframe and engine controller inputs. The partitioned controllers would then be
Airframe Controller:

$$
K^a(s) = \left[\begin{array}{cc} K_{aa}(s) & 0 \end{array} \right]
$$

where "0" is appropriately dimensioned zero matrix, and
Engine Controller :

$$
K^e(s) = \left[\begin{array}{cc} K_{ea}(s) & K_{ee}(s) \end{array} \right]
$$

As the block $K_{ae}(s)$ for properly designed centralized controllers will be negligibly small, compared to the block $K_{ea}(s)$, it was not included in the partitioned control structure [Gar93a] . Also, in the hierarchical structure the propulsion system does not generate any command for the airframe controller.

15.4.4   The Optimization Problem

In order to match the performance and robustness characteristics of the partitioned airframe and engine controllers with those of the centralized controller, an optimization problem was posed by Garg and Schmidt [GS93] as follows:
   Given,

$$
u(s) = K(s) \left[\begin{array}{c} e(s) \\ y(s) \end{array} \right]
$$

where,

$$
u(s) = \left[\begin{array}{c} u_a(s) \\ u_e(s) \end{array} \right], \; e = \left[\begin{array}{c} e_a(s) \\ e_e(s) \end{array} \right],
$$

and

$$
y = \left[\begin{array}{c} y_a(s) \\ y_e(s) \end{array} \right]
$$

and a set of interface variables , z_{ea}

design decentralized hierarchical airframe and propulsion subcontrollers, $K^a(s)$ and $K^e(s)$ with,

$$\begin{bmatrix} u_a(s) \\ z_{ea}(s) \end{bmatrix} = K^a(s) \begin{bmatrix} e_a(s) \\ y_a(s) \end{bmatrix},$$

and

$$u_e(s) = K_e(s) = K^e(s) \begin{bmatrix} e_{ea}(s) \\ e_e(s) \\ y_e(s) \end{bmatrix}$$

such that the closed-loop performance and robustness with the subcontrollers $K^a(s)$ and $K^e(s)$ match those of the centralized K(s) to a desired accuracy. The performance cost to achieve this was defined to be

$$J_{perf}(p) = \int_{\omega_1}^{\omega_2} \sum_k (\sigma_k[K(j\omega) - \tilde{K}(p)(j\omega)])^2 d\omega$$

where $\sigma[.]$ represents the kth singular value of a matrix, "p" represents the parameter vector to be optimized, and $\tilde{K}(p)(s)$ is the assembled partitioned subcontrollers equivalent to the centralized controller i.e.,

$$\tilde{K}(p)(s) = \tilde{C}(p)(sI - \tilde{A}(p))^{-1}\tilde{B}(p) + \tilde{D}(p)$$

The state-space representations of airframe and engine controllers, and the matrices $\tilde{A}(p), \tilde{B}(p), \tilde{C}(p), \tilde{D}(p)$ assembled using these partitioned subcontrollers are listed in Appendix B of [Swa94] .

15.4.5   GA Performance Index

The performance cost is a norm of weighted difference between the loop transfer function matrices of the centralized controller and assembled partitioned controllers. Such a formulation does not guarantee stability of the resulting closed-loop plant obtained using partitioned airframe and engine controllers. Therefore, the performance cost was modified for GA optimization as follows:

GA Performance Index (PI) = $J_{perf}(p) * (1 + Mev)$

where, Mev = Maximum of the real parts of the unstable eigen values of the closed-loop plant.

The above performance index would penalize solutions for partitioned controllers that would result in unstable closed-loop plant and thus ensure stabilizing solutions for partitioned controllers.

15.4.6   GA Results

We have used the GA with stochastic coding to optimize the IFPC problem. Each of the 320 parameters were represented using 5-binary bits leading to a string length of 1600. A GA population of 201 was used and the region of search was designated to be ± 2 times the initial partitioning parameter values. $J_{perf}(p)$ was optimized over the frequency range of $\omega \in [0.01, 100]rad/sec$ with 20 frequency points per decade and trapezoidal rule was used to evaluate the integral. The results of GA evolution is presented in Figure 15.5 along with the value obtained using the initial partitioning controller. Also, a regular GA was used with no success [Swa94]. For this application, the analytical gradient approach performed better than the GA approach. The advantages of the GA approach, as stated earlier, lie in the fact that it could be applied with no major modifications even if the system definitions changed. This will not be true in the case of the analytical gradients. It is also quite obvious that if analytical gradients are available, they are better alternatives to the GA approach. Current research is focused on improving the results obtained using adaptive statistics for the multivariate Gaussian distributions.

15.5   CONCLUSIONS

In this chapter, a modified Genetic Algorithm with stochastic coding was outlined for optimizing problems that have a large number of parameters. These modifications worked well for the test function and the aircraft controller optimization problem. Analytical gradient based methods could be used whenever such gradients are possible to obtain. To solve large parametric space problems for which such gradients are difficult to obtain, the GA with stochastic coding method discussed in this chapter could be applied with reasonable success. Especially, Stochastic GA with adaptive statistics seem to hold lot of promise and its effectiveness in solving large problems needs to be further investigated.

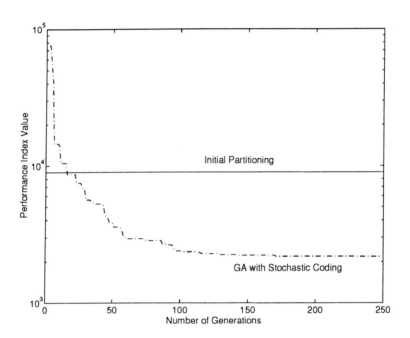

Figure 15.5   Best-so-far Fitness Evolution for the IFPC Problem.

REFERENCES

[Gar93a] Garg S. (June 1993) Partitioning of Centralized Integrated Flight/Propulsion Control Design for Decentralized Implementation. *IEEE Transactions on Control Systems Technology* 1(2).

[Gar93b] Garg S. (1993) Robust Integrated Flight/Propulsion Control Design for a STOVL Aircraft Using H–infinity Design Techniques. *Automatica* 29(1): 129–145.

[Gol89] Goldberg D. E. (1989) *Genetic Algorithms in Search Optimization and Machine Learning*. Addison–Wesley, MA.

[GS93] Garg S. and Schmidt P. H. (August 1993) Application of Controller Partitioning Optimization Procedure to Integrated Flight/Propulsion Control Design for a STOVL Aircraft. In *AIAA Guidance Navigation and Control Conference Proceedings, Monterey, CA*.

[KD92] Kristinsson K. and Dumont G. A. (september 1992) System Identification and Control Using Genetic Algorithms. *IEEE Transactions on Systems, Man and Cybernetics* 22(5): 1033–1046.

[KG92] Krishnakumar K. and Goldberg D. E. (May–June 1992) Genetic Algorithms in Control System Optimization. *AIAA Journal of Guidance, Control and Dynamics* 15(3): 735–740.

[KSGN95] Krishnakumar K., Swaminathan R., Garg S., and Narayanaswamy S. (1995) Solving Large Parameter Optimization Problems Using Genetic Algorithms. In *AIAA Guidance Navigation and Control Conference Proceedings, Baltimore, 1995*, number 95-3223.

[KSM94] Krishnakumar K., Swaminathan R., and Montgomery L. (Sept–Oct 1994) Multiple Optimal Solutions for Structural Control Using Genetic Algorithms With Niching. *AIAA Journal of Guidance, Control and Dynamics* 17(5).

[Sch94] Schwefel H. P. (1994) *Evolution and Optimum Seeking*. John Wiley and Sons, NY.

[Sco93] Scott K. A. (December 1993) Five Ways to a Smart Genetic Algorithm. *AI Expert* pages 35–38.

[Swa94] Swaminathan R. (1994) Controller Partitioning Optimization for a STOVL Aircraft using Genetic Search Methods. Masters thesis, The University of Alabama.

[Tan94] Tang Z. (November 1994) Adaptive Partitioned Random Search Methods. *IEEE Transactions on Automatic Control* 39(1): 2235–2244.

[ZCJ92] Zbigniew M., Cezary J. Z., and Jacek K. B. (1992) A Modified Genetic Algorithm For Optimal Control Problems. *Computers Math Applications* 23(12): 83–94.

16

Evolving Multiple Fuzzy Models and its Application to an Aircraft Control Problem.

K. KrishnaKumar[1] and Antony Satyadas[2]

16.1 INTRODUCTION

Various qualitative and fuzzy modeling techniques have been used by researchers for non-linear system modeling. Evolutionary fuzzy modeling involves near optimal synthesis of fuzzy rules and membership function parameters using the genetic algorithm optimization technique. Several complex problems have multiple, equal or unequal, optimal solutions. This chapter proposes Evolutionary Fuzzy Modeling with Clustered Niches (EFMCN) algorithm to discover hierarchical clusters that may be used to identify these multiple fuzzy models. The niches are identified using a sharing function that modifies the fitness of the individuals in a population based on a normalized distance measure, in conjunction with a centroid clustering algorithm. The availability of alternate solutions provides the user with the much needed flexibility in making cost effective choices. The main components of EFMCN are Genetic Algorithm (GA) module, Niching module, Clustering module, Objective function module, and Fuzzy System (FS) module.

Evolutionary modeling techniques using Genetic Algorithms (GA) was proposed by Satyadas and Krishnakumar [SK94, KGSZ95], following the efforts of Karr et. al. [KG93], Freeman et. al. [FKK90], etc. In reference [SK94], the GA was used to evolve

[1] The University of Alabama, Box 870280,Tuscaloosa, AL 35487-0280,USA;
   e-mail: kkumar@coe.eng.ua.edu
[2] Flexible Intelligence Group, LLC, P. O. Box 1477, Tuscaloosa, AL 35486-1477, USA;
   e-mail: FIGLLC@AOL.COM

near optimum fuzzy membership function parameters and fuzzy rule structure through an iterative procedure using an appropriate performance measure. This concept was successfully applied to the synthesis of a robust fuzzy controller for controlling the attitudes of a space station mathematical model.

It is a known fact that a simple GA converges to a single optimum even though multiple peaks of equal quality may exist. This creates problems in identifying those peaks. In nature, such a predicament is addressed through the formation of stable subpopulations of organisms surrounding separate niches by forcing similar individuals to share their resources. This idea has been introduced in GA literature as Niching [Deb89, GR87, Har94].

Niching helps to maintain subpopulations near global and local optima by introducing a controlled competition among different solutions near every global and local optimal region. Krishnakumar et. al [KSM94], Goldberg and Richardson [GR87] etc. have used GA with niching to identify multimodal solutions. However, when there are several closely lying niches it is difficult to identify them using visual observation as done in reference [KSM94]. It is proposed that a suitable clustering technique be applied to identify the multiple solutions among the evolved niches.

This chapter introduces the combination of three unique ideas to present a new algorithm for discovering multiple fuzzy models: Evolutionary Fuzzy Modeling with Clustered Niches (EFMCN). First we apply the idea of niching in Evolutionary Fuzzy Modeling. Next, we introduce the concept of clustered niches where the population of individuals formed after a number of generations are clustered to discover the multiple fuzzy models. Finally, we put forward the concept of hierarchical alternatives based on secondary cost functions to choose an appropriate fuzzy model.

An outline of this chapter follows. We first give background information on fuzzy concepts, genetic algorithms, GA with niching, clustering, and clustered niches. The Evolutionary Fuzzy Modeling with Clustered Niches (EFMCN) algorithm is then presented. The chapter concludes with an application of this technique to an aircraft wind shear fuzzy controller problem.

16.2   FUZZY SYSTEM CONCEPTS

A fuzzy set is a class of objects (a set of elements) in which there is no sharp boundary between those objects that belong to the class and those that do not. Each object has associated with it a number that represents its grade of membership in the fuzzy set. According to Zadeh [Zad77], the theory of fuzzy sets represents an attempt at constructing a conceptual framework for a systematic treatment of vagueness and uncertainty in both qualitative and quantitative ways.

Parallel to fuzzy set theory, Zadeh proposed a fuzzy propositional logic, based on Lukasiewicz's L Aleph 1 multivalued logic [Zad77, Zad]. Fuzzy logic allows the degree of membership in a fuzzy set to be interpreted as the truth value of a predicate. Thus the truth value of a proposition can be represented by the fuzzy set very true, true, more or less true.

Approximate reasoning is the process by which a possible imprecise conclusion is deduced from a collection of imprecise premises. We use Generalized Modus Ponens (GMP) for reasoning. The rules for the fuzzy system may be expressed as a fuzzy implicational proposition. For example let us consider an aircraft control rule:

IF θ is *Negative* **AND** V is *High* **THEN** δ_e is *Negative*

This rule represents the two states *Pitch Angle* θ and the *Velocity* V, and the action *Elevator Deflection* δ_e. The linguistic terms *Negative* θ, *High* V, *Negative* δ_e can be represented by fuzzy sets based on appropriate membership functions such as a Gaussian function. For a given value of θ and V, we can determine their memberships in the various linguistic categories by applying the corresponding membership functions. These measures of θ and V are combined using fuzzy conjunction, which amounts to taking the minimum of the membership values. In real life, no single rule matches the given input. Therefore any decision about the action (in this case, δ_e) can be taken only through approximate reasoning. In other words, given a set of rules that relates A and B to C, then given $\acute{A}$ and $\acute{B}$, you have to determine $\acute{C}$. According to Zadeh's theory of approximate reasoning [Zad77], this fuzzy implication proposition can be translated into a fuzzy subset using Lukasiewicz's implication. Lukasiewicz's implication is used to get one fuzzy subset per evaluation rule. These series of fuzzy conditions are now combined to draw a conclusion by using the compositional rule of inference. The inference is defuzzified and used as control for the aircraft. In other words, the values of θ and V are adjusted based on their current performance to maintain the flight of the aircraft. The fuzzy system building blocks are given in Figure 16.1.

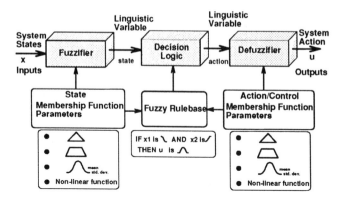

Figure 16.1   Fuzzy System Building Blocks.

In several complex problems, it is not possible to manually generate appropriate rules that describe the relationship between the variables other than by exhaustive search. In this chapter we use Evolutionary Learning of the fuzzy rules and the fuzzy membership function parameters using genetic algorithms.

16.2.1   Design Choices For Fuzzy Systems

A fuzzy system consists of the input and output (including defuzzification and linguistic mapper) modules, rulebase, and decision logic (implication and aggregation) modules. It may be noted that the term input refers to antecedent/state and output

refers to consequent/action/control. The design choices for fuzzy systems include the
following:

1. Type of fuzzy system.
2. Representation scheme.
3. Number of inputs and outputs (antecedents and consequents).
4. Names of inputs and outputs.
5. Units of inputs and outputs.
6. Universe of Discourse of the inputs and outputs.
7. Base set and Granularity of the fuzzy system.
8. Number of linguistic terms (for each input/output).
9. Names of the linguistic terms.
10. Shape of fuzzy membership functions.
11. Fuzzy membership function (fmf) parameters .
12. Learning scheme for the fmf parameters.
13. Linguistic Mapping (if a learning scheme is used).
14. Type of rule.
15. Number of rules.
16. Rule structure.
17. Learning scheme for the rule structure.
18. Fuzzy conjunction and disjunction operators.
19. Fuzzy implication methods.
20. Fuzzy aggregation operators.
21. Defuzzification methods.
22. Fuzzy Reasoning methods.
23. Learning Schemes for Fuzzy Systems.

16.3   MULTIPLE FUZZY MODELS

The problem of system identification, representation, and control for non-linear
systems has been studied using various techniques. The inherent imprecision and
uncertainty of such systems point to the soft computing approach. Soft computing
involves the integrated application of fuzzy systems, artificial neural networks, and
evolutionary computation such as genetic algorithms. Fuzzy models use a fuzzy
rulebase for representing knowledge about the system. These rules specify the
relationship between the input and output variables of the system.

Traditional fuzzy modeling techniques focus on identifying and using a particular
solution for model building. However, it is a known fact that most complex non-linear
problems have multiple solutions. Usually, the criteria (variable parameters) identified
for the model form a sub-set (the most relevant ones) of the complete variable set.
Often, conflicts, time dependent availability of information, and complexity prevents
the designer from specifying the complete variable set. Thus, model building using a
single solution imposes significant restrictions on the functionality and representation.
A single solution approach has the following restrictions.

1. Restricts the user from handling exceptions that might arise in the
   implementation process or in the future.

2. May deprive the user of the system's inherent flexibility.
3. May not be practical from an implementation point of view.
4. May not be robust if designed for performance and vice versa.

We propose and apply the EFMCN algorithm for discovering multiple fuzzy models to overcome this restriction. The following factors may be noted:

1. Discovering multiple solutions is driven by an objective/cost function.
2. Different solution sets may give the same performance.
3. Different solution sets may give different performance, with associated trade off.
4. Multiple fuzzy models can be tested for secondary cost functions to derive hierarchical alternatives.
5. The additional cost in discovering multiple solutions should be considered.

How do we identify and build multiple fuzzy models? This involves the process of deriving multiple fuzzy systems and determining how different the fuzzy systems are - in terms of the rule structure and fmf parameters. If they are different, then multiple solutions exist. Once multiple solutions are identified, the corresponding parameter values may be used to specify the multiple fuzzy models. To determine which of these models to use for defining the hierarchy, secondary cost functions can be employed. EFMCN employs an evolutionary learning technique to evolve these multiple models.

Multiple fuzzy modeling is very useful in problems that require system identification, exception handling, hierarchy of solutions, and crisis management.

16.4 EVOLUTIONARY FUZZY MODELING WITH CLUSTERED NICHES

In this section we present details of the EFMCN algorithm. We first outline the Genetic Algorithm process, the coding used for learning fuzzy models, and present in detail the inclusion of niching in a GA process. A method of clustering based on the Euclidean distance norm is defined for identifying the niches once they are formed. Finally, we present the complete EFMCN algorithm.

16.4.1 Genetic Algorithms

Genetic Algorithm (GA) is a biologically inspired highly parallel mathematical search algorithm pioneered by Holland [Hol75, Gol89]. GA generates entire population of solutions (typically fixed-length chromosome like character strings) each with associated fitness value, tests each solutions independently, and combines qualities from existing solutions to form a new population (next generation) containing improved solutions. The population can be, for example, the fuzzy membership function parameters, fuzzy rules, and so on. The fitness value is computed using the information concerning the quality of the solution produced by the members of the population (objective function values). This adaptive evolutionary learning process relates to the evolutionary selection procedure of genetic chromosomes. They rely on

Darwinian principle of reproduction and survival of the fittest and natural genetic operations such as crossover and mutation. Genetic Algorithm simulates this process, over generations, and identifies the most suitable solution (for example, membership function parameters and fuzzy rules).

16.4.2 Genetic Coding For Learning Fuzzy Models

A typical GA module consists of encoding, evolution, and decoding sub-systems. The population is represented in the GA domain as a character string of 0's and 1's. For the application considered later, we have two inputs (θ, V_A) and one output δ_e. Four linguistic variables (along with their mirror images) are used for each of the inputs and output. We use Gaussian membership functions for all the linguistic variables. For the membership function parameters, we choose four bits each for the mean and standard deviation of the Gaussian functions. These four bits translate to a resolution of 0.1333 and a range of -1 to +1 for the mean and a range of 0 to +1 with a resolution of 0.0667 for the standard deviation. Since each rule consists of two inputs and one output with four possible linguistic variables each, we have $4^3 = 2^6$ possible combinations. This translates to having a 6 bit string to represent each rule. The GA string is arranged with all the rule bits first, mean bits second, and the standard deviation bits last. For an example of 4 rules and 12 linguistic variables, we have a total string length of 120 bits (4 rules * 6 bits/rule + 12 linguistic variables * 2 parameters/variable *4 bits/parameter = 120 bits).

The population of $m - bit$ strings is first randomly generated by the GA. The decoding sub-system generates the actual parameter values, which form the output of the GA module. The FLC module then uses the parameter values set by GA module to define the fuzzy system. For each string in the population, an objective function module generates a fitness value. This sequence is repeated for all the members of the population. The next step is to use this information to breed the next generation. The evolution sub-system breeds the next generation, using reproduction, crossover, and mutation. For crossover, two independent sites are selected: one for the rules and one for the parameters. This is done to achieve good mixing of schemata.

16.4.3 Niching Using Sharing Function

A niche is viewed as an organism's (individual member of the population) environment (fitness function) and a species is a collection of organisms with similar features. A simple GA with no niching converges to a single optimum even though multiple peaks of equal quality may exist. Nature address such a predicament through the formation of stable subpopulations of organisms surrounding separate niches by forcing similar individuals to share their resources [Deb89]. Niching helps to maintain subpopulations near global and local optima by introducing a controlled competition among different solutions near every local optimal region.

Various methods for inducing niches have been cited in Deb's work [Deb89]. It includes Cavicchio's preselection scheme, De Jong's crowding scheme, Booker's bucket brigade sharing scheme, Mauldin's uniqueness operator, Perry's schemata, Stadnyk's sampling method, and Goldberg and Richardson's sharing function [GR87]. More recently, Horn [HNG94], and Harik [Har94] have applied niching in GA for multiple solution problems. In this study, niching is achieved using a sharing function since this method is identified as the best alternative [Deb89]. The sharing function creates subdivisions of the environment by degrading an organism's fitness proportional to the number of other members in its neighborhood. The amount of sharing contributed by each organism x_i into its neighbor x_j is determined by their proximity in the decoded parameter space (phenotypic sharing) based on a distance measure d_{ij}. Given p parameters of unequal boundaries over a parameter range [$x_{min} - x_{max}$] and n uniformly spaced assumed peaks,

$$d_{ij} = \sqrt{\sum_{k=1}^{p}((x_{k,i} - x_{k,j})/(x_{k,max} - x_{k,min}))^2} \qquad (16.1)$$

The limiting distance between the individuals to be shared, σ_{share}, is calculated as the average distance required to identify each niche distinctly in the solution space. This value has to be set carefully.

$$\sigma_{share} = 0.5n^{(-1/p)} \qquad (16.2)$$

For each d_{ij}, we can apply the sharing function $s(d_{ij})$ given by the equation

$$s(d_{ij}) = \begin{cases} 1 - (d_{ij}/\sigma_{share})^\omega & \text{if } d_{ij} < \sigma_{share} \\ 0 & \text{otherwise} \end{cases} \qquad (16.3)$$

We have used the power law function proposed by Goldberg and Richardson here [GR87]. One may choose $\omega=1$ since it allows equal degree of sharing between the neighboring individuals. The following properties hold good for the sharing function.

- $0 \le s(d_{ij}) \le 1$ for all d_{ij}. This condition imposes a fractional contribution to the effect.
- $s(0) = 1$; (When both individuals are identical, they will share a full portion with each other).
- $lim_{d_{ij} \to \infty} s(d_{ij}) \to 0$ (When the individuals are far apart, they produce no effect on each other).

The shared fitness of the ith individual (organism) is given as

$$SharedFitness = \frac{TrueFitness}{\sum_j s(d_{ij})} \qquad (16.4)$$

16.4.4   Clustering

Given any finite data set X of objects, the problem of clustering in X is to assign object labels that identify natural subgroups in the set. This is an unsupervised learning process based on some criteria that ideally strives to maintain homogeneous subsets. Cluster analysis is appropriate in cases where there are finite unknown number of unlabelled patterns. A serial, hierarchical, constructive, deterministic, relational clustering algorithm based on global criteria and Euclidean (hyperspherical) distance is used. The algorithm is as follows: It starts with one data point per cluster, looks for the smallest Euclidean distance between any two clusters (minimum distance pairs), merges the two clusters, thereby generating a branch in the cluster tree (a parent cluster and two sub/child clusters). The merging is repeated until only one cluster remains. This is called the unweighted pair group (centroid) algorithm. The minimum distance pairs are determined by keeping track of the nearest neighbor as well as the distance of each data point. The output of **Clustering** is a hierarchical cluster tree of the population from the final generation. This algorithm is summarized below.

1. Given a set of patterns $(P_1, P_2, .., P_i)$ in which each pattern P_i comprises of j elements. In this study, the fuzzy membership function parameters and the rule structure form the elements of the pattern. Thus each pattern represents a particular fuzzy model (a member of population from the final generation obtained from GA).

2. We compute the Euclidean distance for the ith pattern P_i as follows:

$$D_{ij} = \sum_j (P_i - P_j)^2 \qquad (16.5)$$

3. Find the minimum distance pair from the given set of patterns. The minimum distance (MD_{ij}) is computed as follows:

$$MD_{ij} = \sqrt{min(D_{i1}, D_{i2}, .., D_{ij})} \qquad (16.6)$$

4. Update the hierarchical tree of clusters. The root cluster in the tree will be the super set of all the other clusters in various branches on the tree.

5. Update the nearest neighbor patterns.

6. Repeat for all patterns i.

7. Repeat until there is only one cluster which is the superset of all the other clusters in a hierarchical order, given by the tree.

A suitable level in the hierarchy is decided by the designer. Within this level, Best in the Cluster (BinC) strategy based on the fitness of the individuals is employed to select a representative from each cluster. These representatives form the multiple fuzzy models generated by EFMCN. Best of Best in the Cluster (BoBinC) approach, again based on fitness, could be used as the secondary cost function in choosing a particular fuzzy model.

16.4.5   EFMCN Algorithm

The following steps describe a general procedure for synthesis of multiple fuzzy models using Genetic Algorithms with clustered niches. Reference [FIG95] presents a MATLAB implementation of the EFMCN algorithm.

1. Identify the data set to be modeled or the model of a system to be controlled.

2. Identify the inputs and outputs.

3. Select the type of Fuzzy Membership Function to be used (Gaussian, triangular, etc.) and identify it's parameters. This may be standardized for the system or specific to a particular linguistic variable.

4. Select the number of Fuzzy Membership Functions (FMF) and the number of fuzzy rules desired.

5. Select suitable fuzzy implication, fuzzy aggregation, and defuzzification schemes.

6. Select the coding scheme, Genetic Algorithm function parameters and performance parameters such as population, mutation rate, and crossover rate.

7. Use Genetic Algorithm to evolve the Fuzzy Membership Function Parameter values and the rule structure (Evolutionary Learning). This is an off-line supervised model synthesis process. The learning is driven by a performance measure for the fuzzy model.

8. Use sharing functions to induce niche-like behavior in the genetic search.

9. Stop the search after a fixed number of generations or computing time. Due to the inclusion of niche-like behavior, GA will not converge absolutely.

10. Perform clustering based on **Euclidean Distance** to group the multiple fuzzy models discovered through niching.

11. Select a representative from each cluster using a method such as Best in the Cluster (BinC) fitness criteria.

12. Form a hierarchical choice of fuzzy models using a secondary cost function such as Best of Best in the Cluster (BofBinC). Such secondary cost functions are usually domain dependent.

16.5 FUZZY LOGIC CONTROLLER APPLICATION

To test the EFMCN algorithm on a problem of general interest, the optimal aircraft penetration of a microburst is studied. A microburst [Kri88] is technically defined as a downburst of cool air which interacts with the earth to cause, at low altitudes, a complex wind field that may have high wind shear and dilation. The wind shear causes excessive energy loss leading to aircraft rapidly losing both altitude and airspeed. This has led to several accidents resulting in human loss. A general rule-of-thumb for flying safely through the wind shear is to hold the pitch attitude as high as possible without violating the stall limits. In this EFMCN application, we would like to examine if the EFMCN algorithm can arrive at multiple ways of achieving the end result of flying safely through the wind shear. We use minimization of maximum height loss as a performance measure for evolving the multiple fuzzy models. A secondary cost function that relates to robustness of the controller for varying microburst sizes is used to examine the resulting multiple models.

16.5.1 The Aircraft Model

In this study, a body–fixed axes system is used to define the longitudinal aircraft dynamic equations. The aircraft used for the simulation study is a Boeing B-727 aircraft powered by three JT8D-17 turbofan engines. The kinematic and dynamic body-axis equations of motion (augmented with the linearized short-period dynamics equation) are

$$\dot{X}_I = U \cos\theta + W \sin\theta$$

$$\dot{H}_I = U \sin\theta - W \cos\theta$$

$$m\dot{U} = (T\cos\delta)\delta_t + L\sin\alpha - D\cos\alpha - mg\sin\theta - mWq$$

$$m\dot{W} = (-T\sin\delta)\delta_t - L\cos\alpha + D\sin\alpha + mg\cos\theta - mUq + Z_e\delta_e$$

$$\dot{q} = M_u(U - U_0 - w_x) + M_w(W - W_0 + w_z) + M_q q + M_e\delta_e$$

$$\dot{\theta} = q$$

These equations are supplemented by the functional relations

$$T = T(V_A); \quad L = L(V_A, \alpha); \quad D = D(V_A, \alpha)$$

$$w_x = W_X(X_I); \quad w_z = W_Z(X_I)$$

and the analytical relations

$$V_A = \sqrt{(U - w_x)^2 + (W - w_z)^2}$$

$$\alpha = \arctan\left(\frac{W - w_z}{U - w_x}\right)$$

Approximations for the force terms

The thrust is approximated by a quadratic function [Kri88].

$$T = A0 + A1\,V_A + A2\,V_A^2$$

The coefficients A0, A1, A2 depend on the altitude of the runway, the ambient temperature and the engine power setting. In this study it is assumed that the engines are operating at maximum power setting and the thrust dependence on altitude is neglected. In the above equation, the coefficients are determined for a sea level condition.

The drag coefficient C_D and the total drag is given by

$$C_D = B0 + B1\alpha + B2\,\alpha^2 \qquad \alpha \leq 16\ deg.$$

$$D = \tfrac{1}{2}C_D\rho S V_A^2$$

The coefficients B0, B1 and B2 are determined from the drag curve given in [Kri88]. The lift coefficient is approximated as

$$C_L = C0 + C1\,\alpha \qquad \alpha \leq 16deg.$$

and the total lift is written in the form

$$L = \tfrac{1}{2} C_L \rho S V_A^2$$

The coefficients C0 and C1 are determined from the life curve slope (C_L Vs α curve) given in [Kri88].
The weight is regarded to be a constant.

Data for Take-off Configuration

A0 = 44560.0 lb	$0 \le V_A \le 422$ ft/sec
A1 = -23.98 lb sec/ft	$0 \le V_A \le 422$ ft/sec
A2 = 0.01442 lb sec^2/ft^2	$0 \le V_A \le 422$ ft/sec
$\delta = 2.0$ deg	
$\rho = 0.002203$ lb ft^{-4} sec^2	
S = 1560.0 ft^2	
B0 = 0.07351	$0 \le \alpha \le 16$ $deg.$
B1 = -0.08617	$0 \le \alpha \le 16$ $deg.$
B2 = 1.996	$0 \le \alpha \le 16$ $deg.$
C0 = 0.1667	$0 \le \alpha \le 16$ $deg.$
C1 = 6.231	$0 \le \alpha \le 16$ $deg.$
mg = $180,000$ lb	
$M_u = 0.0$ ft^{-1} sec^{-1}	
$M_w = -0.005$ ft^{-1} sec^{-1}	
$M_q = -0.33$ sec^{-1}	
$M_e = -0.519$ sec^{-2}	
$Z_e = -2.192$ ft sec^{-2}	
$U_0 = 271.69$ ft/sec	
$W_0 = 53.66$ ft/sec	
$\theta_0 = 0.3172$ rad	
$\delta e_0 = -0.51$ ft/sec	

Absolute Limits

Angle of attack : 16 deg;
Positive Elevator : 15 deg;
Negative Elevator: -36 deg;
Maximum thrust ratio = 0.975.

16.5.2   The Wind Shear Model

A simplistic horizontal and vertical wind defined as below was used to simulate the microburst wind shear.

$$wx = -60sin(2\pi x/Width); \quad -Width/2 < x < Width/2 \qquad (16.7)$$

$$wz = --40cos(\pi x/Width); \quad -Width/2 < x < Width/2 \qquad (16.8)$$

In the above equations, wx and wz are the horizontal and vertical wind components in a ground-fixed inertial reference frame, x is the horizontal distance along the take-off path from the center of the microburst, and $Width$ is the width of the microburst. Figure 16.2 shows three different sizes of microburst used in this study.

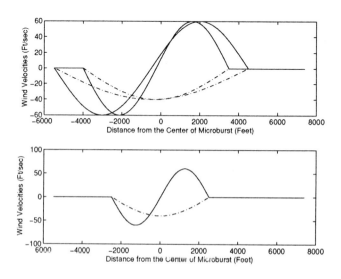

Figure 16.2 Three different simplified wind shear profiles used in this study are presented. The bottom plot shows the Size A microburst and the top plots show Sizes B and C. Solid lines represent horizontal wind and dashed lines represent vertical wind.

16.5.3   The Fuzzy Logic Controller (FLC)

The first step in developing the FLC is to determine which variables will be important in choosing an effective control action. For the wind shear control problem studied in this chapter, airspeed V_A and pitch angle θ are the decision variables and elevator δ_e is the control variable. Once the important decision and control variables have been identified, the linguistic variables that will be used to describe these variables must be defined (fuzzy sets). Four fuzzy sets and their mirror images are used to characterize each of the two decision variables and the control. The **fuzzifier** acts on the system measurements (V_A, θ) and performs the mapping of deterministic numerical data (a crisp set) into fuzzy sets. Given a measurement value for V_A and θ, the fuzzifier interprets it as a fuzzy set Fx with membership function $\mu(V_A$ or $\theta)$ with $\mu(\ ) \in [0,1]$. For the decision logic, we have used rules of the form:

$$IF\ [\theta\ is\ F\theta1\ AND\ V_A\ is\ Fv4]\ THEN\ [\delta_e\ is\ F\delta1] \tag{16.9}$$

where $Fx = [F\theta1, Fv4]$ and $Fu = [F\delta1]$ are fuzzy sets characterizing the respective variables. The mapping from the fuzzy set Fx into the fuzzy set Fu is done using Zadeh's method based on Lukasiewicz's implication. The procedure for computing the fuzzy relation curve for the action is outlined below.

For $i = 1 : N_r$
$\quad Mx = min[(F\theta j), (Fvj)]_i \qquad j \in [1, 2, .., \# of\ fuzzy\ sets]$
$\quad h_i(\delta_e) = min[1, (1 - Mx + F\delta_j(\delta_e)]_i \qquad j \in [1, 2, .., \# of\ fuzzy\ sets]$
end
For $\delta = \delta e_{min} : \delta e_{step} : \delta e_{max}$
$\quad H(\delta) = min[h1(\delta), h2(\delta), h3(\delta), ..., hN_r(\delta)]$
end

In the above procedure, $hi()$ is the fuzzy action curve for the ith rule, $H()$ is the fuzzy relation curve for the control δe, N_r is the number of rules, and δe_{step} is the discretization step size. The defuzzification process is carried out using the centroidal method of defuzzification given as follows:

$$\delta e = \frac{\int_{\delta e_{min}}^{\delta e_{max}} H(x)x\ dx}{\int_{\delta e_{min}}^{\delta e_{max}} H(x)\ dx} \tag{16.10}$$

The feedback elevator perturbation control is computed based on airspeed and pitch angle inputs to the fuzzy controller. A total of 4 rules and their mirror images were used in the FLC. The aircraft model is implemented as a software package (objective function) in the computer. GA optimization is based on a performance index supplied by the objective function is defined as:

$$J = maximum\ height\ loss + penalty \tag{16.11}$$

where

penalty$=100*$maximum $\alpha \qquad \alpha > 16$ degrees
penalty$=0.0 \qquad\qquad \alpha \leq 16$ degrees

The states and actions have the associated linguistic variables. The parameters are supplied by GA. Combination of the states and action with the associated linguistic variables form the IF THEN rules. These rules are selected by the EFMCN algorithm. Typically, a safe flight through a microburst (with throttle held at maximum value) is achieved by flying the aircraft at the maximum allowable (stall) pitch angle. In this study, we examine the capabilities of the EFMCN algorithm in providing multiple strategies for flying through the wind shear.

16.5.4   Controller Results

The genetic algorithm with Niching was executed with the following parameters: population size=77; number of generations=30; crossover rate=0.77; mutation rate=0.0077; number of assumed peaks $-$ 2; total number of parameters = 36; chromosome length = 120; and Size A wind shear (Figure 16.2). We present results for two different fuzzy controllers obtained using the EFMCN algorithm. Controller FLC-A represents the best controller in cluster-1 and FLC-B represents the best controller in cluster-2. Figure 16.3 presents the height, angle-of-attack, and elevator responses generated using the two controllers. From the figures presented, the following are observed: (a) both the fuzzy controllers performed well in maintaining the angle-of-attack within the 16 degree limit and avoiding ground contact; (b) FLC-A performed better than FLC-B in minimizing the height loss; and (c) there is an appreciable difference in the strategy used by the two fuzzy logic controllers. FLC-B uses a much smoother control activity as compared to the control activity of FLC-A.

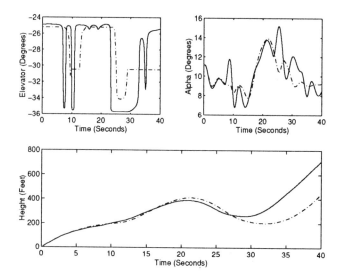

Figure 16.3 Performance comparison of the two controllers evolved by EFMCN. Solid l ines represent response characteristics of FLC-A controller and dashed lines rep resent response curves for FLC-B controller.

To identify the robustness characteristics of the two controllers for varying wind shear sizes, the same controllers were used to fly through Size B and Size C wind shear profiles. The angles-of-attack are plotted for FLC-A and FLC-B in Figure 16.4. It is clearly seen that FLC-A, which did better for Size A wind shear, did poorly for Sizes B and C. On the contrary, FLC-B did better for Size B and C wind shear profiles. Clearly, the EFMCN algorithm found an alternative that is usable for a condition that is different from the one that was used to evolve the FLCs. Here we have used the ability to maintain angle-of-attack limits for varying microburst sizes as the secondary cost function.

16.6   CONCLUSION

This chapter presented the Evolutionary Fuzzy Modeling with Clustered Niches (EFMCN) algorithm that incorporates three unique ideas. First we applied the idea of niching in Evolutionary Fuzzy Modeling. Next, we introduced the concept of clustered niches where the population of individuals formed after a number of generations are clustered to discover the multiple fuzzy models. Finally, we put forward the concept of hierarchical alternatives based on secondary cost functions to choose an appropriate fuzzy model.

The proposed algorithm is unique in that it can be applied to diverse application domains that require reasonable alternatives that permit affordable tradeoff. Typical applications include fault accommodation in aircraft control, financial forecasting, multi criteria multi objective decision making problems, aggregate production planning and scheduling, prospect evaluation in oil industry, managed health care, risk management, and routing problems. EFMCN may also be used to discover implicitly similar fuzzy models.

Future directions include exploring rule interpretation techniques and algorithms for optimal rule pruning from the GA-generated systems.

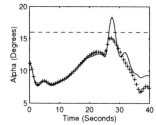 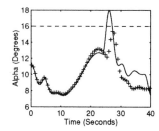

Figure 16.4   Robustness comparison of the two FLCs using Sizes B and C wind Shear profiles. Solid lines represent FLC-A performance and "+" symbols represent FLC-B performance. The dashed line represents the angle-of-attack limit.

REFERENCES

[Deb89] Deb K. (1989) *Genetic Algorithms in Multimodal Function Optimization.* The University of Alabama, Tuscaloosa, AL.

[FIG95] Flexible Intelligence Group L. (1995) *FlexTool(EFM).* Flexible Intelligence Group, LLC, Tuscaloosa, AL 35486-1477.

[FKK90] Freeman L., Krishnakumar K., and Karr C. (1990) Tuning of fuzzy logic controllers using genetic algorithms. *AAAIC'90 Conference (1990)* .

[Gol89] Goldberg D. E. (1989) *Genetic Algorithms in Search Optimization and Machine Learning.* Addison-Wesley, MA.

[GR87] Goldberg D. and Richardson J. (1987) Genetic algorithms with sharing for multimodal function optimization. *Proc. Of the Second International Conference on Genetic Algorithms* pages 41–49.

[Har94] Harik G. (1994) *Finding Multiple Solutions in Problems of Bounded Difficulty.* University of Illinois at Champaign, Champaign, Illinois.

[HNG94] Horn J., Nafpliotis N., and Goldberg D. (1994) A niched pareto genetic algorithm for multiobjective optimization. *Proc. Of the IEEE International Conference on Evolutionary Computation WCCI'94* .

[Hol75] Holland H. J. (1975) *Adaptation in Natural and Artificial Systems.* University of Michigan Press, Ann Arbor, MI.

[KG93] Karr C. and Gentry E. (1993) Fuzzy control of ph using genetic algorithms. *IEEE Transactions on Fuzzy Systems* 1(1): 46–53.

[KGSZ95] Krishnakumar K., Gonsalves P., Satyadas A., and Zacharias G. (1995) Hybrid fuzzy logic controller synthesis via pilot modeling. *AIAA Journal of Guidance, Control, and Dynamics* .

[Kri88] Krishnakumar K. (1988) *Energy Concepts Applied to Control of Airplane Flight in Wind Shear.* University of Alabama, Tuscaloosa, Alabama.

[KSM94] Krishankumar K., Swaminathan R., and Montgomery L. (1994) Multilpe optimal solutions for structural control using genetic algorithms with niching. *AIAA Journal of Guidance, Control, amd Dynamics* .

[SK94] Satyadas A. and Krishnakumar K. (1994) Ga optimized fuzzy controller for spacecraft altitude control. *Proc. Of the IEEE International Conference on Fuzzy Systems WCCI94* pages 1979–1984.

[Zad] Zadeh L.Fuzzy sets. *Information Control* 8: 338–353.

[Zad77] Zadeh L. (1977) *A Theory of Approximate Reasoning.* University of California, Berkeley, Berkeley, California.

17

Genetic Algorithm Modules in MATLAB: Design and Implementation using Software Engineering Practices

Antony Satyadas[1] and K. KrishnaKumar[2]

17.1 INTRODUCTION

The soft computing paradigm combines fuzzy, neural, and evolutionary computation techniques to build complex non-linear systems that exhibit computational intelligence. Genetic Algorithm is a biologically inspired search technique that forms an integral part of the soft computing paradigm. This chapter describes the design and implementation of a genetic algorithm software based on software engineering practices. MATLAB is used as the implementation platform. We follow the 4-stage software development cycle that consists of requirement specification and analysis, system design, object design, and implementation (includes testing and documentation). This emphasizes the interactions between theory, modeling, and application from the perspective of the developer and user. A step by step description of the development cycle, with algorithms and MATLAB code of the basic object modules are provided. The same approach may be employed to develop integrated soft computing software.

The upcoming sections are organized as follows. We first introduce the concepts

---

[1] Flexible Intelligence Group, LLC, P. O. Box 1477, Tuscaloosa, AL 35486-1477, USA;
   e-mail: FIGLLC@AOL.COM
[2] The University of Alabama, Box 870280,Tuscaloosa, AL 35487-0280, USA;
   e-mail: kkumar@coe.eng.ua.edu

of software engineering, soft computing, evolutionary computation and genetic algorithms, the object oriented paradigm, and MATLAB. The next section examines the elements, qualities, and principles of software engineering. This is followed by a description of the software development cycle. The implementation details are provided with MATLAB code for the basic GA modules. A sample problem is used to demonstrate the user interaction with the software. The chapter concludes with a discussion on future directions.

17.1.1   Software Engineering

Software engineering is concerned with building large software systems using teams of programmers [GJM91]. The emerging importance of geographically transparent, large scale application of soft computing emphasizes the need for software engineering practices that provide the required rigor and formalism for reliable and efficient performance. Development of soft computing models have grown from the researchers backyard to professional software houses. Software engineering concepts have their associated development cycles that aim to emphasize qualities such as correctness, reliability, and robustness. They are based on principles of rigor, formality, separation of concerns, modularity, abstraction, anticipation of change, and so on. This includes identification of mappings between algorithms and applications. The product, user, producer, manager, and the process interact with associated trade-off. Soft computing software development strategies require to make use of these software engineering practices.

17.1.2   Uniqueness of Soft Computing Models

Soft computing models [Zad94] are unique in several respects. They solve complex non-linear problems that exhibit traits of uncertainty and imprecision using combinations of fuzzy, neural, and evolutionary computation. The *exploratory* nature of this field makes it difficult to adhere to well defined steps. Specifications evolve with the project. Algorithms evolve with design and testing. Lack of data limits exhaustive verification and testing. Often concerns of time and space complexity are ignored to accommodate for the successful application of an algorithm. Hardware and software platforms and parallel implementation decisions are critical. Correctness, reliability, robustness, and efficiency are of concern. The implicit parallelism and complexity point to an object oriented, modular approach for software development. This chapter emphasizes on genetic algorithms; a subset of the evolutionary computation technique in the soft computing paradigm [FIG95b].

17.1.3   Evolutionary Computation and Genetic Algorithm (GA)

Evolutionary computation techniques include Genetic Algorithm, Evolutionary Strategies, evolutionary programming, and genetic programming. Genetic Algorithm (GA) is a biologically inspired highly parallel mathematical search algorithm pioneered by Holland [Hol75]. Our discussions will focus on GA.

Genetic algorithms are different from normal search methods encountered in engineering optimization in the following ways:

1. GA work with a coding of the parameter set, not the parameters themselves.
2. GA search from a population of points , not a single point.
3. GA uses probabilistic transition rules, not deterministic transition rules.

GA generates entire population of individuals (typically fixed-length chromosome like character strings based on a particular encoding scheme - such as binary, logarithmic, etc; also referred to as string, member, candidate, or point) each with associated fitness value, tests each individual independently, and combines qualities from existing individuals to form a new population (next generation) containing improved individuals. The individual represents a set of parameters (each within a specific range, characterized by their minimum, maximum, and resolution) that corresponds to a particular solution. These parameters could be, for example, fuzzy membership function parameters, fuzzy rule structure, and so on. The fitness value is computed using the information concerning the quality of the solution produced by the members of the population (objective function values). This adaptive evolutionary learning process relates to the evolutionary selection procedure of genetic chromosomes. They rely on Darwinian principle of reproduction and survival of the fittest and genetic operations like mutation [Gol89]. Genetic Algorithm simulates this process, over generations, and identifies the most suitable candidate that meets the criteria specified by the objective function. The GA operation consists of an Encoding, Evolution, and Decoding phase. The encoding phase converts a given parameter value to a sub-string of bits (0 or 1). The evolution phase cycles through the processes of selection, crossover and mutation. The decoding phase maps back the sub-strings to the corresponding parameter values. This is required to evaluate the fitness of the individual using the objective function. Various schemes are available for the operations that occur in each phase. These phases are modular with respect to their functionality and inter-dependencies. It is therefore desirable to examine the object oriented paradigm for the design and implementation of any genetic algorithm software.

17.1.4   The Object Oriented Paradigm

The object oriented paradigm identify software as a collection of discrete objects (an instance of a class) with their associated attributes (data structure) and behavior (operations) [KM90], [Weg87]. Objects with similar behavior are grouped into types. The desired characteristics include identity (in the form of objects), classification/class (objects with the same attributes and operations), polymorphism (Different behavior exhibited by the same operation that permits different methods for implementing it.), inheritance (hierarchy-based sharing of attributes and operations among classes), access controls, and predictability [RBP91]. These characteristics emphasize features such as abstraction, object structure, encapsulation (information hiding), orthogonality, feature sharing, reusability, standardization, concurrency, persistence, and efficiency. This study focuses on genetic algorithm software development. We will now examine the options for designing genetic algorithm objects.

17.1.5   Designing and Implementing GA objects

The software designer can take the application-oriented, algorithm-specific/library, or toolkit approach [FTA94]. The application oriented approach is a black box type implementation specific to a particular application. Algorithm-specific/library approach implement single or a variety of algorithms. The toolkit approach emphasizes the implementation of the software as reusable modules adhering to object oriented software engineering standards.

The notion of an object in genetic algorithms can be addressed at different levels. In the true object oriented sense, objects are instances of classes, each class being identified as a set of data structures and their associated operations. In this case, the GA objects could be instances of classes such as gene, chromosome, individual, population, and environment. The classes may inherit properties from one another. They will have unique characteristics. For example, a chromosome could be haploid or diploid, the coding could be binary, logarithmic, etc. Each class will have an associated set of operations. For example, each individual may undergo selection, crossover and mutation. This requires implementation in regular object oriented programming languages such as C++ [Chr92], Smalltalk, etc.

This chapter focuses on the toolkit approach for implementing Genetic Algorithm modules in MATLAB [FIG95a]. The notion of an object is restrictive here. The GA module can be considered as an object with a set of operations that can be used in association with other objects in MATLAB. We will address GA software development from the software engineering point of view with emphasis on building reusable modules. We follow the 4-prong design cycle that consists of requirement specification and analysis (what the system must do), system design (overall architecture), object design (data structures and algorithms for implementing classes), and implementation (GA in MATLAB). We will now examine the features of MATLAB that can enable the design and development of software engineered GA objects.

17.1.6   MATLAB

MATLAB [TM94] provides an interactive computation intensive environment. The high level, user friendly programming language combined with built-in functions to handle matrix algebra, Fourier Series, and complex valued functions provide the power for large scale number crunching. MATLAB provides a system transparent interoperable environment that allows the developer to build modular, reusable, application packages, while drawing on the modular resources provided by MATLAB. MATLAB provides implicit data type definition facility. The high level language permits writing source code (.m files) that can be directly executed in the MATLAB workspace. The user can define their own functions and script files. The vector and matrix classes with their associated operations are available for immediate use. All the window-based graphic user interface objects come with a defined set of operations. Reusable building blocks are provided for various mathematical computations. This includes sorting and random number generator routines. The toolbox approach allows modular development and addition of diverse features within the same environment. Various, easy to build, data visualization and menu driven GUI interface modules are available for providing user interaction. In short, MATLAB provides a good environment for developing application software using software engineering concepts. Several features of object oriented programming are already available in MATLAB.

17.2   ELEMENTS, QUALITIES, AND PRINCIPLES OF SOFTWARE ENGINEERING ·

Requirement analysis, specification, designing, coding, module and system testing, integration, delivery, maintenance, and risk evaluation form intermediate steps of the software development cycle. The dynamic nature of soft computing models demand the flavor of a spiral model [GJM91]. This applies to genetic algorithms also. The need of global standards for soft computing is critical in a group development environment. Specification approaches may be as simple as just data flow diagrams, charts, and so on or highly descriptive options such as model based or algebraic. The design should ensure robust and reliable performance. The strength of soft computing models is its inherent flexibility, quite typical of human cognition and intelligence. Interoperability is an essential quality for this purpose. Algorithmic efficiency and robustness should be considered when options for membership and activation functions, rules, learning methods, search techniques, interpolations, and matrix manipulations are considered. Choice of programming language, data structure, dynamic and static operations, hardware platform, storage media, and cost are of concern here. From the GA perspective, concerns include random number generators, sorting algorithms, matrix operations, sequential/parallel computation, selecting from a pool of alternatives, and implementation of coding schemes. Module testing, system testing, and verification should satisfy the correctness, reliability, and robustness criteria. Testing choices include specification dependency (black box) and code dependency (white box). The only related area where verification and validation issues have been considered is in Knowledge Base Systems. This is a topic for further research in the soft computing arena.

The desirable qualities of a software engineering based soft computing model include correctness, reliability, robustness, efficiency, user friendliness, verifiability, maintainability, repairability, evolvability, reusability, portability, understandability, interoperability, productivity, timeliness, and visibility. Development of standards in a modular object oriented framework improves reusability. Evolvability is governed by the availability of appropriate software building blocks that accommodates cost effective changes to meet changing requirement specification. This requires a well thought out planning and design phase that gives emphasis to the *anticipation of change* phenomenon. Software evolutions reach the user in the form of software updates, usually patches or new versions. This should be a well documented process that eases the burden off the user. Ease of maintainability and reparability of the software depend upon it's evolvability. Portability concerns different hardware platforms and operating systems. Appropriate Graphic User Interface (GUI) visualization aids enhances user friendliness. Developing understandable software requires consideration of both the user's as well as the designer's and producer's perspective. Interoperability is the ability of the software modules to co- exist and cooperate with other modules. The hybrid nature of soft computing insists on interoperability. For example, in the GA software, links have to be provided to other soft computing modules such as fuzzy systems and artificial neural networks. All these factors influence the productivity and timeliness of software development.

Principles of software engineering include rigor, formality, separation of concerns, modularity, abstraction, anticipation of change, generality, and incrementality [GJM91]. Rigor is an intuitive quality that enhance creativity by raising confidence levels brought about by critical analysis. Formality is the highest degree of rigor that enforces mathematical laws in the software process. Separation of concerns can be augmented by modular development with appropriate abstraction. This abstraction would enhance the productivity of soft computing scientists and engineers. Soft computing in general and GA in particular is a growing discipline. Success of software implementations depend upon the flexibility provided in anticipation to change using generic structures that accommodate for incremental development.

Software engineering practices using object oriented principles, in conjunction with standards, ease the process of creative invention and provide the avenue for quantum growth.

17.3 DESIGN AND DEVELOPMENT CYCLE

We follow the 4-stage design and development cycle: (a) requirement specification and analysis, (b) system design, (c) object design, and (d) implementation.

17.3.1 Requirement Specification

The GA software requirement is addressed in terms of (a) functionality, (b) ease of development, (c) software engineering, (d) user interaction. The following requirements are deemed desirable for developing the software.

1. Functionality

   (a) Various types of GA.
   (b) Options for coding schemes.
   (c) Various selection schemes.
   (d) Options for reproduction operators.
   (e) Facility to evolve multiple solutions [KS95].
   (f) Handle multiple objectives.

2. Software Engineering

   (a) Reusable modules.
   (b) Developer
       i. Ease of development
       ii. Upgradability
       iii. Ease of maintenance
       iv. High level programming language
   (c) User
       i. ease of operation
       ii. accommodate interruptions
       iii. graceful degradation

    iv. handle exceptions
    v. User friendly (user interface, readability, ease of learning).
    vi. Reliable, robust, efficient, upgradable.
    vii. Transparent to a wide range of hardware and operating systems.
    viii. High level programming language

17.3.2 Analysis

Requirement analysis leads us to the following specification of the structural and functional categories, for evolutionary computation in general and genetic algorithms in particular [FIG95b].

1. Functional Description
2. Principles (Theory)
3. Mathematical Procedures
4. Architecture
5. Numerical Values
6. Variable Parameters
7. Data Processing (pre and post) Procedures
8. Coding Procedures
9. Randomization Procedures
10. Selection Procedures
11. Reproduction Procedures
12. Sorting Procedures
13. Decoding Procedures
14. Stopping Procedures
15. Special purpose procedures
16. Objective Function
17. Validation Procedures

A paradigm is properly defined by the specification of a set of such structures and procedures satisfying performance goals. Refer [FIG95b] for more details.

1. Functional Description: Functional description of the EC has a mathematical model or data, simulation/implementation goals, and performance measures. The resulting structured implementation includes mathematical procedures, architecture, numerical values, variable parameters, and various procedures.
2. Principles (Theory): Understanding the underlying principles is critical for ease of implementation, error correction, and upgrades. The following topics are identified as GA principles that are useful.

    (a) Evolution
    (b) Survival of the fittest (Darwinism)
    (c) Schema, Schema theorem, building block hypothesis
    (d) Learning
    (e) Adaptation
    (f) Biological associations
    (g) Searching

   (h) Optimization
   (i) Hill Climbing
   (j) Bias
   (k) Deception
   (l) Hamming Cliff

3. Mathematical Procedures: Mathematical procedures in the GA context include coding, decoding, crossover, mutation, and selection procedures. These procedures require random number generators, sorting algorithms, and several other built-in functions such as exponential and logarithmic functions.

4. Architecture: Architecture is the set of physical or data structures determining the data flow.

   (a) Type:
      i. Genetic Algorithm (GA) Types: Generational, Steady State, Micro; Sequential, Parallel
      ii. Evolutionary Programming (EP)
      iii. Genetic Programming (GP)
      iv. Evolutionary Strategies (ES)
      v. Coding - Binary, Logarithmic, real, floating point, Gray
      vi. Chromosome - Haploid, Diploid

   (b) Number of Chromosomes
   (c) Hybrid links

      i. Fuzzy Systems
      ii. Artificial Neural Networks
      iii. Chaos Dynamics

   (d) Numerical Values: Numerical values in the EC include data to be processed and/or saved and the EC parameters. Input data - parameter values that need to be optimized, ranges, probabilities, EC parameter values, and generational statistics.

   (e) Variable Parameters: Data associated with states and structures may include inputs, coded parameters, parameter values in each generation, and others.
      i. Population Size
      ii. String Length
      iii. Number of generations
      iv. Tournament size (if applicable)
      v. Crossover rate
      vi. Mutation rate

   (f) Data Processing (pre and post) Procedures: This includes normalization procedures, scaling procedures and other techniques for pre-processing and post-processing data.

   (g) Coding Procedures: Coding procedures are algorithms that define the mapping between the input parameters and the chromosome representation.
      i. Binary

  ii. Logarithmic
  iii. Real
  iv. Floating Point
  v. Gray

(h) Randomization Procedures: Required for (a) generating initial population, (2) applying crossover and mutation rates, and (c) applying selection strategies.

(i) Selection Procedures: Selection procedures specify how the mates for next generation are selected.

  i. Tournament
  ii. Roulette Wheel
  iii. Ranking

(j) Reproduction Procedures: Reproduction procedures specify how the selected mates mate to generate their offspring.

  i. Mutation
  ii. Reordering
  iii. Splicing

(k) Sorting Procedures

(l) Decoding Procedures: Decoding procedure specify the mapping between the chromosome representation and the actual parameter value (the final or intermediary output of EC). It depends upon the coding.

(m) Stopping Procedures:

  i. Stop after a number of generations
  ii. Stop after a certain time of computation
  iii. Stop when the average is close to the minimum (or maximum) value
  iv. Stop when there is no improvement in the minimum (or maximum) value
  v. Stop after finding a better solution than the one before (based on some criterion)
  vi. Stop after finding the desired minimum (or maximum)

(n) Special purpose procedures:

  i. Niching for evolving multiple solutions [KS95]
  ii. Multiple Objective Optimization [FIG95a]

(o) Objective Function:

  i. Performance Index
  ii. Penalty functions
  iii. Constraints

(p) Validation Procedures:

  i. for performance
    A. Robustness: Performance, Stability
    B. Consistency
    C. Efficiency
    D. Cost

    E. Speed (time)
   ii. Accuracy
  iii. Reliability
  iv. Global and local optimum
   v. Allelic Loss

17.3.3 System Design

The following system design categories are identified.

(a) Objective function
(b) Function parameters: Parameter ranges (minimum, maximum), parameter resolution, number of parameters.
(c) GA parameters: Type of GA, coding, etc.
(d) System operation: decode and evaluate, select, reproduce, cycle through
(e) Data Outputs: Optimized parameter values, objective function values, etc.
(f) Compute and save statistics
(g) Graphical User Interface (GUI)
(h) Plot and print results
(i) Load and save data

The analysis of requirement specification brings into focus two separate aspects of GA: (a) Function specific (b) Parameter specific. The functional aspects include details such as (1) Name of an initialization function, (2) Name of the objective function, (3) Number of objectives, (4) Type of optimization (minimization or maximization), (5) input (parameter) details (number of parameters, maximum value, minimum value, resolution). The parameter specific factors include (1) type of GA, (2) population size, (3) stopping criteria, and (4) crossover and mutation rates.

17.3.4 Module Design

Modules consists of data structures and operations. We have employed the vector and matrix class provided by MATLAB. The following module designs are considered here.

(a) FTcode: Defines the binary coding based on the minimum, maximum, and resolution desired
(b) FTinitpop: Initializes the population at random
(c) FTgen: Generates a new population
(d) FTselect: Selection module
(e) FTcross: Crossover module
(f) FTmutate: Mutate module
(g) FTdecode: Converts Genotype to Phenotype (binary string to real parameters)

Algorithms for these modules follow.

```
00 ============================================================
10 CODING (FTcode)
11 Compute range based on maximum and minimum
12 Compute binary coding for each parameter based on range and resolution
13 Concatenate to form a single string representation
99 ============================================================


00 ============================================================
10 GENERATION (FTgen)
11 Select new individuals based on selection scheme
        from the old population using PI (Objective Function)
        values as a measure
12 For each pair of individuals
     {
13        Perform crossover
14        Perform mutation
     }
15 Return the newpop (new population) as the next generation
99 ============================================================


00 ============================================================
10 TOURNAMENT SELECTION (FTselect)
11 Calculate fitness for all members of the population
      (using objective function).
12  Given Tsize, the tournament size
        Randomly pick Tsize members from the population with
        no replacement.
13  Pick the best member among the Tsize members
14  Repeat till the population list is empty.
15  Repeat steps 12-14 till the new population size equals the
        old population size
99 ============================================================


00 ============================================================
10 DETERMINISTIC ROULETTE WHEEL SELECTION (FTselect)
11 Calculate fitness for all members of the population
            (using objective function).
12 Calculate average fitness, avg.
13 for i=1 to population size,
            number of copies NC(i) = int[f(i)/avg]
99 ============================================================


00 ============================================================
10 n POINT CROSSOVER (FTcross)
```

```
11    Pick two strings (chromosome) at random
12    Generate a random number
13    if the random value is within crossover rate
        {
14        randomly select a crossover point (site)
15        exchange string bits between 2 chromosomes
16        save new chromosomes
        }
17    Repeat till all strings are picked once
18 Repeat steps 11 through 17 "n" times
99 =============================================================


00 =============================================================
10 MUTATION (FTmutate)
11 Take each bit in the chromosome
12 Generate a random value
13   if random value is within mutation rate
       {
14   flip the bit (if 0 put 1 and vice versa)
       }
99 =============================================================
```

17.3.5 Implementation

This includes coding the software using a computer language, testing
and validation, and document generation. In this chapter, we use
the MATLAB high level programming language environment. Several
test functions were generated to test the software for functionality,
bugs, performance, user interface, and robustness. The software was
tested for validation purposes. Internal and external documentation are
important. The following format (template) may be used for the internal
documentation.

(a) Internal Documentation

```
NAME of the Function
PURPOSE
GIVEN
      GLOBAL VARIABLES
        PARAMETERS PASSED
OPERATION
RETURNS
APPLICATION
SYNTAX
FUNCTION DECLARATION
GLOBAL
CODE
```

EXAMPLE
REFERENCE

(b) External Documentation: It is desirable to provide software features, software layout, installation requirements (hardware and software), technical support, quick start instructions, commonly used commands, frequently asked questions, GA overview, tutorial, variable dictionary, users reference, bibliography, glossary, index, and license agreement in the manual (external documentation).

The implementation details of GA modules follow in the next section.

17.4   IMPLEMENTING GA MODULES IN MATLAB

The following algorithm is followed to implement the GA software.

```
00 =================================================
01 GA (FTga)
10 User Interaction
      {
11    Choose objective function.
12    Choose objective function parameters
13    Choose type of GA
14    Choose GA parameters
      }
20 Define coding scheme
30 GA Initialization
      {
31    Define coding scheme
32    Initialize a population at random
33    Evaluate the population (test the fitness of each member of
          this population using the objective function.
      }
40  Main GA Loop
41  While stopping criteria has not been satisfied
      (example: for G generations)
      {
42    Generation (FTgen)
        {
43      Apply selection (FTselect).
44      Perform Crossover (FTcross).
45      Perform Mutation (FTmutate).
        }
46    Decode parameters (FTdecode)
47    Evaluate the fitness of the population.
48    Update Statistics.
49    Plot Evolution.
```

```
50      Save Results
        }
99  ========================================================
```

Codes for the following GA module in MATLAB are provided in this section.

(a) FTcode: Coding module (step 31)
(b) FTinitpop: Initialization module (step 32)
(c) FTgen: Generation module (step 42)
(d) FTselect: Selection module (Step 43)
(e) FTcross: Crossover module (Step 44)
(f) FTmutate: Mutation module (Step 45)
(g) FTdecode: Decode module (part of step 46)
(h) FTshuffl: Shuffle module

```
%=============================================================
% FTcode
%---------
%  PURPOSE
%            Generates Phenotype to Genotype mapping.
%  GIVEN
%       nop :  Number of parameters
%       pmin :  Minimum for the parameters, pmin(1:nop)
%       pmax :  Maximum for the parameters, pmax(1:nop)
%       pres :  Resolution for the parameters, pres(1:nop)
%  OPERATION
%            Generate coding definitions
%  RETURNS
%            lchrom     :  Length of the chromosome
%            nbits         : Coded bits for each parameter, nbits(1:nop)
%  APPLICATION
%            In GA applications, part of initialization.
%  FUNCTION DECLARATION
function [nbits,lchrom]=FTcode(pmin,pmax,pres)
% CODE
%...............................................................
% Resolution, Range, Nbits, Lchrom
%...............................................................
range   = (pmax-pmin)./pres;
nbits   = round(log(range+1.0)./log(2.0));
lchrom = sum(nbits);
%
% end FTcode
%
```

```
%===========================================================
% FTinitpop
%---------
%   PURPOSE
%           Generate Initial random population of individuals.
%   GIVEN
%       GLOBAL
%           npop :  Number of population
%           lchrom :  Length of the chromosome
%   OPERATION
%           Generates random population of individuals
%   RETURNS
%           oldpop :  Population of chromosomes
%                           oldpop(1:npop,1:lchrom)
%   APPLICATION
%           In GA applications, part of initialization.
%   FUNCTION DECLARATION
function oldpop=ftinipop
% CODE
x=rand(npop,lchrom);
x=x*2.0;
oldpop=fix(x);
%
%END FTinipop
%


%===========================================================
% FTgen
%------
%   PURPOSE
%           Generation Function. An iteration of the creation
%           of a new population using reproduction operators.
%           Use a measure of fitness (PI).
%   GIVEN
%           oldpop:  Old Population of chromosomes
%                        oldpop(1:npop,1:lchrom)
%           PI :  Performance Index PI(1:npop)
%       GLOBAL
%           npop :  Number of population
%           sspop :  This Population will not be replaced or
%                        modified (steady state GA)
%   OPERATION
%           performs, selection, crossover, mutation, reproduce
%           new generation of population
%   RETURNS
%           newpop :  New Population of chromosomes
%                           newpop(1:npop,1:lchrom)
```

```
%  APPLICATION
%          In GA applications, for reproducing the next
%          generation of population.
%  FUNCTION DECLARATION
function [newpop]=ftgen(oldpop,PI)
%  CODE
%--------------------
%  CALL Selection
%--------------------
[xpop]=ftselect(oldpop,PI);
newpop=xpop;
for i=1:2:(npop-sspop)
    %--------------------
    %  CALL Crossover
    %--------------------
    x1o=xpop(i,:);
    x2o=xpop((i+1),:);
    [x1n,x2n]=ftcross(x1o,x2o);
    %--------------------
    %  CALL Mutation
    %--------------------
    x1n=ftmutate(x1n);
    x2n=ftmutate(x2n);
    %--------------------
    %  NEW Population
    %--------------------
    newpop(i,:)=x1n;
    newpop((i+1),:)=x2n;
end %  i for loop
%
%  end ftgen.m
%


%==========================================================
%  FTselect
%  --------
%  PURPOSE
%          Selection Function. Individuals from the old
%          population are copied to a mating pool. Fit
%          individuals will have more copies in the pool.
%          Models nature's survival of the fittest mechanism.
%  GIVEN
%          oldpop:  Old Population of chromosomes
%                      oldpop(1:npop,1:lchrom)
%          PI :  Performance Index PI(1:npop)
%      GLOBAL
%          npop :  Number of population
```

```
%             sspop :  This Population will not be replaced
%             nselect: Selection scheme choice
%                         1. Tournament Selection
%                         2. Roulette-Wheel Selection
%                         3. Ranking Selection
%             mm :  Maximization (mm=-1) or Minimization (mm==1)
%             Noofobj:  Number of objectives (1-4 allowed for
%                         Tournament selection)
%             Tsize :  Tournament Size (for tournament selection)
%   OPERATION
%             Performs, selection of population based on the value
%             of nselect.(only roulette wheel selection is shown here)
%   RETURNS
%             newpop :  New Population of chromosomes
%                         newpop(1:npop,1:lchrom)
%   APPLICATION
%             In GA applications, for selecting suitable parents
%             to mate and reproduce to form the next generation
%             (of population).
%   FUNCTION DECLARATION
function [newpop]=ftselect(oldpop,PI)
%   CODE
PIN=PI(:,1);
j=0; i=0;
%-------------------
% CALL Shuffle
%-------------------
index=ftshuffl;
%---------------------------
% Tournament Selection
%---------------------------
if nselect == 1
  if Noofobj ==1, J1=npop; J2=npop; J3=npop; J4=npop; end
  if Noofobj ==2, J1=fix(npop/2); J2=npop; J3=npop; J4=npop; end
  if Noofobj ==3, J1=fix(npop/3); J2=2*J1; J3=npop; J4=npop; end
  if Noofobj ==4, J1=fix(npop/4); J2=2*J1; J3=3*J1; J4=npop; end
  while j < npop-sspop
    j=j+1;
    for k=1:Tsize
      pin(k)=PIN(index(k+i));
      INDEX(k)=index(k+i);
    end % k for loop
    [Y,I]=min(pin);
    newpop(j,:)=oldpop(INDEX(I),:);
    i=i+Tsize;
    %-------------------
    % CALL FTshuffl.m
```

```
      %-------------------
      if i > (npop-Tsize), i=1; index=ftshuffl; end
      if j > J1, PIN=PI(:,2); end
      if j > J2, PIN=PI(:,3); end
      if j > J3, PIN=PI(:,4); end
    end  % while loop
  end
  %-------------------------------
  % Roulette Wheel Selection
  %-------------------------------
  if nselect == 1
    copy=zeros(npop,1);
    PIN=-PI(:,1);
    tf=mean(PIN);
    for i=1:npop
      copy(i)=round(PIN(i)/tf);
    end
    k=0;   i=0;
    while k <= npop
      i=i+1;
      if i > npop, i=1; end
      for j=1:copy(i)
        k=k+1;
        newpop(k,:)=oldpop(i,:);
      end % end j loop
    end % end k loop
  end


%
% end ftselect.m
%


%=========================================================
%  FTcross
%  -------
%  PURPOSE
%           Crossover Function.
%           Do the 1 or multi-point crossover operation for GA
%  GIVEN
%        x1o :  Old Chromosome x1o(1:lchrom)
%        x2o :  Old Chromosome x2o(1:lchrom)
%      GLOBAL
%        npop :  Number of population
%        nop :  Number of parameters
%        pcross :  Probability of Crossover
%                        (crossover rate)
%        ncross :  Number of crossover points
```

```
%            lchrom :  Length of the chromosome
%  OPERATION
%        Based on crossover rate, randomly selects crossover
%        points performs crossover, repeats ncross times,
%        determines crossover points, perform crossover based on
%         crossover rate, repeats ncross times
%  RETURNS
%            x1n(1:lchrom), and x2n(1:lchrom), the newly
%            formed chromosomes
%  APPLICATION
%  In GA applications, for providing random information
%        exchange.(recombination of building blocks)
%  FUNCTION DECLARATION
function [x1n,x2n]=ftcross(x1o,x2o)
%  CODE
for i=1:ncross
   x1s=x1o; x2s=x2o;
   if(rand(1) < pcross)
       site=fix(rand(1)*lchrom);
       if(site > 1)
           x1s(site:lchrom)=x2s(site:lchrom);
           x2o(site:lchrom)=x1o(site:lchrom);
       end % if
   end % if
   x1n=x1s; x2n=x2o;
   x1o=x1n; x2o=x2n;
end % for loop

%
% end ftcross.m
%


%===========================================================
% FTmutate
% --------
%  PURPOSE
%            Mutation Function.
%  GIVEN
%            xold :  Old Chromosome xold(1:lchrom)
%        GLOBAL
%            npop :  Number of population
%            pmutate :  Probability of Crossover
%                        (crossover rate),< 0.01
%            lchrom :  Length of the chromosome
%  OPERATION
%            Based on mutation rate, performs mutation
%  RETURNS
```

```
%               xnew :   New Chromosome xnew(1:lchrom)
% APPLICATION
%          In GA applications, for restoring lost genetic
%          material.  Usually after the crossover
%          operation. Improves global search.
% FUNCTION DECLARATION
function xnew=ftmutate(xold)
% CODE
for i=1:lchrom
   if(rand(1) < pmutate)
      xold(i)=abs(xold(i)-1);
   end% if
end% for i loop
xnew=xold;


%
% end ftmutate.m
%


%===========================================================
% FTdecode
% --------
%          Decode a given chromosome into the actual parameters
% GIVEN
%          x :  chromosome x(1:lchrom)
%      GLOBAL
%          nop :  Number of parameters
%          lchrom:  Length of the chromosome
%          nbits :  number of bits per parameter nbits(1:nop,1)
%          pmin :  Minimum for the parameters, pmin(1:nop)
%          pres :  Resolution for the parameters, pres(1:nop)
%          coding:  type: (1)binary,(2)logarithmic, (3) Gray
% OPERATION
%          Decodes the string based on the max, min,
%          resolution and coding
% RETURNS
%          P :  Decoded Parameters P(1:nop)
% APPLICATION
%          In GA applications, part of initialization.
% FUNCTION DECLARATION
function P=ftdecode(x)
% CODE
sbit=1;
for i=1:nop
    ebit=sbit+nbits(i)-1;
    sum=0;
    for j=sbit:1:ebit
```

```
        k=j-sbit;
        sum=sum+2^k * x(j);
    end % j for loop
    sbit=ebit+1;
    P(i)=sum*pres(i)+pmin(i);
end % i for loop

%
% end ftdecode.m
%


%==========================================================
% FTshuffl
% --------
%  PURPOSE
%           Perform random shuffling.
%  GIVEN
%      GLOBAL
%         npop           :  Number of population
%  OPERATION
%           performs, selection of population based on the
%           value of nselect.
%  RETURNS
%           index        :  index(1:npop)
%  APPLICATION
%           In GA applications, for selecting suitable parents,
%           used in select()
%  FUNCTION DECLARATION
function index=ftshuffl
%  CODE
x=rand(npop,1);
[y,i]=sort(x);
index=i;

%
% end ftshuffl.m
%
```

It may be noted that the modular environment in MATLAB permits the user to easily integrate these modules to their application by carefully choosing the Global and Local variables.

17.5   SAMPLE APPLICATION AND GUIDELINES

This section provides some basic guidelines for using genetic algorithms.
These are not specific to the software implementation. This is followed
by a sample application of the software.

17.5.1   Basic Guidelines for using GA

The following items are useful to bear in mind for efficient use of GA.

(a) Initial population needs to be large and random.
(b) Mutation needs to be low.
(c) Crossover disrupts high defining length schema–but brings in new
    high order, high length schema.
(d) Mutation disrupts all schema–but brings in fresh set of schema.
(e) Selection biases towards schemata with high fitness.
(f) Effective schema processing is proportional to population cubed.
(g) Low order, short defining length, highly fit schemata are the building
    blocks of the search. When the building blocks are pieced together,
    similar to solving the JIGSAW PUZZLE, the solution emerges.
(h) From complex design solutions to human thinking (creative) process
    to biological systems (such as immune systems), the building block
    approach seems to be the main theme.
(i) A coding should be selected such that short, low-order schemata
    are relevant to the underlying problem and relatively unrelated to
    schemata over other fixed positions.
(j) The user should select the smallest possible alphabet that permits a
    natural expression of the problem.

17.5.2   Sample Application

A sample application is examined here.

(a) Function
    i. Function Name: Himmelblau Function
    ii. Function : $f(x_1, x_2) = (x_1^2 + x_2 - 11)^2 + (x_1 + x_2^2 - 7)^2$
(b) MATLAB code

```
function PI=FThimmel(x)
PI=(x(1)^2+x(2)-11)^2+(x(1)+x(2)^2-7)^2;
```

(c) Function Parameters
    i. Coding: binary
    ii. Objective Function file Name: FThimmel.m
    iii. Initialization function: None
    iv. Number of Objectives: 1
    v. Type of optimization: Minimization
    vi. Number of parameters: 2
    vii. Minimum value of parameters (1 and 2) : $[-10.24 - 10.24]^T$

    viii. Maximum value of parameters (1 and 2) : $[-10.24 \; -10.24]^T$
    ix. Resolution of the parameters (1 and 2) : $[0.01 \; 0.01]^T$

(d) GA parameters

    i. Type of GA: Regular GA
    ii. Population size (Number of individuals) : 77
    iii. Stopping Criteria: Number of generations
    iv. Number of Generations: 77
    v. Crossover Rate: 0.77
    vi. Number of crossover points: 2
    vii. Mutation Rate: 0.0077
    viii. Type of Selection: Tournament
    ix. Tournament Size: 2

(e) Final Generation Results

    i. Best parameter set: $p = [-2.81 \; 3.14]$
    ii. Fitness of the first parameter set: PI=0.0038

17.6 CONCLUSION

Application of object oriented software engineering practices enhance the
productivity of software implementations of soft computing models like
genetic algorithms. These ensure correct, reliable, robust, and efficient
performance. This chapter demonstrated how a GA software can be
designed and implemented using software engineering techniques. The
importance of interactions between theory, model, and application is
of significance here. This mutual interaction and growth helps towards
building large scale hybrid intelligent systems that can address real world
problems.

REFERENCES

[Chr92] Christian K. (1992) *The Microsoft Guide to C++ Programming.* Microsoft Press.

[FIG95a] Flexible Intelligence Group L. (1995) *FlexTool(GA).* Flexible Intelligence Group, LLC, Tuscaloosa, AL 35486-1477.

[FIG95b] Flexible Intelligence Group L. (1995) *Guidelines for Evolutionary Fuzzy Modeling.* Flexible Intelligence Group, LLC, Tuscaloosa, AL 35486-1477.

[FTA94] Filho J. L. R., Treleaven P. C., and Alippi C. (June 1994) Genetic-algorithm programming environments. *Computer* 27(6): 29–43.

[GJM91] Ghezzi C., Jazayeri M., and Mandrioli D. (1991) *Fundamentals of Software Engineering.* Prentice-Hall, NJ.

[Gol89] Goldberg D. E. (1989) *Genetic Algorithms in Search Optimization and Machine Learning.* Addison-Wesley, MA.

[Hol75] Holland H. J. (1975) *Adaptation in Natural and Artificial Systems.* University of Michigan Press, Ann Arbor, MI.

[KM90] Korson T. and McGregor J. D. (Semptember 1990) Understanding object-oriented: A unifying paradigm. *Communications of the ACM* 33(9).

[KS95] KrishnaKumar K. and Satyadas A. (1995) Discovering multiple fuzzy models using efmcn algorithm. In *Proceedings of ACM Computing Week Symposium on Applied Computing (SAC), Nashville, TN*, pages 537–541.

[RBP91] Rumbaugh J., Blaha M., and Premerlani W. (1991) *Object-Oriented Modeling and Design.* Prentice-Hall, NJ.

[TM94] The MathWorks I. (1994) *MATLAB.* The MathWorks, Inc.

[Weg87] Wegner P. (October 1987) Dimensions of object-based language design. In *Proceedings of ACM OOPSLA*, pages 168–182.

[Zad94] Zadeh L. A. (1994) Fuzzy logic, neural networks, and soft computing. *Communications of the ACM* 37(3): 77–84.

18

ELECTROMAGNETIC SYSTEM DESIGN USING GENETIC ALGORITHMS

ERIC MICHIELSSEN AND DANIEL S. WEILE

18.1 INTRODUCTION

Electromagnetic phenomena permeate many aspects of our lives. Electromagnetic theory governs the propagation of high-speed signals in computer chips, the guidance of microwaves and millimeter waves in monolithic millimeter integrated circuits, the transmission of gigabit signals in optical fibers, the propagation of atmospheric and ionospheric waves, and the principles of antenna, waveguide, and radar design for air traffic control, navigation, and target identification. In addition, electromagnetic fields are used for microwave and magnetic resonance imaging, nondestructive testing, microwave heating, remote sensing, and geophysical probing.

The basic physics describing electromagnetic phenomena has been understood since James Clerk Maxwell published his famous 1873 treatise [Max93]. The analysis of electromagnetic phenomena involves the solution of coupled partial differential equations known as Maxwell's Equations, which have no known analytical solution for most problems of practical interest. During the past two decades, tremendous research efforts have been directed towards the development of algorithms for numerically solving Maxwell's equations. Today, a wide range of integral and differential equation-based numerical techniques exists that permit the accurate analysis of a large class of electromagnetic scattering, radiation, and guidance problems; computer-aided design (CAD) tools for many industries rely on these schemes. As technological frontiers advance, the demand for more sophisticated CAD tools grows, as they are expected to tackle increasingly complex problems within ever shorter turnaround times. As a consequence, computational electromagnetics (CEM) remains a wide-open research field that is intensively pursued by engineers, computer scientists, and mathematicians alike.

Unfortunately, research on algorithms for the systematic *design* of electromagnetic devices has historically received scant attention, and tedious trial and error procedures often prevail over systematic approaches in the design of electromagnetic components. The lack of progress in the area of systematic synthesis of electromagnetic devices can be attributed to the nature of the objective functions that arise in the formulation of electromagnetic optimization problems. These objective functions are often highly nonlinear, stiff, multiextremal, and nondifferentiable. In addition, their evaluation is almost always computationally expensive. Historically, the vast majority of research efforts related to the design of electromagnetic systems has relied on Deterministic Optimization Methods (DOMs) [Fle80]. DOMs have proven to be powerful tools for solving optimization problems with smooth, unimodal objective functions, as they typically require few objective function evaluations when compared to stochastic optimization methods. However, DOMs have important drawbacks when applied to multiextremal and stiff optimization problems [Dix78]. Therefore, their use often leads to highly interactive and expensive design procedures.

Genetic Algorithms (GAs) [Gol88] and Evolution Strategies (ESs) have recently been introduced as powerful stochastic alternatives to DOMs. GAs and ESs are known to be computationally expensive when compared to DOMs but overcome many of their drawbacks. GAs are stochastic, highly robust, and inherently parallel optimization techniques that mimic natural evolution. Whereas GAs have been successfully applied across a wide range of engineering disciplines [Gol88], their application in electromagnetic design and optimization is still in an embryonic stage. It is expected that many recent CEM developments will help induce a more widespread use of GAs and ESs in electromagnetic system design, including:

- The increasingly stringent design specifications for electromagnetic systems. Global optimization techniques will gradually supersede local optimizers and DOMs, which are often incapable of reaching strong objective function extrema. GAs and ESs offer opportunities for global optimization.
- The development of fast algorithms for the analysis of electromagnetic wave phenomena. Historically, the applicability of stochastic optimizers to electromagnetic design problems has been hampered by the significant computational cost associated with the evaluation of objective functions arising in this context. A host of recently developed fast algorithms, e.g., the fast multipole method [Rok90], fast recursive algorithms [Che93], the adaptive integral method, the multilevel domain decomposition algorithms, and wavelet transform techniques, will enable the application of GAs and ESs to CEM design problems because these algorithms drastically accelerate the evaluation of the pertinent objective functions.
- The increasing availability of massively parallel computers and technology that permits the effective utilization of workstation clusters. GAs and ESs are inherently parallel and greatly benefit from these platforms.

In this chapter, different applications of GAs to problems in electromagnetics are examined. Design problems from widely different fields of electromagnetics have been studied using GAs; this chapter will show that standard GAs are a very powerful tool for solving all of them. The success of the standard GA notwithstanding, this chapter will also demonstrate the use of more advanced GA techniques such as real coding and Pareto optimization methods, which have been proven useful in the solution of various electromagnetic problems.

This chapter is organized as follows. Section 18.2 summarizes the current state of the art of GA-based design tools for electromagnetic and optical systems by reviewing currently available literature. Section 18.3 demonstrates the effectiveness of GAs within the context of electromagnetic system design by describing GA-based optimizers for specific electromagnetic systems, namely multilayered filter systems, absorbers, frequency selective surfaces, and antennas.

18.2 REVIEW OF GENETIC ALGORITHM-BASED TECHNIQUES FOR ELECTROMAGNETIC SYSTEM DESIGN

Applications of GAs to electromagnetics fall into four main categories: magnetostatic device design and inverse problem solution, optical system synthesis, microwave absorber design, and antenna design. The section on magnetostatics details the application of GAs to electric motors and nondestructive testing. A discussion of the design of optical devices such as lenses and layered filters follows the section on magnetostatics. Closely related to the optical filter design problem is the design of stratified microwave absorbers, which is discussed next. Lastly, the applications of genetic algorithms to the solution of radiation and scattering problems such as antenna and array design are reviewed.

18.2.1 Applications of GAs in Magnetostatics

GAs were first used in electromagnetic design to optimize magnetostatic devices. The goal of many such studies is to design a system of electromagnetic coils and magnetic pole pieces to produce a desired magnetic field configuration for the operation of a motor or similar electromechanical device. Although the numerical analysis of such problems is well-understood, design methods are scant because of the high dimensionality and multimodality of the search spaces involved. Thus, these problems are natural candidates for the application of GAs. For example, Uler, Mohammed and Koh [UMK95] use a GA to optimize the shape of a magnetic pole face in a motor to give a desired magnetic field. The geometry of the coil and the position of the pole piece are fixed, and the GA is used to shape one face of the pole piece. The vertical positions of four points evenly spaced in the horizontal direction across the pole face are encoded into a chromosome, and the shape of the pole face is then determined by interpolation with a cubic function through those points. The magnetic field created by each pole configuration is then determined by solving the pertinent differential equations using the finite element method. The sum of the square differences between the magnetic field at several different points produced by the pole piece being evaluated and the desired field constitutes the objective function and is minimized by the GA.

In a similar vein, Vasconcelos et al. [VKNN94] apply a field-shaping GA to a problem in electrostatics. Parameters describing the surface of an insulator are encoded into a chromosome. A GA using the three basic operators and Holland's inversion operator, which reverses the order of randomly selected bits in the chromosome with a preset probability [Hol75], was used to minimize the maximum difference between the normal component of the electric field on the surface of the insulator and its desired value.

Mohammed and Uler also apply GAs to the design of a pot core transformer [MU95]. In this problem, instead of just optimizing a device's shape to produce a specific field pattern, the physical size of the device is optimized as well. Specifically, Mohammed and Uler Gray encode into a chromosome the total width and height of the device itself, as well as the width, height and radius of the coil and the thickness of the air gap, and attempt to achieve a desired magnetic flux while simultaneously minimizing the size of the transformer. To accomplish both goals, an objective function is maximized, which is the weighted sum of a reciprocal function of the transformer's area and the logarithm of the reciprocal of the absolute difference between the desired and calculated magnetic fields. Note that the absence of a physical method for combining design goals into one overall objective function results in an empirical combination of objectives based solely on intuition or experience. In fact, Mohammed and Uler [MU95] explain the use of the logarithm as a method to make the ranges of the two objectives commensurable and experiment to find the best weighting constants for the sum. This heuristic/experimental approach is quite common in most electromagnetic design problems using GAs; eradicating it would make GAs more useful design tools.

Other studies focus instead on the use of GAs in the nondestructive testing of magnetostatic devices. Some of these studies are especially interesting in that, unlike almost all of the other GA work in electromagnetics, they do not rely on GAs for design purposes, but instead employ GAs for the solution of inverse problems. In an electromagnetic inverse problem, the unknown is some physical object which alters a given field distribution due to a known source of excitation. The problem is termed *inverse* in contrast to the *forward* or *direct* problem in which the unknown is the field generated by a known excitation in the presence of a known physical layout. GAs can be applied to the solution of inverse problems by coding possible characteristics of the unknown object into chromosomes and then evaluating how closely the field produced by the known excitation of the object described by the chromosome matches the actual field measurements. A prime example of the application of GAs to solve inverse problems can be found in the work of Arkadan, Sareen and Subramaniam [ASS94] in which GAs are used to locate surface cracks on a steel vessel subjected to a magnetic field. A crack in a vessel is described by the location of its center as well as its width, length and orientation, and this description is then encoded into a chromosome. The magnetic field is calculated for each chromosome assuming that the gap may be described as a region in the steel vessel with the same constitutive parameters as air. The GA locates the crack by minimizing the difference between the field generated in the presence of the crack and the measured field.

Another application of GAs to the solution of an inverse problem may be found in the work of Enokizono and Akinari [EA95], who use a GA to find currents in an object from knowledge of the magnetic fields on its surface. The sampled pattern matching method is employed to estimate the direction and magnitude of currents inside an object. These estimates are then used to construct an initial population for the GA, which then further optimizes the magnitude of current flow in each element by minimizing the difference between the observed field and the field created by the current distributions being optimized in the population.

A different application of GAs to nondestructive testing may be found in Thollon and Burais [TB95], who use GAs to optimize the design of an electromagnet for nondestructive testing. The electromagnet's coil radius, width and current frequency are encoded into chromosomes, and the finite element method is used to evaluate the field around the structure with and without the crack. A crack geometry is assumed and the GA is used to maximize the difference of the field at a given point in the presence of the crack and without the crack.

18.2.2   Applications of GAs in Optics

Beyond their various uses in static problems, GAs have also been applied to the synthesis of optical equipment. Betensky [Bet93] uses a GA to optimize the design of compound lenses used in a cathode ray tube. Unlike most GA optimization schemes that start from a random population of initial designs, Betensky's method assumes a single initial design. The chromosomes in Betensky's GA are composed of single-bit genes, which decode to inform the GA whether or not to apply a particular "zero power" operator to the initial design to change the shape or placement of a simple lens in the compound system. In other words, instead of directly encoding the physical parameters of the compound lenses into chromosomes, Betensky encodes perturbations to an assumed design. The GA then optimizes the sequence of operators contained in the chromosome to achieve a design with low astigmatism.

More often than in lens design, GAs are used in optics for the design of thin film filters. In such systems, the designer layers different materials so that incident electromagnetic waves are either reflected or transmitted depending on their frequencies. In the first of such studies, Michielssen, Ranjithan and Mittra [MRM92] use a real-coded GA to optimize the thicknesses of the layers that constitute the filter (see section 18.3.1). The objective function used in this study measures how closely the filter's reflection characteristic approaches the desired characteristic over a prescribed frequency band. The results of the study showed that the GA-based method surpass the best results attained by traditional methods on all problems considered.

Along similar lines, Martin, Rivory and Schoenauer [MRS94] also design multilayered filters with real coded GAs using a similar objective function. The major distinction between this study and the one reported in [MRM92] is in its assumptions regarding available materials. Instead of assuming two available materials and only encoding the electrical lengths of the layers, Martin, Rivory and Schoenaur encode both the electrical length and the refractive index of each layer into a chromosome. Thus, the material properties of each layer can assume continuous values within a given range. This approach has the advantage of being very general, but on the other hand, there is no guarantee that a material with the parameters arrived at by the GA will be available.

Instead of concentrating on the filter characteristic of layered dielectrics, Eisenhammer et al. [ELL+94] optimize interference filters constructed of both dielectric layers and thin layers of silver based on their ability to trap heat from the sun. Though this problem amounts to creating a multilayer which transmits visible light and reflects infrared, Eisenhammer et al. use a more direct objective function which measures the heat trapping efficiency of the device. Also unlike the above two studies, the work by Eisenhammer et al. employs a binary coding scheme. The discrete nature of the binary encoding permits the selection of the material for each layer from a database of available materials instead of assuming that materials with continuously varying parameters are available.

18.2.3 Applications of GAs to Microwave Absorber Design

Stacked microwave absorbers are very similar to multilayered optical filters. In fact, the physics governing the operation of both systems is identical; the only difference is that, in contrast to optical filters that are designed to selectively reflect certain frequency bands while transmitting others, absorbers are designed to suppress reflection over large bandwidths by using materials that efficiently absorb electromagnetic energy. Michielssen et al. [MRM93] use a binary encoded GA employing crowding [DeJ75] to solve this problem. Much like the previous study discussed [ELL+94], materials for the multilayer are chosen from a database of available materials. The objective function for this study, however, is more complex than those used in the studies on multilayered filters for two reasons. First, the objective function optimizes the absorption over both a wide band of frequencies and many different angles of incidence of the illuminating wave. The inclusion of waves that are not normally incident on the face of the absorber also forces the consideration of two different polarizations; in other words, the reflection from the face of the absorber is different depending on whether the electric or magnetic field is perpendicular to a normal to the absorber. In addition to being more complex because of the inclusion of the multiple incident angles and two polarizations, the objective function for this study includes a term which rewards thin designs. Because the relative importance of thinness and absorptivity may vary depending on the application, results are presented for different weightings of both goals, thus demonstrating how the objective function may be easily tailored to a designer's needs.

Another study on the GA-based design of microwave absorbers was undertaken by Chambers and Tennant [CT94], who used an entirely different approach to evaluating designs. Unlike in the previous study [MRM93], the frequency band in [CT94] is not fixed; instead, the bandwidth between frequencies having a given level of reflectance is optimized. Of course, both polarizations must be taken into account, so the objective function is constructed to simultaneously make the sum of the bandwidths for the two polarizations large while making their difference small. Additionally, computational time is saved by optimizing the structures at only one oblique angle. Chambers and Tennant found that this is sufficient to ensure satisfactory results for all angles less oblique than the design angle.

18.2.4 Applications of GAs to Antenna Array Design and Scattering Problems

Other examples of GA-based applications in electromagnetics occur in the design of radiating systems and scatterers. These problems involve the generation of electromagnetic waves by antennas, or the scattering of such waves by different physical objects. One problem that has received considerable attention is that of antenna array optimization by GAs. For example, Tennant, Dawoud and Anderson [TDA94] use a GA to perturb the positions of antennas in a uniformly spaced linear array so that the resulting array has reception nulls at angles where there is known to be interference. The position perturbations are encoded into a chromosome, and an objective function that is proportional to the antenna gain in the desired direction of transmission and inversely proportional to the gain in positions of desired nulls is optimized. Using this method, the GA places nulls in the antenna pattern by perturbing the positions of its elements without shifting the position of the main lobe.

To optimize linear arrays according to different criteria Haupt [Hau94] uses GAs to reduce the maximum sidelobe level of array patterns. Instead of placing nulls in the pattern, Haupt attempts to reduce the maximum radiation in all secondary directions. To accomplish this, Haupt "thins" the array by switching off some of its elements as opposed to perturbing their positions. Thus, the chromosomes for this study contain one bit corresponding to each element in the array which tell the GA if the element is on or off. The objective function is constructed to minimize the maximum sidelobe level produced by the array. Haupt not only optimizes linear arrays but also synthesizes rectangular arrays using two-dimensional chromosomes and several different crossover schemes.

Finally Haupt and Ali [HA94, Hau95] use GAs for optimizing the scattering from different structures. First, in a manner analogous to that employed in the sidelobe level reduction problem, Haupt thins a linear array of perfectly conducting strips to reduce the sidelobe level associated with the scattered wave. The details of the operation of the GA in this problem are identical to those in the previous study on thinning antenna arrays. Second, Haupt reduces the sidelobe level of the field scattered by a single perfectly conducting strip by placing resistive strips next to the conductor and optimizing their thicknesses and resistivities. The objective function is the same as that used for the previous two problems, and the encoding is not unlike that used in the optical coating or microwave absorber problems.

18.3   SPECIFIC DESIGN EXAMPLES

This section details the application of GAs to the design of a variety of electromagnetic systems. Section 3.1 describes a synthesis procedure for optical filters based upon a real coded GA. Section 3.2 outlines a procedure for designing Pareto-optimal absorbers using Pareto GAs, which employ advanced techniques such as fitness sharing and nondomination ranking. Section 3.3 describes a GA-driven procedure for synthesizing complex frequency selective surfaces. Finally, Section 3.4 presents a GA-based method for synthesizing loaded monopole antennas that reside on a complex platform.

18.3.1   Optical Filter Design

Many electromagnetic applications require devices that exhibit specific frequency-dependent properties. Perhaps the simplest electromagnetic and optical filter is the multilayered stack shown in Figure 1. The multilayer consists of N alternating layers

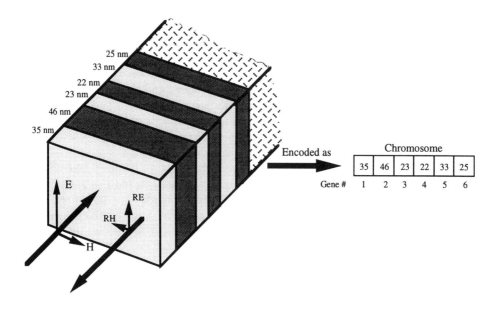

Figure 18.1   Six-layer optical filter and its chromosome representation

of two materials with fixed effective refractive indices. The structure is bounded by air on one side and a substrate medium with known refractive index on the opposite side. It is well-known that such multilayers partially reflect incident electromagnetic energy and that the reflection and transmission characteristics of the multilayer vary with the frequency of the incident electromagnetic wave. The structure is characterized by a reflection coefficient R, and the fractions of the incident energy that are reflected from and transmitted through the multilayer are termed reflectance and transmittance. The reflectance equals $|R^2|$, and assuming that the multilayer is lossless, the transmittance equals $1 - |R^2|$. Given the electrical properties of the two-layer materials and the substrate medium, the design of a multilayer involves a search for the optimal thicknesses of the layers, such that the frequency response of the resulting multilayer closely matches a desired response, which may correspond to either a low-pass, a high-pass or a band-pass characteristic. Unlike most CEM problems, the analysis of reflection from a multilayered stack can be carried out analytically in a straightforward manner [Che90c].

In the past, multilayered filters have been designed using a variety of approaches including circuit-analog methods [See83, SLW85, See65, SLC73], coefficient matching techniques [Che88c, Che88b, Che88a], and, most recently, the turning point method [Che90b, Che90a]. The latter technique seems to overcome certain drawbacks of the circuit-analog and coefficient matching techniques, since the filters designed using this technique have a smaller passband ripple, larger cutoff and cuton rates, and a higher rejection than those designed using the classical methods. The turning point method is an iterative technique, which starts from an initial design which can either be a classical stack or a filter obtained using the coefficient matching method. The turning point technique iteratively refines the design by altering the thicknesses of the layers, its goal being the design of a filter with equiripple response.

In [MRM92], a novel iterative technique based on a real coded GA for designing multilayer filters is presented. Given the total number of layers in the filter and the electrical properties of the materials constituting each layer, the algorithm iteratively constructs multilayers, whose frequency response closely matches a desired response. The objective function used in this study measures the quality of a design by densely sampling the filter's frequency response and integrating the difference with respect to that of the desired characteristic. The chromosome for this problem is just a real array of layer thicknesses.

Because the GA is real coded, there is some ambiguity in the formulation of crossover and mutation rules. For crossover, two designs are picked at random from the population created by the selection operator and a crossover point is chosen at random. This crossover point, unlike that used in binary GAs, is not picked between in two alleles, but is actually chosen to be one of the thicknesses in the array. The gene at the crossover site is inherited not directly from the one parent or the other, but is hybridized—that is, the children receive a randomly weighted average of their parents values for that gene. The other genes in the chromosome are inherited in the usual fashion: Those to the left of the crossover gene are inherited from one parent; those to its right are inherited from the other. Mutation is then implemented by adding or subtracting a random fraction of a gene's range to that gene with a predetermined mutation probability.

There are three major advantages to the GA-based algorithm when compared to classical design procedures.

- First, compared to classical iterative techniques, including the turning point method, the GA technique does not require a crude preliminary design to ensure convergence. This is mainly due to the fact that the genetic algorithm is not a gradient-based search procedure, and, therefore, is not trapped easily in local maxima.
- Second, the design procedure is independent of the nature of the multilayer as well as the characteristics of the incident and substrate media, which implies that the same implementation can be used for directly designing low-pass, high-pass, and band-pass filters that operate between practical terminal media.
- Third, the design objective can be changed easily by manipulating the fitness function. For instance, given a database of available materials, the algorithm could be used to simultaneously optimize the material choice for each layer as well as its thickness.

This algorithm was applied to the design of 19-layer low-pass and high-pass filters composed of alternating layers of ZnS, with a refractive index of 2.20, and PbTe with a refractive index of 5.10, and operating between air and a germanium substrate. Both filters outperform those obtained using the turning point method (Figs. 2 and 3). For example, the low-pass filter designed using a GA has a higher transmittance than the filter designed using the turning point method, which is considered the most powerful classical algorithm for the synthesis of multilayered filters. The versatility of the GA technique is evident in its ability to design filters of high performance for varying design

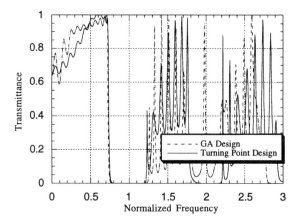

Figure 18.2   Transmittance $(1 - R^2)$ versus normalized frequency of 19-layer
low-pass filters composed of alternating layers of ZnS and PbTe on a Ge
substrate, designed using a GA and the turning pointing method

conditions without human interference. The current implementation of the design
technique is certainly more computer intensive than for the classical methods and
the turning point technique. However, this drawback is offset by the above-mentioned
advantages of the GA over the classical design techniques. In addition, the full potential
of the new technique could be achieved through parallel implementations of the GA.

18.3.2   Pareto Optimal Absorber Design

This section considers the design of electromagnetically absorptive coatings
constructed by layering materials on a perfectly conducting object to suppress the
reflection of incident electromagnetic waves. A designer, given a database describing
the electromagnetic properties of available materials, synthesizes an absorber by
choosing a material and thickness for each layer (Fig. 4). The resulting structure
should not only absorb as much energy as possible over a given frequency band, but
should also be thin and light to be cost-effective.

Like most of the design problems encountered in electromagnetics, the absorber
synthesis problem is not only multimodal, but also involves the simultaneous
consideration of multiple and possibly conflicting objectives. In many electromagnetic
design problems, various goals can be combined to arrive at one objective function for
the entire system. Unfortunately, unlike those problems in which different goals are
physically related, the absorber design goals are completely incommensurable. This
implies that not only is there no obvious mathematical method for combining design
goals, but that there is no single optimum as various tradeoffs between thickness and

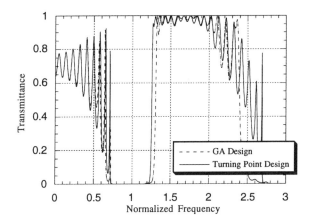

Figure 18.3 Transmittance $(1 - R^2)$ versus normalized frequency of 19-layer high-pass filters composed of alternating layers of ZnS and PbTe on a Ge substrate, designed using a GA and the turning pointing method

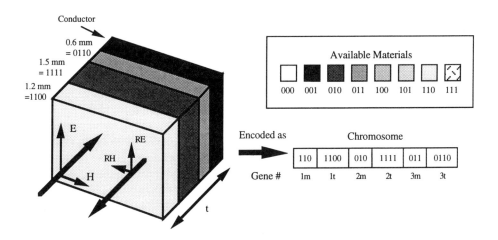

Figure 18.4 Three-layer absorber design and its chromosome representation assuming a four-bit coding for the thickness and an eight-material database

absorptivity may be considered optimal in different applications. Therefore, the goal of an algorithm for designing microwave absorbers should not be to return one design as in classical optimization problems, but should instead be to generate a set of designs representing all optimal tradeoffs.

To define what constitutes an optimal tradeoff, the concept of *Pareto optimality* is used. Though the problem at hand only involves the minimization of two parameters, the concept of Pareto optimality can be applied to design problems with any number of criteria. Consider a design problem with a vector $\mathbf{f} = (f^1, \cdots, f^G)$ of G objectives, each of which we wish to minimize, and two candidate designs with objective function vectors $\mathbf{f}_1$ and $\mathbf{f}_2$, respectively. Design 1 is said to *dominate* design 2 (or $\mathbf{f}_2$ is said to be *inferior* to $\mathbf{f}_1$) if for all $i \in \{1, 2, \ldots, G\}$, $f_1^i \leq f_2^i$, and there exists at least one i such that $f_1^i < f_2^i$. A design is said to be *nondominated* if no feasible design exists in the entire solution space which dominates it. The *Pareto front* is the set of all such nondominated designs [FF93, Gol88, CM75, Coh78, Ste89]. Therefore, roughly speaking, a Pareto optimal design is optimal in the sense that no other design exists which performs better or equally well with respect to all design goals—in other words, it is a design in which improvement with respect to one goal implies degradation with respect to another. In the absorber design problem, given a material database and a frequency band of interest, the Pareto front consists of all those designs for which no design in the solution space is simultaneously thinner and more absorptive over that frequency band. The Pareto front thus provides the designer of microwave absorbers all optimal alternatives: Designs on the Pareto front reflect as little as possible for a given thinness of coating; points not on the front are either physically unrealizable given the material constraints, or represent an inferior tradeoff to a design on the curve.

To study trade-offs in absorber design using GAs, the absorber structure has to be encoded into a chromosome, and the standard GA modified to meet the special requirements of Pareto optimization. In this application, binary coded GAs are used and chromosomes constructed as illustrated in Figure 4: each layer in the stack is described by two genes relating the material choice for each layer and its thickness, respectively, and it is assumed that the number of layers is fixed.

Pareto GAs differ from standard GAs in two respects. First, the objective function used to determine which designs are reproduced favorably must be some direct measure of the relative Pareto optimality of different designs as opposed to an algebraic combination of objectives. Second, the GA cannot be permitted to converge to a single design but must instead return the entire Pareto optimal set. This requires the maintenance of populational diversity. Though several Pareto Genetic Algorithms have been suggested to accomplish this [HNG94, FF93, SD95], prolonged experimentation has shown that for the design of broad-band microwave absorbers the Nondominated Sorting Genetic Algorithm (NSGA) of Srinivas and Deb [SD95] outperforms other schemes.

The NSGA, based on a suggestion by Goldberg [Gol88], handles the above two problems separately. To select population members according to Pareto dominance, Goldberg suggests the use of *nondomination ranking*. Specifically, Goldberg delineates a scheme in which all nondominated population members are assigned a rank of one and removed from the population. The nondominated members in the remaining part of the population are then assigned a rank of two, and this process is continued until the entire population is ranked. Each design is then assigned a fitness value based on its nondomination ranking.

Maintaining population diversity is accomplished using the concept of fitness sharing, which is based on the competition between similar organisms for limited resources in the environment. In a fitness sharing GA, in addition to calculating the objective function value f_i for each chromosome i, the GA calculates a niche count m_i, which measures the density of the population surrounding that chromosome. The niche count is calculated according to

$$m_i = \sum_{i=1}^{N_p} \text{Sh}(d_{i,j})$$

where $\text{Sh}(d)$ is a monotonically decreasing function of x such that $\text{Sh}(0) = 1$, and $d_{i,j}$ is a distance measure between designs i and j. In designing microwave absorbers, the triangular sharing function

$$Sh(d) = \begin{cases} 1 - \frac{d}{\sigma_{share}} & d < \sigma_{share} \\ 0 & d \geq \sigma_{share} \end{cases}$$

is used where σ_{share} is a designer-chosen niche radius . The distance $d_{i,j}$ itself may be measured with respect to a metric in either *genotypic* or *phenotypic* space. A genotypic sharing scheme defines $d_{i,j}$ as the interchromosonal Hamming distance. Phenotypic sharing, on the other hand, measures the distance between the designs' objective function values. In Pareto GAs, phenotypic sharing is usually preferred because we seek a large variety of different optimal tradeoffs in *objective function space*, but we care little whether or not different points on the Pareto front represent large differences in the design of the absorber itself [HNG94, SD95]. To ensure that the metric does not weigh objectives inequitably, the metric is normalized so that all objectives are effectively scaled to range between 0 and 1 [HNG94, SD95]. Once the m_i have been calculated, chromosomes are reproduced assuming an objective function value of $\frac{f_i}{m_i}$ instead of f_i so that the GA will not allow all chromosomes to converge to one solution.

The NSGA combines the above described nondomination ranking and sharing schemes with one more rule added to maintain diversity in the population and find the Pareto front. To assign fitness values in the nondomination scheme, rank-one population members are assigned a "before sharing" fitness of 1.0. After all of the "shared" fitness values for the first rank are calculated by dividing the "before sharing" fitness by the niche count, the lowest fitness value "after sharing" becomes the "before sharing" fitness value for the second rank. This process is continued until all ranks have been considered.

In the microwave absorber study, the NSGA was employed to minimize both the reflection and thickness of a five-layer absorber over the frequency band ranging from 2-8 GHz using a sixteen-material database from [MRM93] containing representative materials, e.g., lossy and lossless dielectrics, and those with a relaxation-type characteristic. Figure 5 shows the Pareto front obtained by the NSGA and another scheme known as the CTPGA, which uses crowding [DeJ75] to maintain diversity and tournament selection based on nondomination ranking for selection. Note that the NSGA returns a superior front. The frequency response of two very different designs on that front is shown; design #1 is very absorptive, and design #2 is very thick.

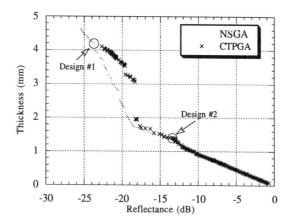

Figure 18.5 The Pareto front (thickness vs. reflectance (R^2)) obtained over 2-8 GHz by the NSGA compared to that obtained by the CTPGA, the NSGA front is denser and dominates the CTPGA front

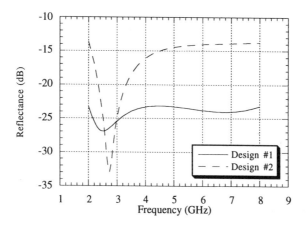

Figure 18.6 The frequency responses (reflectance (R^2) vs. frequency) of two different Pareto optimal designs, design # 1 could be used when thickness is a relatively minor issue; design #2 is useful where a design may have to be very thin

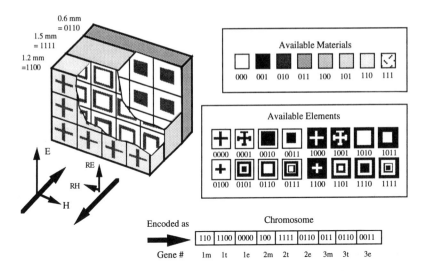

Figure 18.7   Three layer FSS and its chromosome representation assuming an 8-material database, a 4-bit coding for the layer thicknesses, and a 16-element database

18.3.3   Design of Frequency Selective Surfaces

Another popular filter structure, often used in electromagnetic applications, is the Frequency Selective Surface (FSS). FSSs find widespread applications as spatial filters over a wide range of the microwave and optical spectra. These surfaces consist of a periodic array of perfect electrically conducting (p.e.c.) elements, which are printed on a dielectric substrate [MCC88, Cwi86, CM87]. The periodicity of the array extends along two axes parallel to the plane of the substrate (Fig. 7). The surfaces exhibit total reflection/transmission for the incident electromagnetic wave in the neighborhood of the resonant frequency of the patch/aperture element. The resonant frequency depends upon the size and shape of the p.e.c. elements in the array, as well as upon the material properties of the substrate layer. Often, multiple FSSs are cascaded to form a superstructure with enhanced frequency-dependent characteristics, e.g., filters with a very narrow transition region between the reflection and transmission bands of the structure [CM87]. It should be noted that multilayered FSSs are very similar to the optical filters discussed in Section 18.3.1, with the exception of the periodic arrays of metallic elements that reside between substrate layers. Applications of FSSs are omnipresent. In the microwave frequency range, the frequency-dependent properties of the screen can be exploited for separating feeds in reflector antenna systems [OM66] and enhancing the characteristics of antenna radomes [Lee79]. In the far-infrared spectrum, FSSs find applications as polarizers, beam splitters, and mirrors [DD81], and in sensor design. In the near-infrared and visible portions of the electromagnetic spectrum, FSSs have been proposed in the design of solar energy systems [Hor74].

To synthesize a filter with a desired frequency response, one typically resorts to a tedious trial and error procedure, since few systematic alternatives are currently available for this purpose. The objective of this section is to describe a novel technique, based on a genetic algorithm, for designing multilayered FSSs. The present method does not search for the optimal design by optimizing the dimensions of a given element shape. Rather, such a design is obtained by cascading elementary building blocks of predefined element shapes and dielectric layers.

Key to the successful application of the proposed technique is the compilation of general purpose databases comprised of element shapes (square and rectangular patches, simple and Jerusalem crosses, single and double loops, apertures, etc.) and dielectric materials with permittivities that fall within an adequate range. The chromosome describing an FSS then consists of a binary string describing the material choice, discretized thickness and element shape for each layer. Given a desired frequency response, the algorithm iteratively selects, from these databases, the optimal sequence of components for construction of the filter by minimizing the difference between the actual filter response and the desired characteristic, much like in the optical filter discussed in Section 18.3.1. As a preamble to the iterative search procedure, the individual building blocks are analyzed and their respective scattering matrices stored in a data file [Cwi86, CM87]. Given the scattering matrices of the various building blocks, the frequency response of an arbitrary multilayered design can be computed efficiently using a cascading procedure. It is important to note that the choice of the element shapes in the database should not be guided by the design objective, since element shapes whose characteristics are incompatible with the design objective are automatically eliminated by the genetic search algorithm. On the contrary, the element database should contain a wide variety of element types, so as to be useful in the design of a large class of filters.

For the purpose of illustrating the GA-based design technique, a 3 layer high-pass FSS was synthesized using the 16-element database shown in Figure 7. In addition, a material database containing 8 different materials with permittivities ranging from 2 to 18 was provided to the algorithm. The genetic algorithm starts with a population of 100 arbitrarily selected designs, and converges, after approximately 150 iterations, to a design, the frequency response of which is shown in Figure 8, together with that of the desired high-pass response. It should be noted that, without the use of the GA, the design of this FSS would be difficult, if not impossible. Using the GA, the design was completed in approximately 15 minutes of CPU time on a low-end workstation.

18.3.4   Antenna Optimization

One of the more recent applications of GAs in the field of electromagnetics is the computer-aided design of loaded broad-band wire antennas. Though wire antennas have been in use since the inception of electromagnetic theory and the experiments of Heinrich Hertz, they continue to be used in a myriad of applications today. The radiation from these antennas was understood fairly well even in 1897 by Pocklington [Poc97]; today they may be analyzed easily with a plethora of computer codes implementing any one of various numerical solution techniques, such as the Method of Moments (MoM) [Har68, Wan93]. Unfortunately, even though the analysis of radiating wire structures is well-understood, the synthesis of such structures still presents some difficulty, and fully automated methods for the design of wire antennas are rare.

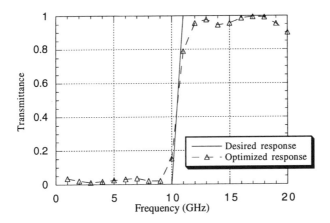

Figure 18.8 Frequency response (transmittance vs. frequency) of a 3-layer
FSS optimized using the GA technique. The optimized high-pass characteristic
closely matches the desired characteristic

The optimization of antenna characteristics is intrinsically complicated, as the design goals often conflict, and analysis methods for such structures are computationally intensive. To further complicate matters, any objective function used to compare different antenna designs is likely to be highly multimodal. Thus, as Popovic [Pop91] notes, "Although the CAD design of wire antennas is of course desirable, for various reasons it can not be implemented in all cases. Quite frequently, we are compelled to use an interactive type of CAD ... ."

The goal of any antenna design algorithm is to synthesize a structure that efficiently radiates energy in desired directions over a given frequency band. The algorithm discussed here uses a GA to design loaded broad-band monopoles (i.e., wire antennas excited at their base) that may be situated in complex environments such as on a vehicle (Fig. 9) The synthesis of such an antenna involves optimizing two facets of the design. First, a given number of loads consisting of a parallel resistor/capacitor/inductor network are to be placed at various positions along the antenna to maximize its radiation in any number of given directions. Second, a matching network composed of a sequence of series or shunt connected capacitors and inductors must be designed so that the signal generator efficiently couples energy into the antenna terminals. To synthesize a broad-band monopole antenna using a GA, a scheme for encoding the loading and matching network design parameters into a chromosome must be devised, and an objective function measuring the efficiency by which the transmitter couples energy onto the antenna, must be constructed.

Encoding the antenna loading data into a binary string is quite straightforward; each load can be described by its location, resistance, inductance and capacitance

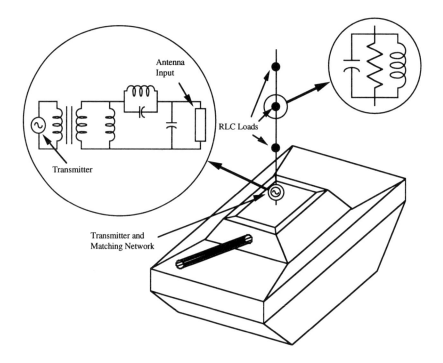

Figure 18.9 Broad-band loaded antenna situated on a tank. The monopole is 5 m high, and the tank is 2.9 m high, 3.7 m wide and 8 m long

values, which, upon discretization, can be directly encoded on the chromosome. The description of the matching network poses more of a problem. While optimizing component values for the capacitors and inductors situated in a given matching network topology would be straightforward, finding the optimal matching network topology to optimize is by no means trivial. The most obvious method for encoding the matching network topology into a chromosome is by concatenating binary coded representations of the values of the matching network components in some logical order and including extra bits that describe how the component fits into the topology of the network. For example, one could start describing components from the antenna terminals and move towards the transmitter, adding two extra bits after those describing each component value to determine (i) whether the element in question is a capacitor or an inductor and (ii) if it is shunt or series connected. This naïve approach fails, however, because it violates one of the necessary conditions for successful GA optimization: designs characterized by similar strings should behave similarly.

To circumvent this problem, it is assumed that the matching network consists of an integral number of four component sections as shown in Figure 10. The total number of components in the matching network equals four times the number of sections. The inductors and capacitors are allowed to take on values through a binary representation of their inductance or capacitance values. However, unlike in typical GA codings, the matching network components described by these bits are also permitted to take on values of infinity or zero in addition to the finite values in between. Infinity and zero are coded in the GA as the largest or smallest few binary numbers in the binary representation of the component value, respectively. Since an infinite capacitor acts like a short circuit and a zero capacitor acts like an open circuit (and vice versa for inductors) this coding allows the topology itself to evolve. Unlike the previous scheme mentioned, however, this coding permits the matching network topology to evolve slowly, with similar "genotypes" producing similar "phenotypes." Instead of a capacitor "suddenly" changing into an inductor after a mutation or crossover as would happen in the "two-bit" method mentioned above, a one-bit change in this scheme represents a slow evolution toward a design.

With the coding in place, an objective function must be designed that takes into account both how well the antenna radiates energy in the desired directions as well as how efficiently the matching network couples energy into the antenna terminals. To accomplish this, an objective function composed of two parts is used. The first part of the objective function measures the overall efficiency of the antenna-matching network system by calculating how much of the signal generator's power is coupled into the antenna and radiated in the desired direction. To prevent the GA from prematurely converging to a suboptimal design, a penalty function was added to help steer the GA toward the optimal design. This function penalizes antennas which exhibit especially low radiation in the desired directions, which have highly reflective matching networks or whose radiative properties vary wildly with frequency (making them hard to match).

Of course, the antenna's radiation patterns must be computed to evaluate this objective function. This analysis is usually accomplished by setting up an integral equation that describes the scattering of electromagnetic fields and by solving it using the MoM [Har68, Wan93]. For electromagnetic problems involving objects of considerable size such as those considered here, however, the MoM procedure requires the solution of extremely large systems of linear equations. Direct solution of these

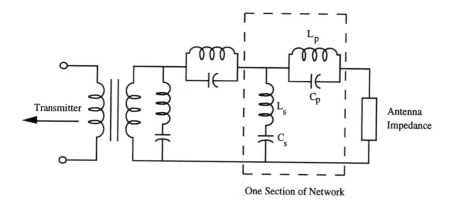

Figure 18.10   Generalized matching network topology

equations for each design in the GA's population would take far too long to make the technique practical. Fortunately, since the only differences between the various antennas in the GA's population are their load locations and values and matching networks, the entire MoM matrix need not be inverted for each design. A perturbation analysis shows that this fact can be exploited and that the only matrix equation that has to be inverted for each specific design is of the size of the number of loads. Thus, the full antenna analysis only has to be done once in a preprocessor to the GA, and the GA optimization of antennas becomes a practical possibility.

As a practical example of the antenna optimization technique, a tank (Fig. 9) antenna was optimized for operation between 15 and 60 MHz in the direction of its gun, perpendicular to the gun, and in the direction opposite the gun. Each load and matching network component value was described by seven bits, antenna load locations were described by six bits, and five loads and two matching network sections were used. The objective function was set up to maximize the average radiation in all three directions. The resulting matching network is shown in Figure 11 and the response of the system in Figure 12 The antenna load information in summarized in Table 18.1. Figure 12 shows that the system exhibits positive gain at all angles of interest over the whole frequency band even in this difficult problem where without loading some part of the gain characteristic will drop below -40 dB! In addition, the results obtained using methods such as simulated annealing do not even compare, and are therefore not shown.

18.4   CONCLUSION

Research on the application of global optimization techniques for electromagnetic synthesis is regarded as a high priority by members of the electromagnetics research and industrial communities. Such techniques should be sufficiently reliable to consistently find either global or strong local optima in their domain of application, yet

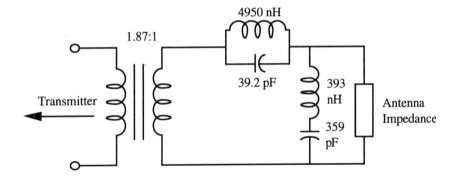

Figure 18.11 Matching network for the tank antenna

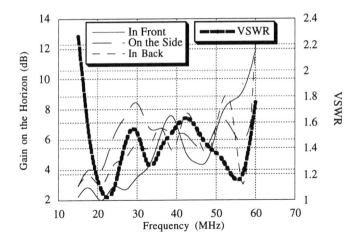

Figure 18.12 Frequency response of the tank antenna

Table 18.1 Load locations and values for the tank antenna

Load No.	Location (cm)	Resistance (Ω)	Capacitance (pF)	Inductance (nH)
1	226.6	1500.0	706.0	36.7
2	257.8	1500.0	76.6	244.0
3	273.4	238.0	59.1	3930.0
4	289.1	1500.0	76.6	753.0
5	320.3	1500.0	107.0	2420.0

fast enough to be economical. The discussion presented in this chapter underscores the power of GAs to accomplish these goals for electromagnetic device design and the solution of inverse problems. Overall, GAs are robust, simple and powerful optimization tools which already have resulted in novel methodologies that largely eliminate the need for trial-and-error-based synthesis paradigms, and which drastically shorten the design cycle for many electromagnetic components. Furthermore, many such GA-based techniques have advanced the state of the art in their respective applications, greatly improving the designs of devices that were previously the best in existence.

Regardless of these successes, however, many issues regarding the application of GAs to electromagnetic engineering problems remain to be explored. On one hand, there is a plethora of applications yet untried. For example, the applicability of GAs to radome design and waveguide filter synthesis needs to be investigated. More importantly, the use of GAs for solving inverse problems in electrodynamics has not even been attempted.

On the other hand, many advanced GA techniques and operators remain largely unapplied to electromagnetic synthesis problems. The Pareto GA technique shows enormous promise for studying choice trade-offs in many applications; it awaits application to the design of filters, antennas and frequency selective surfaces. In addition, since GAs are inherently parallel as they simultaneously analyze entire populations, they are implemented quite naturally on parallel machines. Parallel GAs are especially promising for electromagnetic design problems because of the computationally expensive nature of the objective functions encountered in this area.

GAs have introduced the power of stochastic algorithms into the realm of electromagnetic design, and their use has resulted in novel designs for devices heretofore unimagined. There is little doubt that the future will bring many more GA-based design tools to the electromagnetic community and that these tools will become commonplace in many electromagnetic engineering environments.

18.5 ACKNOWLEDGMENTS

This work was supported by the US National Science Foundation under grant ECS-9502138.

REFERENCES

[ASS94] Arkadan A. A., Sareen T., and Subramaniam S. (1994) Genetic algorithms for nondestructive testing in crack evaluation. *IEEE Transactions on Magnetics* 30(6): 4320–4322.

[Bet93] Betensky E. (1993) Postmodern lens design. *Optical Engineering* 38(2): 1750–1756.

[Che88a] Chen T. C. (1988) Analysis and design of optical low-pass multilayer filters via noncommensurate to pseudocommensurate system transformation. *IEE Proc. J. Optoelectron.* 135: 359–371.

[Che88b] Chen T. C. (1988) Optimized design of even-order low-pass and high-pass multilayer filters by coefficient matching. *IEE Proc. J. Optoelectron.* 135: 359–371.

[Che88c] Chen T. C. (1988) Optimized design of odd-order low-pass and high-pass multilayer filters by coefficient matching. *IEE Proc. J. Optoelectron.* 135: 166–177.

[Che90a] Chen T. C. (1990) Towards optimum multilayer filter design. *IEE Proc. J. Optoelectron.* 138: 241–248.

[Che90b] Chen T. C. (1990) Turning point method of optimized filter design. *IEE Proc. J. Optoelectron.* 137: 102–107.

[Che90c] Chew W. C. (1990) *Waves and Fields in Inhomogeneous Media*. Van Nostrand Reinhold, New York.

[Che93] Chew W. (1993) Fast algorithms for wave scattering developed at the university of illinois' electromagnetics laboratory. *IEEE Antennas and Propagation Magazine* 35: 22–32.

[CM75] Cohon J. L. and Marks D. H. (1975) A review and evaluation of multiobjective programming techniques. *Water Resources Research* 11(2): 208–219.

[CM87] Cwik T. and Mittra R. (1987) The cascade connection of planar periodic surfaces and lossy dielectric layers to form an arbitrary periodic screen. *IEEE Trans. Antennas Propagat.* 72: 1397–1405.

[Coh78] Cohon J. L. (1978) *Multiobjective Programming and Planning*. Mathematics in Science and Engineering. Academic Press, New York.

[CT94] Chambers B. and Tennant A. (1994) Design of wideband jaumann radar absorbers with optimum oblique incidence perfomance. *Electronics Letters* 30(18): 1530–1531.

[Cwi86] Cwik T. (1986) *Scattering From General Periodic Screens*. PhD dissertation, University of Illinois at Urbana-Champaign, Urbana, IL.

[DD81] Durschlag M. S. and DeTemple T. A. (1981) Far-ir optical properties of freestanding and dielectrically backed metal meshes. *Appl. Opt.* 20: 1245–1253.

[DeJ75] DeJong K. A. (1975) *An Analysis of the Behavior of a Class of Genetic Adaptive Systems*. PhD dissertation, University of Michigan, Ann Arbor, MI.

[Dix78] Dixon L. C. W. (1978) *Design and Implementation of Optimization Software*, chapter Global Optima Without Convexity. Sijnoof and Noordhof, Aalphen aan den Rijn.

[EA95] Enokizono M. and Akinari Y. (1995) Estimation of current distribution by a hybrid of genetic algorithms and sampled pattern matching method. *IEEE Transactions on Magnetics* 31(3): 2012–2015.

[ELL+94] Eisenhammer T., Lazarov M., Leutbecher M., Schoeffel U., and Sizmann R. (1994) Optimization of interference filters with genetic algorithms applied to silver based mirrors. *Applied Optics* 32(32): 6310–6315.

[FF93] Fonseca C. M. and Fleming P. J. (1993) Genetic algorithms for multiobjective optimization: Formulation, discussion and generalization. In *Proc. Fifth Annual Conference and Their Applications Symposium on the Theory of Computing*.

[Fle80] Fletcher R. (1980) *Practical Methods of Optimization*. Wiley Interscience, Chichester, UK, first edition.

[Gol88] Goldberg D. E. (1988) *Genetic Algorithms in Search, Optimization. and Machine Learning.* Addison-Wesley, Reading, MA.

[HA94] Haupt R. L. and Ali A. S. (1994) Optimized backscattering sidelobes from an array of strips using a genetic algorithm. In *Proc. Applied Computational Electromagnetics Conference Symposium on the Theory of Computing.*

[Har68] Harrington R. F. (10 January 1968) *Field Computation by Moment Methods.* Krieger, Malabar, FL.

[Hau94] Haupt R. L. (1994) Thinned arrays using genetic algorithms. *IEEE Transactions on Antennas and Propagation* 42(7): 993–999.

[Hau95] Haupt R. L. (1995) An introduction to genetic algorithms for electromagnetics. *IEEE Antennas and Propagation Magazine* 37(2).

[HNG94] Horn J., Nafpliotis N., and Goldberg D. E. (1994) A niched pareto genetic algorithm for multiobjective optimization. In *Proc. First IEEE Conference on Evolutionary Computation Symposium on the Theory of Computing.*

[Hol75] Holland J. H. (1975) *Adaption in Natural and Artificial Systems.* University of Michigan, Ann Arbor, MI.

[Hor74] Horwitz C. M. (1974) A new solar selective surface. *Opt. Commun.* 11: 210–212.

[Lee79] Lee S. W. (1979) Scattering by dielectric loaded screen. *IEEE Trans. Antennas Propagat* 19: 478–486.

[Max93] Maxwell J. C. (1893) *A Treatise on Electricity and Magnetism.* Oxford, Oxford, UK, third edition.

[MCC88] Mittra R., Chan C. H., and Cwik T. (1988) Techniques for analyzing frequency selective surfaces, a review. *Proc. IEEE* 72: 1593–1615.

[MRM92] Michielssen E., Ranjithan S., and Mittra R. (1992) Optimal multilayer filter design using real coded genetic algorithms. *IEE Proceedings J* 139(12): 413–420.

[MRM93] Michielssen E., Ranjithan J.-M. S. S., and Mittra R. (1993) Design of lightweight, broad-band microwave absorbers using genetic algorithms. *IEEE Transactions on Microwave Theory and Techniques* 41(6): 1024–1031.

[MRS94] Martin S., Rivory J., and Shoenauer M. (1994) Simulated darwinian evolution of homogeneous multilayer systems: A new method for optical coatings design. *Optics Communications* 110: 503–506.

[MU95] Mohammed O. and Uler G. F. (1995) Genetic algorithms for the optimal design of electromagnetic devices. In *Eleventh Annual Review of Progress in Applied Computaional Electromagnetics Symposium on the Theory of Computing.*

[OM66] O'Neans F. and Matson J. (1966) Antenna feed system utilizing polarization independent frequency selective intermediate reflector. *U.S. Patent 3-231-892* .

[Poc97] Pocklington H. E. (1897) Electrical oscillations in wires. *Cambridge Philosophical Society Proceedings* 9: 324–332.

[Pop91] Popovic B. (1991) *CAD of Wire Antennas.* Research Studies Press, Somerset, UK.

[Rok90] Rokhlin V. (1990) Rapid solution of integral equations of scattering in two dimensions. *J. Comput. Phys.* 86(7): 414–439.

[SD95] Srinivas N. and Deb K. (1995) Multiobjective optimization using nondominated sorting in genetic algorithms. *Evolutionary Computation* 2(3): 221–248.

[See65] Seeley J. S. (1965) Lc ladder used as broadband prototype for distributed components. *Electron. Lett.* 1: 265–266.

[See83] Seeley J. S. (1983) Synthesised design of optical filters assisted by microcomputer. *Proc. Soc. Photo–Opt. Instrum. Eng.* 369: 118–129.

[SLC73] Seeley J. S., Liddell H. M., and Chen T. C. (1973) Extraction of chebycheff design data for the lowpass dielectric multilayer. *Optica Acta* 20: 641–661.

[SLW85] Seeley J. S., Lim Y. K., and Wu S. Y. (1985) New algorithms for synthesized design of optical filters. *Appl. Opt.* 24: 875–878.

[Ste89] Steuer R. E. (1989) *Multiple Criterial Optimization: Theory, Computation and Application.* Krieger, Malabar, FL.

[TB95] Thollon F. and Burais N. (1995) Geometrical optimization of sensors for eddy currents nondestructive testing and evaluation. *IEEE Transactions on Magnetics* 31(3): 2026–2031.

[TDA94] Tennant A., Dawoud M. M., and Anderson A. P. (1994) Array pattern nulling by element position perturbations using a genetic algorithm. *Electronics Letters* 30.

[UMK95] Uler G. F., Mohammed O. A., and Koh C.-S. (1995) Utilizing genetic algorithms for the optimal design of electromagnetic devices. *IEEE Transactions on Magnetics* 30(6): 4296–4298.

[VKNN94] Vasconcelos J., Karahenbuhl L., Nicholas L., and Nicholas A. (1994) Design optimization using the bem coupled with genetic algorithms. In *Second International Conference on Computation in Electromagnetics Symposium on the Theory of Computing.*

[Wan93] Wang J. J. H. (1993) *Generalized Moment Methods in Electromagnetics.* John Wiley and Sons, New York.

19

Robust Genetic Algorithms for Optimization Problems in Aerodynamic Design

J. Périaux [1]   M. Sefrioui [1][2]   B. Stoufflet [1]   B. Mantel [1]

E. Laporte [1]

Abstract

The purpose of this paper is to discuss the use of Genetic Algorithms as a part of a complex design process devoted to the solution of problems originating from Aerospace Engineering. It first presents the role of genetic algorithms in such a process and their interaction with other ones such as Finite Elements mesh generator and Euler flow analysis solvers. A classical deterministic approach based upon a conjuguate gradient method is then presented providing genetic algorithms a reference for comparison. The application part presents the results obtained in inverse and optimization problems which aims are to rebuild or optimize an airfoil shape embedded in transonic Euler flows. Results of both methods are then compared in terms of robustness and convergence speed.

[1] DASSAULT AVIATION, 78 Quai Marcel Dassault, 92214 Saint-Cloud. France
[2] LAFORIA-IBP-CNRS Université Paris VI, 4 place Jussieu, 75252 Paris Cedex 05. France

19.1   Introduction

There has been in the last few years a growing industrial interest in the numerical solution of constrained optimization problems governed by large scale non linear P.D.E.'s. This interest is motivated both by the improvment of distributed computer technology and the high priority given to minimum cost target in aircraft manufacturing.

The most challenging problems in a Aerodynamics, involving complex flow solvers modelized by the Euler or the Navier-Stokes equations for shock and/or viscous drag reduction purpose, require robust non-convex optimization tools for the design of optimal shapes.

In a very classical sense the mathematical problem is formulated as a control one using the calculus of variatiopn in order to derive gradient information algorithms. An efficient gradient computation requires the introduction of adjoint equations following the optimum control theory path developed by O. Pironneau [Pironneau84] and more recently by A. Jameson [Jameson88].

Different variants of gradient methods like reduced gradient methods [Young et al.94] or one shot methods proposed by [Taasan94] have been recently proposed with acceleration of the optimization process.

On the other hand, Genetic Algorithms are derivative free search procedures based on the mechanics of natural selection, and Darwin's main principle : *survival of the fittest.* They have been introduced by J. Holland [Holland75] who explained the adaptative process of natural systems and laid down the two main principles of genetic algorithms : the ability of simple bit string representation to encode complicated structures and the power of simple transformations to improve such structures. More recently D. Goldberg [Goldberg89a] brought GAs in non-convex optimization theory for quantitative study of optima and introduced a decisive thrust in the GAs research field.

A major advantage for GAs in complex industrial environment is robustness : they are computationaly simple and powerful in their search for improvment, not limited by restrictive assumptions about the search space (continuity, existence of derivatives, unimodality) and accomadate well to discontinuous environments compared to classical optimization methods.

Indifference to problem specifics, codings of decision variables , process of population and randomized operators are the main characteristics which contribute to the robustness of GAs.

The main purpose of this paper is to compare results of deterministic and evolutive approaches in selected optimization problems from industrial aerodynamic design.

Section 19.2 is concerned with the GAs integration in a CFD industrial environment system and the chain access to the fitness function is described. Section 19.3 describes the remeshing strategy with adaptation used for each candidate airfoil. The accuracy of the computed shape relies on the quality of the mesh all along the design process. In Section 19.4 different optimization strategies using control theory with the discrete state equations and an automated differentiation process are presented for an inverse problem and the associated Euler flow analysis solver described. Section 19.5 is mainly concerned with the evolutive optimization by GAs, including the Bezier Spline parametrization, adopting a floating-point represenation of design variables and describing the mechanisms of selection, cross-over and mutation with small populations. In Section 19.6 we discuss and compare results of design airfoils shapes obtained by both determistic and evolutive approaches in term of accuracy and efficiency. The computerized optimizations are performed for inverse or optimization problems in transonic aerodynamics proposed in a recent workshop within the european ECARP project. Hybridization of the two methods is then clearly defined as the next step of the design process.

19.2 GAs Integration in a CFD system

When using classical benchmark functions to evaluate genetic algorithms performances, the computation time of the fitness function is generally low. But real-world problems are much more complicated, and even when the fitness function is clearly defined it might involve sophisticated solvers that require the solutions of large-sized Partial Differential systems of equations. In inviscid aerodynamic design, the fitness associated to an airfoil shape — for instance the shock-drag— depends on the solution of Euler equations (Euler state equations represent the environment in which the genetic algorithms shapes evolve). But this solving can be done only after having properly generated a mesh corresponding to the shape by a finite element technique. The interaction of the various processes is shown in Figure 19.1. The shape is represented by a Bezier Spline and the design variables of the genetic algorithms based search correspond to the Bezier control points. After a discretization step, the shape is then passed to a mesh generator that use finite element technique to adapt the mesh to the profile. When the mesh is adapted to the shape, the state evaluation can begin. The state evaluation is an iterative process whose objective is to approximate with a given accuracy the solutions of Euler equations for compressible inviscid fluids in a transonic regime. This solver is used by both genetic algorithms and conjugate gradient approaches. But an important difference arises because of the iterative character of the solver. On the one hand, a gradient method needs a high accuracy stopping criteria, since it is supposed to work with exact solutions, so there must be many iterations before reaching the good accuracy. On the other hand, genetic algorithms are very robust with respect to noise and they accomodate well with less precise solutions, since the convergence of the algorithm does not depend on a single individual but on a whole population. That robustness makes it possible to consider only approximate solutions at the beginning of the process before refining them during the course of the evolution.

19.3 Mesh Generation and Parametrization

This section is devoted to an automatic mesh generator/adaptation whose data is the solution of a previous computation. A directional error estimator leads to the definition of a control space including some anisotropic information, or equivalently a Riemannian metric for mesh re-generation. The mesh fineness is controlled by creating points and its stretch by swapping edges. The obtained meshes are adapted to the given solution and can be constituted on stretch elements.

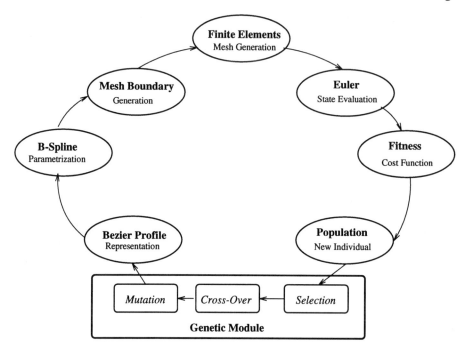

Figure 19.1   GA interaction with CFD solvers

19.3.1   Control spaces

The mesh adaptation process discussed below is guided by a control space. This contains information about the size and the shape of the expected elements depending on their location.

General background

Let Ω be the bounded domain of R^2 or R^3 we want to triangulate. According to [George et al.91], we define a control space as follows:

Definition: (O,H) is a control space for the domain Ω under consideration if:

- Ω is an open subset of R^2 or R^3 containing Ω,
- H is a real valued function defined at each point P of O and in each direction d, $H(P,d)$ represents the desirable step size of the mesh at P in d direction and must always be positive.

If $H(P,d)$ is independent of d, the control is said to be *isotropic*. In this case, a good triangulation of Ω will be constituted of nearly equilateral triangles of a given size. On the contrary if $H(P,d)$ is d-dependent, the control is *anisotropic* and stretched elements are desirable. The problem is to create a mesh such that H is optimally satisfied for all points and all directions.

A particular class of control spaces

Here, a particular class of control spaces is considered, including some anisotropic properties. We assume that O is partitioned by a mesh whose vertices are $(S_i, \ i = 1, .., N)$. A n x n-matrix M_i symmetric and positive definite is associated with each vertex S_i (n = 2 or 3 is the dimension of the space). A new metric is defined by the following norm

$$\| X \|_{S_i} = \sqrt{X^t M_i X} \tag{19.1}$$

Then, for each direction d, the function H is given at point S_i by

$$H(S_i, d) = \frac{\| X_d \|}{\| X_d \|_{S_i}} \tag{19.2}$$

where X_d is a vector in the d direction. With this notation, a vector X_d has a length of $H(S_i, d)$ if and only if it is a unit vector in the metric associated with M_i. In this way the mesh control becomes isotropic from the point of view of the local metric. Note that, except if M_i is a multiple of identity matrix, $H(S_i, d)$ is d-dependent. The control space (O, H) is now fully defined by interpolating the matrices on each element of the partition. Only continuous interpolations are considered to avoid jumps of the control function over O. The simplest one is linear on each element.

Riemannian metric

The open set O can be considered as a C^∞-manifold and matrices $M(P)$, $P \in O$ as bilinear forms. For a control space of the considered class, these forms define a *Riemannian metric* on O. The length of a parametric curve $\gamma(\tau)$, $\tau \in [0, 1]$ in such a metric is the value of

$$\int_0^1 \sqrt{\gamma(\tau)^t \ M(\gamma(\tau)) \ \gamma(\tau) \ d\tau)}$$

Its interest is expressed in the following proposition.

Proposition 1:
Let $M(P)$, $P \in O$ be a set of symmetric and positive definite n x n-matrices with continuous variation of M versus P. The control function defined by

$$H(P, d) = \frac{\| X_d \|}{\sqrt{X_d^t M(P) X_d}}$$

becomes in the Riemannian metric associated with $M(P)$, $P \in O$:

$$H(P, d) = 1.$$

Consequently, the change from Euclidean metric to Riemannian metric reduces mesh control to an isotropic and constant step size control. From the point of view of this metric, a mesh satisfying the function H all over the domain $\Omega \subset O$ is constituted of equilateral elements of diameter equal to 1.

The next section is devoted to the mesh adaptation controlled by space (O, H) where function H is everywhere nearly 1. With a suitable change of metric, it takes into account any control space of the considered class.

19.3.2   Adaptation

The mesh generator needs a fully defined control space. For this, a function H can be given over O from a priori knowledge of the physical solution. With a view to generating adaptive meshes, H is determined from a computed solution obtained on a previous (coarse) mesh T_0. This mesh becomes a natural partition of Ω to support the definition of the control function H. We base the mesh adaption on an error estimation for a selected key variable σ. More precisely H is determined in order to equi-distribute the interpolation error. The linear interpolation error depends on second derivatives of σ. This leads to computing the desirable directional step size H at each vertex S_i according to

$$H(S_i, d)^2 \mid \frac{\partial^2 \sigma}{\partial d^2}(S_i) \mid = c_0 \tag{19.3}$$

where c_0 is the acceptable error chosen by the user. We then have the proposition
Proposition 2
Control function H defined by (19.3) can be expressed as (19.1) - (19.2).

The previous algorithm can be applied with one of these metrics instead of the Euclidian one. As a consequence of **proposition 2**, the obtained mesh is adapted to variable σ in the sens that the interpolation error becomes equally distributed all over the domain and in each direction.

(3927 Nodes ; 7671 Elements)

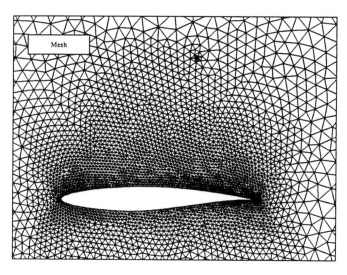

Mesh

Figure 19.2   Initial Mesh Generation

The initial mesh (Figure 19.2) is uniformly generated according to a given airfoil shape. Afterwards, the adaptation process is applied until the mesh fits the profile in order to provide a very fine tuning in the critical regions whereas the mesh is coarse in the less interesting areas (Figure 19.3).

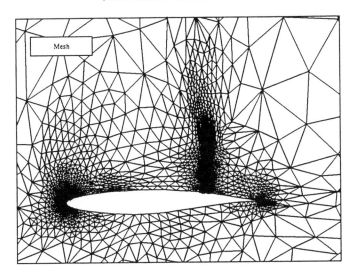

(2828 Nodes ; 5484 Elements)

Mesh

Figure 19.3 Adapted Mesh

The mesh generation step shows another advantage that can be used by genetic algorithms . When the mesh is very well adapted , the state evaluation process converges in a very fast way. For the gradient conjugate method, since it is a step by step approach, the initial mesh has to be uniform because there is no prior knowledge that could be used for the meshing. But for genetic algorithms , it is entirely different since we do have previous meshes : we can take advantage of the meshes that were generated for the previous populations instead of beginning a blind-search at each fitness evaluation.

19.4 Control Theory Approach

19.4.1 Schemes on unstructured meshes

From these last years, design principles of numerical schemes for compressible flows tend to emerge in a frame shared by the whole Computationnal Fluid Dynamics (CFD) community based on schemes satisfying discrete maximum principles for scalar equations. This frame dictates the way followed by various teams to construct high-order non-oscillatory schemes.

For conservation laws, a convenient approach to design discretization schemes is based on two steps:

- Monotonicity principle for non-oscillatory low-order scheme   (1)
- High-order construction preserving property (1)   (2)

On this basis (to satisfy (2)) are constructed a large number of family of schemes (at the top stand MUSCL type reconstruction or introduction of anti-diffusive term). Successful attemps avoiding step (2) are reported comprising Finite Element approaches based on Petrov-Galerkin formulations ([Chalot et al.94]), distributive schemes balancing fluxes on the element ([Struijs et al.91]) and recently mixed Finite Volume approximations considering the gradient as a degree of freedom [Berde et al.95]. In this section we present a classical MUSCL reconstruction or diffusive term addition.

Concerning the point (1), for a large number of years, total variation diminushing (TVD) schemes concept has dominated. This notion is difficult to extend in more than one-dimension and specially when dealing with unstructured meshes. Moreover, this is confirmed by recent theoretical investigations on the convergence of schemes on unstructured meshes ([Perthame et al.94]) where the functional space BV is not the appropriate one for weak convergence. As mentionned before, a more incisive frame is to search for a discrete maximum principle. This has motivated to construct discretizations with positive coefficients ([Dervieux et al.93], [Reuther et al.95], [Barth et al.95]).

Consider a semi-discrete in time discretization of the conservation law equation

$$w_t + \vec{\nabla} \cdot (\vec{f}(w)) = 0 \tag{19.4}$$

of the following form

$$\frac{dw_j}{dt} = \sum_k c_{kj} (w_k - w_j) \tag{19.5}$$

The aforementionned monotonicity principle is satisfied with positive coefficients $c_{kj} \geq 0$. Considering a Lagrange-Galerkin formulation of the system of d conservation laws

$$W_t + \vec{\nabla} \cdot (\vec{F}(W)) = 0 \tag{19.6}$$

and introducing V_h^d a set of piecewise polynomial functions of degree k (P_k), written as

$$\int_{\Omega_h} W_t \, \phi \, dv + \int_{\Omega_h} \phi \, \vec{\nabla} \cdot \vec{F}(W) \, dv = 0 \quad , \forall \phi \in V_h \tag{19.7}$$

If $\vec{F}(W)$ is taken as element of V_h^d (group representation), one obtains the following form of the scheme

$$\int_{\Omega_h} W_t \, \phi_i \, dv + \int_{\Omega_h} \phi_i \, \vec{\nabla} \cdot \vec{F}(W) \, dv = 0$$

For steady-state solutions, mass-lumped variant is often considered as

$$S_i \frac{dW_i}{dt} + \int_{\Omega_h} \phi_i \vec{\nabla} \cdot \vec{F}(W) \, dv = 0 \tag{19.8}$$

A simple integration by part of the second term leads to the following identity

$$\int_{\Omega_h} \phi_i \, \vec{\nabla} \cdot \vec{F}(W) \, dv = \sum_j \int_{\Omega_h} \phi_i \vec{\nabla} \phi_j \cdot \vec{F}_j(W) \, dv = - \sum_j \int_{\Omega_h} \phi_j \vec{\nabla} \phi_i \cdot \vec{F}_j(W) \, dv + \int_{\Gamma_h} \phi_i \vec{F}(W) \cdot \vec{v}_T \, d\sigma \tag{19.9}$$

We can introduce the following vector which has a geometrical interpretation (construction of a **dual mesh** with cells C_i) for P1 interpolation:

$$\vec{\eta}_{ij} = \int_{supp(i) \cap supp(j)} (\phi_i \vec{\nabla}\phi_j - \phi_j \vec{\nabla}\phi_i) dv \; \left(= \int_{\partial C_i \cap \partial C_j} \vec{\eta} \, d\sigma\right)$$

Considering the following identity based on a close contour

$$\sum_j \vec{\eta}_{ij} \vec{F}_i(W) + \int_{\Gamma_h} \phi_i \vec{F}_i(W) \cdot \vec{\nu} \, d\sigma = 0 \tag{19.10}$$

gives the centered approximation of the Finite Volume Galerkin formulation

$$\sum_j \int_{\Omega_h} \vec{\eta}_{ij} \cdot \frac{\vec{F}_i(W) + \vec{F}_j(W)}{2} + \int_{\Gamma_h} \frac{\vec{F}_i(W) + \vec{F}(W)}{2} \cdot \vec{\nu}_\Gamma \, d\sigma = 0 \tag{19.11}$$

Introducing a numerical flux function Φ_F and as numerical conditions are expressed in the boundary integrals, the previous equation reduces to

$$S_i \frac{dW_i}{dt} \, dv + \sum_{j \in K(i)} \Phi_F(W_i, W_j, \vec{\eta}_{ij}) + \int_{\partial C_i \cap \Gamma} \frac{\vec{F}_i(\bar{W}) + \vec{F}(\bar{W})}{2} \cdot \vec{\nu}_\Gamma \, d\sigma = 0 \tag{19.12}$$

A modified finite volume Galerkin is of interest when the median dual mesh can become highly distorted in the case of large aspect ratio elements. A first modification is given by minimizing the angle $(\vec{\eta}_{ij}^T, N_i \vec{N}_j)$ in each element T where $\vec{\eta}_{ij}^T$ denoted the contribution to $\vec{\eta}_{ij}$ of the element T. In [Barth et al.95], T. Barth proposes to define cells based on the center of the smallest circle which contains the triangle.

The second step of reconstruction is an extension of MUSCL approach where the flux is replaced in (1.9) by $\Phi_F^{first, \, upwind}(W_{ij}, W_{ji}, \vec{\eta}_{ij})$.

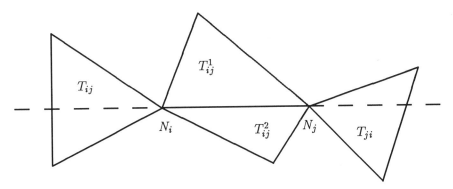

Figure 19.4   Edge $[N_i, N_j]$ and adjacent elements

Introducing limiter functions $l(u, v)$ such that $l(u, v) = \psi(\frac{u}{v})v = \psi(\frac{v}{u})u$, $\psi(r) = r\psi(\frac{1}{r})$ and $\psi(r) = 0$, $\forall r \leq 0$, one can now define the reconstruction for an edge

$E = [N_i, N_j]$ as

$$W_{ij} = W_i + \Pi_E \, Lc(\nabla W \mid_{T_{ij}} \cdot \overrightarrow{N_i N_j}, \nabla W_E \cdot \overrightarrow{N_i N_j}) \, \Pi_E^{-1}$$

$$\nabla W_E = \nabla W \mid_{T_{ij}^1} \cdot \overrightarrow{N_i N_j} = \nabla W \mid_{T_{ij}^2} \cdot \overrightarrow{N_i N_j}$$

where Lc is the diagonal matrix of limiters l and Π_E is a characteristic matrix constructed at some mean.

A variant (where the reconstruction acts on the diffusion terms only) will be of the form

$$(V_4) \quad S_i \frac{dw_i}{dt} + \sum_{j \in K(i)} \left(\frac{\vec{f}_j \cdot \vec{\eta}_{ij} - \vec{f}_i \cdot \vec{\eta}_{ij}}{2} - \alpha_{ij}(w_{ji} - w_{ij}) \right) = 0$$

which is a centered FVG scheme with scalar or matrix diffusion with a parameter $\tilde{\alpha}$ controlling the fourth order diffusion term.

Most of the ingredients described in this section are used to solve the Euler equations that represent the flow analysis solver of our optimum design process.

19.4.2 Optimization strategies

The general formulation of a shape optimization problem has been introduced by many authors [Cea81], [Pironneau84], [Dervieux et al.94]. It has been applied in many situations and recently by [Reuther et al.95]. He advocates the application of the control theory directly to the system of partial differential equations. The adjoint equations are formed as a system of differential equations. These differential equations are then discretized and solved in the same manner as the original flow equations. This approach can be called as continuous sensitivity analysis. Here, a different approach, called discrete sensitivity analysis is applied consisting in applying the control theory to the discrete equations itselves. Two advantages make it attractive: one manipulates exact gradient and the adjoint operator is easily derived from the implicit matrix. The general formulation may be written as

$$\text{Find a shape } \gamma \text{ such that } \gamma^* = \arg \min_{\gamma \in \mathcal{O}} j(\gamma)$$

where $j(\gamma)$ denotes a cost function given by

$$j(\gamma) = J(\gamma, W(\gamma)) = \int_\gamma C(W(\gamma)) d\sigma$$

$$J : \mathcal{O} * V^d \to I\!R$$

$W(\gamma) \in V^d$ state vector solution of steady Euler equations

$$E(\gamma, W(\gamma)) = 0.$$

The optimality condition satisfied by γ^* is $\frac{dj(\gamma)}{d\gamma} = 0$ given by

$$\frac{dj(\gamma)}{d\gamma}\delta\gamma = \frac{\partial J}{\partial\gamma}\delta\gamma + <\frac{\partial J}{\partial W}, \frac{dW}{d\gamma}\delta\gamma>$$

$$\frac{\partial E}{\partial\gamma}\delta\gamma + \frac{\partial E}{\partial W}\frac{dW}{d\gamma}\delta\gamma = 0$$

Introducing an adjoint state Ψ (Lagrange multiplier) satisfying the adjoint equation

$$\frac{\partial E}{\partial W}^*\Psi = \frac{\partial J}{\partial W}.$$

Then the gradient of the cost function is

$$\frac{dj(\gamma)}{d\gamma}\delta\gamma = \frac{\partial J}{\partial\gamma}\delta\gamma - <\Psi, \frac{\partial E}{\partial\gamma}\delta\gamma>$$

We will treat the case of two-dimensionnal shape optimization for (multi) - airfoil profiles including the capacity of multi-point optimization. As a first step, we will limit to inviscid flow as a model for compressible flow. In a first study, we applied the theoretical approach to treat full potential equation in [Dinh et al.93]. In this present paper, we will detail the construction of the optimization algorithm when using numerical approximations of the Euler equations described in the first section.

As numerical methods considered here are based on unstructured meshes in order to treat general geometries, a key point of the method is the relation between optimization parameters and geometrical characteristics of the mesh. This relation is decomposed in elementary functions; the overall gradient will be the product of these elementary gradients for each of which an appropriate evaluation can be used. One can anticipate to use finite differences restricted to some specific elementary operations to calculate the necessary information. In the present work, this manner has been avoided by using analytical dependance between intervening variables. The strategy chosen here can be easily described. Although an initial mesh is constructed using a mesh generator starting from prescribed boundary points, perturbed meshes arising in the optimization process are determined from variations of the boundary mesh points (of coordinates $(x_i^{(b)}, y_i^{(b)})$) by solving an elasticity equation which will provide a new position of the interior mesh points (of coordinates $(x_i^{(m)}, y_i^{(m)})$). As the topology of the mesh is preserved by this approach, the relation between boundary and interior mesh points is differentiable and the corresponding gradient is easily computed. We have used to build directly the subroutines evaluating the gradient an automatic differentiator developped by INRIA [Rostaing et al.93]. The optimization parameters z_i chosen here are explicit functions of the ordinates v_i of the control points (of coordinates (u_i, v_i)) of the cubic splines defining the boundary and the angle of attack α of the profile. Their relations with boundary points is then analytical, thus the gradient is trivially evaluated.

We have now sufficient information to apply the control theory which sets up the following system of equations

$$
\begin{cases}
E(\gamma, W(\gamma)) = 0 & \text{(state)} \\
\dfrac{\partial E}{\partial W}(\gamma, W(\gamma))^* \ \Pi(\gamma) = \dfrac{\partial J}{\partial W}(\gamma, W(\gamma)) & \text{(adjoint)} \\
j'(\gamma) = \dfrac{\partial J}{\partial \gamma}(\gamma, W(\gamma)) - \ <\Pi(\gamma), \dfrac{\partial E}{\partial \gamma}(\gamma, W(\gamma)) > & \text{(optimality)}
\end{cases}
\qquad (19.13)
$$

The introduction of the adjoint problem resulted from the application of the control theory on the continuous system where the flow equations are viewed as a constraint. This analysis can be directly performed on the discretized system. This discrete sensitivity analysis is detailed hereafter.

Firstly, we give expressions of considered cost functions. A first class of problems addressed are the inverse problems aiming to control a prescribed boundary pressure distribution. The cost function can be formulated as

$$
j(\gamma) = \int_{\gamma} (p - p^{obj})^2 d\sigma \quad \text{or in discrete form}
$$

$$
j(\gamma) = \sum_{l \in \mathcal{L}} \alpha_l \, (p_l - p_l^{obj})^2
$$

where $card(\mathcal{L}) = N$ is the number of pressure points given on the profile.

The problem of minimizing the drag coefficient can be addressed in defining the cost function as the pressure drag coefficient (augmented with viscous drag if a viscous correction is considered) submitted to constraints as given target pressure distribution or given lift. It can be noticed that any modification of the cost function (which will be currently envisaged in using such design tools) only affects the right-hand side of the adjoint equation and the derivative of the cost function with respect to the optimization parameters in the whole system of equations.

The discretized state equations - solving the two-dimensional Euler equations - are given by the formulation described in the first section.

$$
S_i \frac{dW_i}{dt} + \sum_{j \in K(i)} \Phi_F(W_{ij}, W_{ji}, \vec{\eta}_{ij}) + \int_{\Gamma_h} \Phi_i \frac{\vec{F}_i(\tilde{W}) + \vec{F}(\tilde{W})}{2} \cdot \vec{\nu}_\Gamma \, d\sigma = 0 \qquad (19.14)
$$

which includes boundary conditions expressed in $\tilde{W}$.

A linearized implicit solution of the system involves the Jacobian of the first-order approximation (1.9).

$$
\frac{Id}{\Delta t} \, \delta W_i + \sum_j A_{ij} \, \delta W_j = RHS_i \qquad (19.15)
$$

where Id is the identity matrix of $\mathcal{L}(I\!R^d, I\!R^d)$ and $A_{ij} \in \mathcal{L}(I\!R^d, I\!R^d)$.

Let perform the control theory directly on the discretized system. Consider the optimization parameters z_m $(m = 1, M)$ introduced before. To fix ideas, let take a cost function of the form (what follows is general and is not restricted to a specific problem)

$$j(z) = \frac{1}{2} \sum_{l \in \mathcal{L}} \alpha_l(z) \, (p_l - p_l^{obj})^2 \tag{19.16}$$

The corresponding partial derivatives can be written

$$\frac{\partial J}{\partial W_k} = \sum_{l \in \mathcal{L}} \alpha_l \, (p_l - p_l^{obj}) \, \frac{\partial p_l}{\partial W_k} \tag{19.17}$$

$$\frac{\partial j}{\partial z_m} = \frac{\partial J}{\partial z_m} + \sum_{k} \sum_{l \in \mathcal{L}} << \alpha_l \, (p_l - p_l^{obj}) \, \frac{\partial p_l}{\partial W_k} \, , \, \frac{\partial W_k}{\partial z_m} >> \tag{19.18}$$

where $<< >>$ is the duality product in IR^d. For any operator of $\mathcal{L}(IR^d, IR^d)$, we denote by B^T the adjoint which verifies $<< B^T U, V >> = << U, B \, V >>$.

Let introduce the state equation

$$E_i = \sum_{j \in K(i)} \Phi_F(W_i, W_j, \vec{\eta}_{ij}) + B.C. \tag{19.19}$$

Its derivative with respect to parameters is given by

$$\frac{\partial E_i}{\partial z_m} + \sum_{j} A_{ij} \frac{\partial W_j}{\partial z_m} = 0 \tag{19.20}$$

Let introduce the adjoint state Π solution of the following equation

$$\sum_{j} A_{ji}^T \, \Pi_j = \sum_{l \in \mathcal{L}} \alpha_l \, (p_l - p_l^{obj}) \, \frac{\partial p_l}{\partial W_i} \tag{19.21}$$

By substitution, we get:

$$\frac{dj}{dz_m} = \frac{\partial J}{\partial z_m} + \sum_{k} << \sum_{j} A_{jk}^T \Pi_j \, , \, \frac{\partial W_k}{\partial z_m} >> = \frac{\partial J}{\partial z_m} + \sum_{j} << \Pi_j \, , \, \sum_{k} A_{jk} \frac{\partial W_k}{\partial z_m} >> \tag{19.22}$$

The gradient of the cost function is then finally given by

$$\frac{dj}{dz_m} = \frac{\partial J}{\partial z_m} - \sum_{j} << \Pi_j \, , \, \frac{\partial E_j}{\partial z_m} >> \tag{19.23}$$

Where Π is the adjoint state solution of equation 19.21.

In the above conditions, the whole chain can be exactly differenciated, the gradient of j is expressed with an adjoint state and a conjugate gradient algorithm can be applied to the mininmization of j. In the study of this section the Euler system 19.19 is discretized via a triangulation by means of unstructured, MUSCL or central-differenced, finite-volume methods described above. The application of the ODYSSEE automated differenciator in order to derive the exact adjoint system is carried out in [Rostaing et al.93]. A classic Fletcher-Reeves conjugate gradient algorithm is used in the optimization loop.

19.5   Evolutive Optimization by GAs

Binary-coded genetic algorithms are very well suited to some problems and they facilitate theoretical analysis, but the power of genetic algorithms does not depend on the binary representation [Goldberg90]. There are various domains where the binary bias is no longer necessary [Janikow et al.93]. When we deal with optimum design problems in aerodynamics, there are several parameters taking values in different intervals that are to be optimized with a high accuracy. In this context, another crucial point emerges : the time needed for each fitness evaluation. We can no longer afford having an algorithm that converges after thousands of evaluations because every fitness evaluation requires quite a long time to be computed. The objective is to get a fast-convergence algorithm that is still robust enough to avoid local minima, i.e. a balance between exploration and fast convergence.

In the sequel, the GA's terminology applied to design optimization is described in the following table.

Population	Set of airfoils
Individual	Airfoil shape
Gene	Bezier Spline control point
Fitness	Drag or Pressure target
Social success	Best airfoil

19.5.1   Bezier Spline Representation

The first and most natural representation of a shape that is generally considered is a point by point representation. It is quite interesting since in engineering most of airfoils shapes are given in the form of table functions or fixed points. Thus, the optimization problem parameters could be those very points that define the shape. But the drawbacks of this approach soon appear :

- To have a fine local tuning, the shape might be defined by more than a hundred points. And since the convergence of genetic algorithms depends on the number of parameters (i.e. the size of the search space), it is obvious that having such a representation would lead to slow population convergence.
- The most important point is that this representation is not robust toward standard genetic operators. The crossing-over of two very different individuals might lead to individuals that are non-feasible aerodynamically and which are impossible to evaluate in terms of fitness.

The above reasons favored the choice of a Bezier Spline representation because it deals very well with the two main drawbacks of a point by point representation:

- There is no need to have hundred of parameters: a few control points are enough to represent with a very good accuracy any kind of airfoil shape.
- The genetic algorithms operators can not create obviously non-feasible solutions: the new individuals are always candidate solutions because of the underlying smoothness properties of Bezier curves.

A Bezier curve of order n is defined by the Bernstein polynomes $B_{n,i}$

$$Q(t) = \sum_{i=0}^{n} B_{n,i} P_i \text{ with } B_{n,i} = C_n^i t^i (1-t)^{n-i}$$

Where t denotes the parameter of the curve taking values in $[0,1]$, P_i are the coordinates of the control points and $C_n^i = \frac{n!}{i!.(n-i)!}$.

We use a 7 order Bezier Spline Representation, which corresponds to 2 fixed points end 6 control points.

$$\begin{cases} x(t) = \sum_{i=0}^{7} C_7^i t^i (1-t)^{7-i} x_i \\ y(t) = \sum_{i=0}^{7} C_7^i t^i (1-t)^{7-i} y_i \end{cases}$$

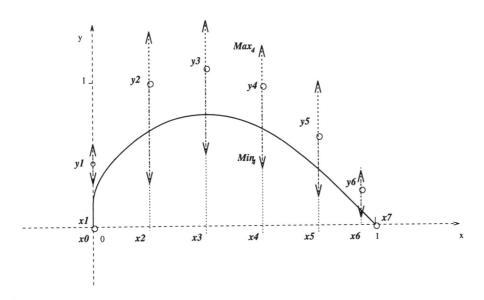

Figure 19.5 Extrados Bezier Spline Parametrization

We define a Bezier curve for the extrados (Figure 19.5) and another Bezier curve for the intrados, the whole airfoil shape is then defined by the merge of the two curves.

A Bezier curve is completely determined by the coordinates (x_i, y_i) of the P_i. In the case we treat, the shape is normalized, so the 2 fixed points are $P_0 = (0,0)$ and $P_7 = (1,0)$ that correspond to the leading and trailing edges of the airfoil shape. We also fix all the x_i of the control points $P_1 \ldots P_6$: we fix $x_1 = 0$ to respect an aerodynamic constraint that corresponds to the fact that the curve is to be tangent to the y axis for the leading edge, $x_1, \ldots, x_6$ are uniformly distributed in $[0,1]$. The only parameters that are taken into account are the ordinates y_i of those points. We proceed in the same way with the intrados curve and then we obtain 12 real numbers that are the ordinates of the airfoil shape control points. A chromosome C is a vector of $\Re^{12}$ whose genes are the ordinates of the control points.

$$C = (\underbrace{y_1, y_2, y_3, y_4, y_5, y_6,}_{Extrados} \underbrace{y_7, y_8, y_9, y_{10}, y_{11}, y_{12}}_{Intrados})$$

$$\text{with } \forall i \in [1..12], y_i \in \Re \cap [Min_i, Max_i]$$

Min_i and Max_i are lower and upper bounds of the variable y_i. The values of the bounds are problem dependent and can be used in order to apply a limited form of constraint (i.e. a geometrical constraint).

19.5.2 Mechanisms

Selection

The selection process is a very classical one, since it is based upon a Roulette-Wheel approach [Goldberg89b] with an elitist strategy. During our experiments, the reproduction rate was fixed to 0.2 whereas the cross-over rate was 0.8.

Cross-Over

The cross-over for a real-coded genetic algorithms is very close to the one defined for binary-coded genetic algorithms . Let A and A' be the parents chosen by the selection process. They produce the 2 following offsprings B_1 and B_2 if the randomly determined cutting point falls after the 4th gene.

$$A = \begin{pmatrix} y_1 \\ y_2 \\ y_3 \\ y_4 \\ \cdots \\ y_5 \\ y_6 \\ y_7 \\ y_8 \\ y_9 \\ y_{10} \\ y_{11} \\ y_{12} \end{pmatrix} \quad A' = \begin{pmatrix} y'_1 \\ y'_2 \\ y'_3 \\ y'_4 \\ \cdots \\ y'_5 \\ y'_6 \\ y'_7 \\ y'_8 \\ y'_9 \\ y'_{10} \\ y'_{11} \\ y'_{12} \end{pmatrix} \quad \xrightarrow{Cross-Over} \quad B_1 = \begin{pmatrix} y_1 \\ y_2 \\ y_3 \\ y_4 \\ \cdots \\ y'_5 \\ y'_6 \\ y'_7 \\ y'_8 \\ y'_9 \\ y'_{10} \\ y'_{11} \\ y'_{12} \end{pmatrix} \quad B_2 = \begin{pmatrix} y'_1 \\ y'_2 \\ y'_3 \\ y'_4 \\ \cdots \\ y_5 \\ y_6 \\ y_7 \\ y_8 \\ y_9 \\ y_{10} \\ y_{11} \\ y_{12} \end{pmatrix}$$

Mutation

The mutation operator is different from the one used in binary-code genetic algorithms (i.e. a bit inversion of a gene) because we deal with floating-points representations. We used the Non-Uniform Mutation defined by Michalewicz [Michalewicz92]. If a gene y_i is to be mutated, the new value y'_i is chosen at random within the interval $[Min_i, Max_i]$.

$$y'_i = \begin{cases} y_i + (Max_i - y_i).r.(1 - \frac{t}{T})^b & \text{if a random digit is 0} \\ y_i + (y_i - Min_i).r.(1 - \frac{t}{T})^b & \text{if a random digit is 1} \end{cases}$$

r is a random number from $[0,1]$

t is the generation number

T is the maximal generation number

b generation-dependancy parameter

We took T=30 (i.e. we fixed a maximum of 300 evaluations since the population size was 10) and b=5. We needed diversity among the population, so the mutation rate was fixed to 0.1.

Population

The algorithm we implemented uses small populations (from 10 to 20 individuals) in order to get the oppportunity to let them evolve during a few hundred generations. The choice consisting in taking larger populations has not been retained since it would have drastically limited the maximum number of generations that can be computed in a reasonnable time [Goldberg89a].

19.6   Applications

19.6.1   GAs fit of the database shape

All the runs we performed dealt with transonic aerodynamic test cases. Those test cases consist in optimization problems taking into account several parameters, and , for sake of comparison, they always consider a given initial shape which is precisely defined by the coordinates of the shape points. The first step is then to find the Bezier representation of that shape. To do so, we use a GA approach that use a least square fitness function. Let S be a shape and $f(S)$ its fitness. Let n be the number of points that define the target shape, $(x_i^{target}, y_i^{target})$ the coordinates of this points and (x_i^{target}, y_i) the coordinates of the points that define the shape S (after a discretization process that transforms the Bezier spline representation in a point by point representation).
The fitness function is then given by :

$$f(S) = \frac{1}{n} \cdot \sum_{i=1}^{n} (y_i - y_i^{target})^2$$

We obtain the target shape with a 10^{-4} accuracy on each point after 150 generations.

19.6.2   Inverse Problem : A general formulation

The first problem we deal with is an inverse problem that consists in finding the shape (denoted γ) of an airfoil which realizes a surfacic target pressure distribution for a given Euler flow condition. This problem has the following formulation:

$$minimize\ J(\gamma)\ with\ J(\gamma) = \frac{1}{2} \int_{\gamma} |p_\gamma - p_t|^2 d\gamma \qquad (19.24)$$

p_t is a given target pressure and p_γ is the actual flow pressure on γ.

We do compute the pressure distribution for a given shape and then we start the optimization process with another shape and try to find the first shape.Let n be the number of discretization points of the profile. The following discretized cost function f_n (fitness) is used both by the gradient approach and the genetic approach :

$$f_n(\gamma) = \frac{1}{2} \sum_{i=1}^{n} (P_{\gamma_i} - P_i^{target})^2$$

Where P_γ is the pressure of the evaluated shape via the Euler flow analysis solver described in Section 19.3 whereas P^{target} denotes the pressure distribution of the target shape.

Deterministic Approach : NACA0012 airfoil reconstruction (case : $M_\infty = 0.8, \alpha = 0^o$)

We first present the deterministic approach results.The determist optimizer has been applied to airfoil design problems using the Fletcher-Reeves conjugate gradient algorithm. State equations are solved by the centered scheme (V_4) described in the second section whereas the adjoint operator is based on the first-order implicit matrix. Parametrization of the profile is based on piecewise B cubic spline (74 nodes). The initial airfoil guess to start the computation was a NACA63215. After 120 evaluations (30 iterations) we almost found the target shape (Figure 19.6), which was the optimal one with a 10^{-2} accuracy for the cost function defined in equation (19.24).

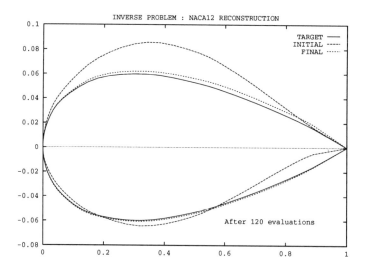

Figure 19.6   Inverse Reconstruction Problem by Deterministic method

Genetic Approach : RAE2822 airfoil reconstruction (case : $M_\infty = 0.73, \alpha = 2^\circ$)

In this computation, the reconstruction the RAE2822 airfoil is achieved with the GA optimizer described in section 19.4. Parametrization of the profile is based on Bezier Splines (16 control points, and 60 nodes on the profile). The initial guess was a NACA64a410 airfoil. The initial population was randomly generated except one individual that was a NACA64a410 airfoil. The parameters of the GA are the ones described in section 19.4. The genetic approach needed more function evaluations to converge, but the result (i.e. the best individual) was quite better (Figure 19.7).

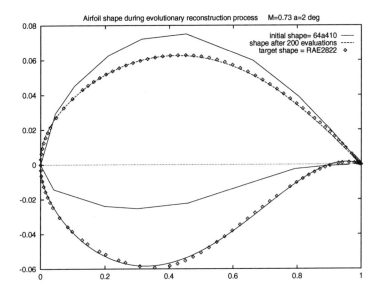

Figure 19.7   GAs : Shape Evolution during the Inverse Reconstruction Process

The pressure distribution (Figure 19.8) is almost the same than the target one. It is possible to notice the quasi-absence of shock in this case due to the Van-Leer first order scheme used in the Euler flow analysis solver and the small number of nodes on the profile.

The logarithmic convergence evolution (Figure 19.9) shows that the accuracy of the solution reaches 10^{-4} after only 140 evaluations (which corresponds approximatively to 14 generations)

19.6.3   Optimization Problem

The second test case aim is to reduce the shock-drag on a RAE2822 airfoil (with a Mach number $M_\infty = 0.73$ and an incidence angle $\alpha = 2^\circ$) for a given lift value (here, the value obtained on the initial profile). This optimization problem corresponds to the workshop test case TE3 in the ECARP european program. The general form of

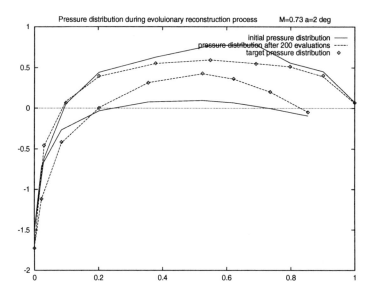

Figure 19.8   GAs : Pressure Distribution Evolution for the inverse problem

the cost function in such a problem is

$$j(\gamma) = \omega_1 \, I_{tar} \; + \; \omega_2 \, C_D \; + \; \omega_3 \, (C_L \; - \; C_L^{target})^2$$

The aim is to minimize the drag C_D while maintaining the same lift C_L^{target}. That constraint is taken into account by fixing the weights $\omega_1 = 0, \omega_2 = 1, \omega_3 = 10$. The cost function becomes

$$j(\gamma) = C_D \; + \; 10 \, (C_L \; - \; C_L^{target})^2$$

Deterministic Approach

The following results were found for the conjugate gradient approach

Shape	Drag	Lift	Number of Evaluations
Initial shape : RAE2822	0.0070	0.69	1
Optimized shape	0.0055	0.684	10

Table 1 : Deterministic shock-drag reduction

The aerodynamic solver used for this optimization is the one described in section 19.3 without farfield conditions. As the average radius of the outer boundary of the mesh is less than 8 chords. The computed lift is lower than the one obtained with this correction (checked to be around 0.89)

The shape is very slightly changed (Figure 19.10) but it is enough to decrease the drag from 0.0070 to 0.0055 while the lift remains almost constant. We can observe after convergence that the maximum thickness location moves toward the trailing edge.

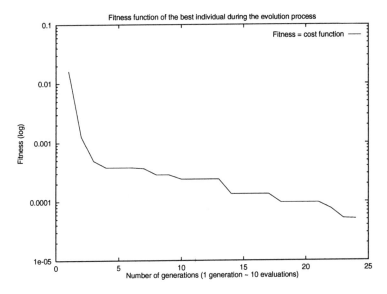

Figure 19.9   Convergence Evolution

On Figure 19.11 we can see the reduction of the shock strenght due to the drag reduction.

Genetic Approach

The genetic algorithms work well on that kind of problem, since we obtained the results described on Table 2. We can observe that GAs provide a better minimum with a thicker intrados leading edge.

Shape	Drag	Lift	Number of Evaluations
Initial shape : RAE2822	0.0070	0.69	1
Gradient method Optimized Shape	0.0055	0.684	10
GAs Optimized Shape	0.0033	0.696	120

Table 2 : GA's shock-drag reduction

The shape is a lot more modified than in the deterministic approach (Figure 19.12) and it made the drag decrease to 0.0033. But there was 120 fitness evaluations instead of the 10 needed by the gradient method.

The solution found by genetic algorithms is better, but further investigations including the continuation of the optimization process and the mesh refinement dependency for stability are considered in order to qualify the shape obtained by GAs.

Furthermore, the GA optimization process needs a lot more evaluations than the conjugate gradient one. So the next step consists in hybridizing the two algorithms in order to get the robustness of genetic algorithms and the fast convergence and accuracy of deterministic methods.

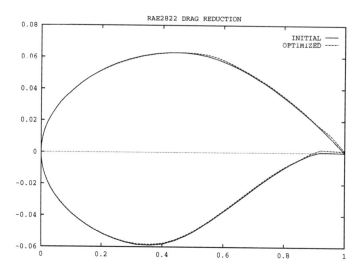

Figure 19.10   Gradient Method : Shape Optimization

Conclusion

Two different optimization approaches have been considered capable of treating
(multi) shape design optimization in an industrial environment and applied to selected
test cases of an Optimum Design Workshop. Both techniques rely on the same
aerodynamics — finite element Euler flow analysis solver — , finite element mesh
regenerator and cost or fitness functions modules.

The two approaches differ from their optimizer characteristics. The first
deterministic approach is based on control theory applied to the discrete system using
an approximate discrete sensitivity analysis and a conjugate gradient method has
been selected as descent method. The second approach is based on genetic algorithms
using real coding of the design variables. The selection process is based on an elitist
roulette-wheel whereas mutation acts directly on the floating-point numbers of the
representation by an adaptative refined gaussian function and the mating of two airfoil
individuals is achieved by cross-over on the genes representing a string of control nodes
of the parametrized shapes.

From the results of the optimization of the shape RAE2822 in term of shock-
drag reduction, different shapes have been obtained with the deterministic (CG) and
evolutive (GAs) optimizers. Point wise (CG) is definitely more efficient but (GAs)
provided a better minimum corresponding to a very different shape exhibiting a
significant modification of the intrados of the airfoil.

The design problems considered in this paper were quite simple. In a near future
CG and GAs will be tested on a more difficult constrained optimization problems such
as transonic multi-point design with strong viscous effects and the two approaches
hybridized in order to get robustness and a faster convergence to a global minimum.
On the one hand, much more efficiency can be gained from a possible combination

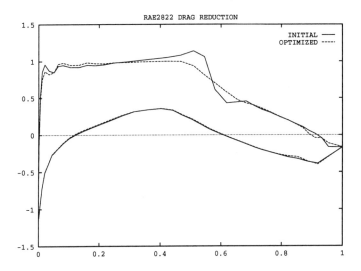

Figure 19.11   Gradient Method : Pressure Distribution

of conjugate gradient methods with hierarchical optimization [Beux et al.94].On the other hand, due to the costly repetitive solutions of large non-linear systems for each fitness evaluation of the individuals of the population in this aerodynamic context, the use of Parallel GAs on distributed architectures is investigated for cost-efficiency [Muhlenbein et al.91].

Acknowledgements

The results and methods described in this paper have been obtained in cooperation with many colleagues in France and abroad through fruitful discussions. We would like to mention G. Bugeda, H. Q. Chen, P. Chariavopulos, M. Galante, M. J. Galan, W. Haase, F. Hecht, T. Labrujere, R. A. E. Makinen, G. Montero, C. Poloni, D. Quagliarella, P. Questa, V. Selmin, J. Toivonen, G. Winter and most particularly A. Dervieux, J. M. Malé, N. Marco, who have implemented many deterministic optimization methods used in this paper.

The support of the CEC, DG XII under european contract ECARP and for CNRS for access to the IBM SP2 of CNUSC-Montpellier is also aknowledged.

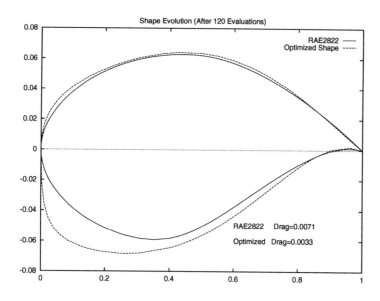

Figure 19.12   GAs : Shape Optimization for the direct problem

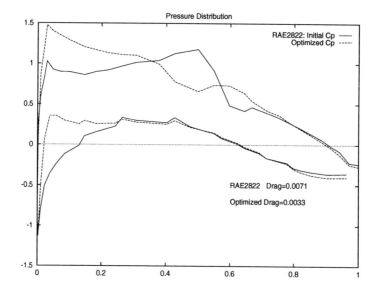

Figure 19.13   GAs : Pressure Distribution for the direct problem

REFERENCES

[Barth et al.95] Barth (T.J.) et Linton (S.W.). – An unstructured mesh newton solver for compressible fluid flow and its parallel implementation. *In : 33rd Aerospace Sciences Meeting and Exhibit.* AIAA. – Reno, 1995.

[Berde et al.95] Berde (B.) et Borel (M.). – Moment approach for the navier-stokes equations. *In : 12th Computational Fluid Dynamics.* AIAA.

[Beux et al.94] Beux (F.) et Dervieux (A.). – A hierarchical approach for shape optimization. *Engineering Computation,* no11, 1994.

[Cea81] Cea (J.). – Optimization of distributed parameter structures. *In : Numerical Methods of Shape Optimal Design,* éd. par Haug (E.J.) et Cea (J.). – The Netherland, 1981.

[Chalot et al.94] Chalot (F.), Mallet (M.) et Ravachol (M.). – A comprehensive finite element navier-stokes solver for low and high-speed aircraft design. *In : 32nd AIAA Aerospace Sciences Meeting and Exhibit.* AIAA. – Reno, 1994.

[Dervieux et al.93] Dervieux (A.), Fezoui (L.), Leclercq (M.P.) et Stoufflet (B.). – A general upwind formulation for compressible flows on multi-element meshes. *In : Numerical Methods for Fluid Dynamics 4,* éd. par Baines (M.J.) et Morton (K.W.). – Oxford Science Publications.

[Dervieux et al.94] Dervieux (A.), Male (J.M.), Marco (N.), Periaux (J.), Stoufflet (B.) et Chen (H.Q.). – Some recent advances in optimal shape design for aeronautical flows. *In : Computational Fluid Dynamics '94,* éd. par Wiley (J.).

[Dinh et al.93] Dinh (Q.V.), Stoufflet (B.) et Vossinis (A.). – *In : ICIAM Conference.* – Philadelphia, 1993.

[George et al.91] George (P.L.), Hecht (F.) et Saltel (E.). – Automatic mesh generator with specified boundary. *Computer methods in applied mechanics and engineering,* vol. 92, 1991, pp. 269–288.

[Goldberg89a] Goldberg (D. E.). – Sizing populations for serial and parallel genetic algorithms. *In : Proceedings of third International Conference on Genetic Algorithms,* éd. par Schaffer (J. D.). – Los ALtos, CA, 1989.

[Goldberg89b] Goldberg (D.E.). – *Genetic Algorithms in Search, Optimization, and Machine Learning.* – Reading, Mass., Addison-Wesley, 1989.

[Goldberg90] Goldberg (D.E.). – *Real-coded Genetic Algorithms, Virtual Alphabets, and Blocking.* – Technical Report 9001, University of Illinois at Urbana-Champain, September 1990.

[Holland75] Holland (J. H.). – *Adaptation in Natural and Artificial Systems.* – Ann Arbor, University of Michigan Press, 1975. *An early treatment of the genetic algorithm and classifier systems.*

[Jameson88] Jameson (A.). – Aerodynamics design via control theory. *Journal of Scientific Computing,* vol. 21, Nov. 1988.

[Janikow et al.93] Janikow (C. Z.) et Michalewicz (Z.). – An experimental comparison of binary and floating point representations in genetic algorithms. *In : Proceedings of the 5th Internations Conference on Genetic Algorithms.*

[Michalewicz92] Michalewicz (Z.). – *Genetic algorithms + data structures = evolution programs.* – New York, Springer-Verlag, 1992, *Artificial Intelligence.*

Genetic Algorithms in Engineering and Computer Science
Editor J. Périaux and G. Winter

[Muhlenbein et al.91] Muhlenbein (H.), Schomisch (M.) et Born (J.). – The parallel genetic algorithm as function optimizer. *Parallel Computing*, vol. 17 (6-7), September 1991, pp. 619–632.

[Perthame et al.94] Perthame (B.), Qiu (Y.) et Stoufflet (B.). – Sur la convergence des schémas fluctuation-splitting pour l'advection et leur utilisation en dynamique des gaz. *C.R. Académie des Sciences*, vol. 319 (Série I), 1994, pp. 283–28.

[Pironneau84] Pironneau (O.). – *Optimal Shape Design for Elliptic Systems.* – New-York, Springer-Verlag, 1984.

[Reuther et al.95] Reuther (J.) et Jameson (A.). – Aerodynamic shape optimization of wing and wing-body configurations using control theory. *In: 33rd Aerospace Sciences Meeting and Exhibit.* AIAA. – Reno, 1995.

[Rostaing et al.93] Rostaing (N.), Dalmas (S.) et Galligo (A.). – *Automatic Differentiation in Odyssée.* – Tellus, 1993.

[Struijs et al.91] Struijs (R.), Roe (P.L.) et Deconinck (H.). – *Fluctuation splitting schemes for the 2D Euler Equations.* – Technical Report LS 1991-01 in CFD, Von Karman Institute, 1991.

[Taasan94] Ta'asan (A. S.). – *Multigrid one shot methods for optimal control problems : Infinite dimensional control.* – Technical Report 94-52, ICASE, 1994.

[Young et al.94] Young (D.P), Huffman (W.P), Melvin (R.G.), Biterman (M. B.), Hilmes (C. L.) et Johnson (F. T.). – Inexactness and global convergence in design optimization. *In: 5th AIAA/NASA/USAF/ISSMO Symposium on Multidisciplinary and Optimization.* AIAA 94-4386. – Pananma City Beach, FL, September 1994.

20

Hybrid GA for Multi Objective Aerodynamic Shape Optimisation

CARLO POLONI

20.1 INTRODUCTION

In recent years much interest has been addressed to the use of Genetic Algorithms as general purpose optimisers and a large number of examples of engineering application can be found in the literature [Goldberg, 1992]. Recently this technique has been introduced even in the case of aerodynamic design where GA has been used for aerodynamic shape optimisation in the case of viscous incompressible flows [Poloni et al., 1993] and in the case of inviscid compressible flows [Quagliarella et al., 1994]. However the main concern related to the use of Genetic Algorithm for aerodynamic design is the computational effort needed for the accurate evaluation of a design configuration that, in the case of a crude application of the technique, might leads to unacceptable computer time if compared with other more classical algorithms [AGARD, 1994].

Three main issues makes however GA more than attractive and maybe unique among the aerodynamic design optimisation methods: GA are usually much more robust than gradient based algorithm and can tolerate even approximate design objectives evaluation, GA can be efficiently parallelised and can therefore take full advantages of the oncoming massively parallel computer architecture, GA can directly approach a multi objective optimisation problem [Hajela et al., 1993], [Horn et al., 1993], [Belegundu et al., 1994].

Most real-life design procedures are complex tasks that have to deal with multi disciplinary environments, not always clearly defined targets, constraints to be satisfied. In this sense even tough the target of the optimisation could be expressed with a single expression like: "do the best possible design", the optimisation process must consider several different usually conflicting objectives and the compromise obtainable might not be a-priori known. The possibility of looking not only for a single good solution but for a set of solutions (the Pareto Set) [Eschenauer et al., 1990] that satisfy different levels of compromise might be of great help to the decision maker that must select the most suitable one.

In this paper a Parallel Genetic Algorithm expressively designed for Pareto Frontier detection is applied to a single objective aerodynamic test problem and to a two objectives optimisation problem related to the pressure reconstruction problem and to a multi-point airfoil design.

20.2   GENETIC ALGORITHM FOR MULTI OBJECTIVE OPTIMISATION

The Genetic Algorithm used in this paper is similar to classical GA with the exception of the selection process and the alphabet that uses integers instead of binary. A fixed number of individuals is evolved for a given number of generation by means of selection, cross-over and mutation.

While traditional selection schemata allow to select an individual within the whole population, *local geographic selection* is based on the idea that the population has a particular spatial structure. It is divided in demes or semi-isolated sub-populations, with relatively thorough gene mixing within the same deme, but restricted gene flow between different demes. One way in which demes can be created in a continuous population and environment is isolation by distance: the probability that two individuals mate is a fast declining function of their geographical distance [Collins et al., 1991].

To simulate this schema, individuals are placed on a toroidal grid with one individual per grid location. Selection takes place locally on this grid. Each individual competes with its nearby neighbours. Specifically, an individual finds his mate during a random walk starting from his location: the individual with the best fitness value is selected. One dimensional and two dimensional grids have been used in the case here presented and even though in the preliminary tests no clear evidence of the advantages of one or the other grid, it seems that a correlation between the number of objectives and the cardinality of the grid might exist.

Local selection has been mainly adopted because of its applicability to multi-objective optimisation. It represents, in fact, a niching technique, whose aim is to maintain a useful form of diversity in the population [Harik, 1994]. In this sense, it is an alternative to the fitness sharing techniques [Goldberg, 1989]. Local selection has been preferred to the use of sharing techniques as it should naturally create niches without the need of problem dependent parameter tuning.

As for the crossover, a two-point crossover operator has been used to improve GA's search, as suggested by Booker [1987]. Elitism over on the whole population is also applied. Finally, the mutation operator acts by randomly changing the value of a gene.

Instead of combining the two objectives in a scalar fitness function, each individual is evaluated by means of a vector function, i. e. the value related to each objective is explicitly and separately considered. In [Horn et al., 1993], the multi-objective optimisation is carried on by the use of a modified tournament selection scheme and fitness sharing techniques, while in this approach the sub-population construction is secured by the use of the local selection for reproduction based on crossover operator. The selection schemata that has been used is based on a set of comparisons between two individuals; the winner is the individual that dominates the other one. The meaning of the dominance relation is the following: an individual dominates another one if all its fitness values are not worse, and at least one of them is strictly better. If neither individual dominates the other, one of them is randomly selected.

20.2.1   Parallelisation of GA

In the case of costly fitness evaluation the computing time used by the evolution process is usually negligible if compared even with a single design configuration evaluation. The first and most intuitive parallelisation is to distribute on all the available processors the fitness evaluations. However this approach is efficient only in the case of fitness evaluation of almost identical CPU requirements.

Quite often the CPU time needed by a single CFD simulation might be highly different for different geometry causing consequent synchronisation problems. Taking full advantage of the distributed memory capabilities offered by the CRAY T3D machine, it has been decided to use shared arrays as database of the active individuals in the population that is updated when the new individuals are evaluated. Using the local selection schemata and the shared individuals database the whole *selection / cross-over / mutation / fitness evaluation* can be viewed as a local process and consequently be done in parallel. With this algorithm no synchronisation exist at the end of each generation and all the processors are maintained busy all the time while the communications between the processors is still limited to the updating of the database.

20.2.2   Powell Direction Set Local Optimiser

To be applicable, an aerodynamic optimisation procedure should be both efficient as well as robust. Most of the time the designer needs first of all a feasible solution and secondly the optimum one. Two different perspectives can then be underlined: it should be possible to find innovative design and it should be possible to improve them. While Genetic Algorithm has been proved to be capable of finding unconventional solution to a design problem [Tong, S.S., 1994], it usually performs poorly in terms of computational effort for finer tuning of an existing solution. A hybrid optimisation technique should offer a compromise between efficiency and robustness requirements.

Most of the efficient optmisation algorithms are based on the assumptions that the objective function is convex and the domain is simply-connected. These assumptions can be considered applicable if a "local" optimisation in the neighbourhood of a good feasible solution is done.

The use of the derivatives of the objective function with respect to the problem variables is a questionable matter. If existing analysis codes, usually not available in their source form are used, the correct computation of derivatives by finite differencing becomes costly and inaccurate making more attractive methods that do not need the explicit evaluation of the derivatives. For all these reasons it has been decided to use a modified version of Direction Set Powell algorithm [Acton,F.S., (1990)] as a "local" operator and Genetic Algorithm as a "broad scale" operator.

Being $f(x)$ a continuous function with n variables, let us consider the Taylor expansion series:

$$f(\overline{x}) = f(\overline{x}_0) + \sum_{i=1}^{n} \frac{\partial f}{\partial x_i} x_i + \frac{1}{2} \sum_{i,j=1}^{n} \frac{\partial^2 f}{\partial x_i x_j} x_i x_j + \cdots \approx c - b \cdot \overline{x} \cdot A \cdot \overline{x}$$

where: $c \equiv f(\overline{x}_0)$ $b \equiv -\nabla f \mid_{\overline{x}_0}$

and $[A]_{i,j} \equiv \frac{\partial^2 f}{\partial x_i x_j} x_i x_j \mid_{\overline{x}_0}$

with this approximation the gradient of the function can easily be computed:

$$\nabla f \approx A \cdot \overline{x} - b$$

and its variation: $\delta(\nabla f) \approx A \cdot \delta \overline{x}$

Suppose that a minimum point is reached along the generic direction $\overline{u}$ and that a new one-dimensional minimisation is to be performed along $\overline{v}$. In order to maintain a minimum condition along $\overline{u}$ it is necessary that $\overline{u}$ and $\overline{v}$ are orthogonal:

$$\overline{u} \cdot \delta(\nabla f) = \overline{u} \cdot A \cdot \overline{v} = 0$$

Successive one-dimensional minimisation along conjugated direction will bring to the minimum of the quadratic form. The Powell here summarised produce in k iteration k direction mutually conjugated and therefore with $n(n+1)$ one dimensional minimisation the minimum is found:

- Initialize the direction set with the basis vectors
- Save the starting point as P_0
- For $i = 1...N$ move P_{i-1} to the minimum along direction u_i and call the point P_i
- For $i = 1...N-1$ set $u_i = u_{i+1}$
- Set $u_N = P_N - P_0$
- Move P_N to the minimum along u_N and call the point P_0

In practice some heuristic rules are needed in order not to produce linearly dependent conjugated directions. Minor modifications to the algorithm reported in [Press, W.H., et al. 1992] have been done in order to allow the use of penalty functions to treat constrained optimisation.

20.3 MULTI POINT PRESSURE RECONSTRUCTION PROBLEM

The optimisation problem that is here addressed deals with the construction of a compromise airfoil that should deals with two incompatible working condition: high lift in subsonic flight and low drag in transonic flight. The testcase considered is a

Table 20.1   Two point airfoil design

'airfoil'	case	incidence angle ($\alpha°$)	Mach number
high lift	1	10.8	0.20
low drag	2	1.0	0.77

modified version of the testcase shown by Laburjere [AGARD, 1994]. The contour geometry of two existing airfoils one for high lift and the other for low drag are given in a tabulated form. The pressure distribution is therefore to be computed by the same flow solver that is used for the following pressure reconstruction. No assumption is made on the flow solver to be used that in the case that is here presented is a classical inviscid full-potential transonic solver that has been chosen to show the applicability of this methodology. The two design point at which a specific profile is given are summarised in table 20.1.

Once the reference pressure distribution are computed, the two-point design problem is defined as the minimisation of the objective function:

$$F\left(\alpha_1, \alpha_2, x(s)\right) = \sum_{i=1}^{2} \left[W_i \int_0^1 \left(C_p^i(s) - C_{p_{tar}}^i(s) \right)^2 ds \right] \quad (1)$$

where s is the fractional arc length measured along the airfoil contour to be defined, where $\alpha_1, \alpha_2, x(s)$ and $y(s)$ are the design variables and where the weight factors W_i are given constants.

C_p^i represents the actual pressure distribution at design condition i (given by α_i and Mach number), $C_{p_{tar}}^i$ represents the target pressure distribution at design condition.

20.3.1   Airfoil Shape Parametrisation

Any optimisation approach needs a parametric representation of the problem and this task is usually matter of studies by itself. In the case of aerodynamic design by means of direct flow solvers one has always to cope with computationally expensive objective function evaluation. Efficiency of the solvers and of the optimisers might in fact vanish if an efficient geometrical representation is not adopted. The number of degree of freedom given to the parametrised shape must be large enough to explore a wide range of shapes but small enough in order to limit as much as possible the number of variables involved in the problem.

Satisfactory results have been obtained using an assembly of Bezier curves [Arcilla, A.S. et Al., (1991)] for the definition of parametrised shapes and therefore even in the case of the airfoil a similar approach have been used. A Bezier polynomial of order n is defined by:

$$b(t) = \sum_{i=0}^{n} P_i \Phi_i(t)$$

where:
$$\Phi_i(t) = \left(\frac{n!}{i!(n-i)!} \right) t^i (1-t)^{n-1}$$

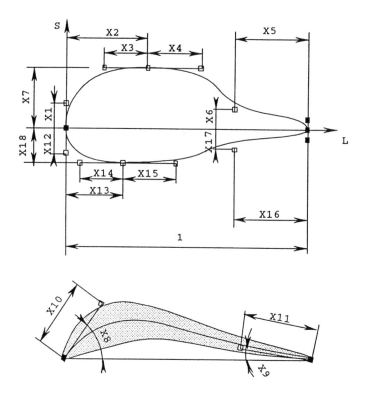

Figure 20.1   Airfoil shape parametrisation

and P_i's are the vertices of the Bezier control polygon. In matrix form the Bezier curve can be expressed by $b(t) = TBP$ where T is a primitive polynomial basis of order n, B is a symmetric basis change matrix made of constants and P is the vector of the $n + 1$ control points. The main properties that make these curves suitable for the geometry parametrisation are the following:

- the curve is independent of the choice of the coordinate system
- the curve will lie within the convex generated by the vertices of its control points
- the first two and the last two control points define the tangent to the curve at the start and at the end of it

In the example here shown 18 parameters define the airfoil shape, four of them determines the camber line defined by means of one cubic Bezier curve while the remaining 12 determines the location of the control points related to the upper and lower side of the thickness distribution that is superimposed to the camber line. Each side of the thickness distribution is given by two Bezier polynomials one of order three for the leading edge and one of order four for the rear part. A schematic representation of the variables' meaning is given in the figure 20.1.

String Representation of the Shape

The eighteen variables needs to be converted into a genetic string when GA is applied and into a vector of variables when Powell is used. In order to have a coherent representation all the variables are adimensionalized within a given feasible interval.

In the case presented it has been preferred to use an integer alphabet for the genetic string with a basis equal to number of intervals n_{int} that each variable interval is divided. The optimisation is then performed in a transformed variable space where a given variable $X(i)$, that can span from $min_{X(i)}$ to $max_{X(i)}$, is translated into the adimensionalized interval for the variable $X_{tr}(i)$:

$$X(i) = min_{X(i)} + n_{int} \left(\frac{X_{tr}(i)}{max_{X(i)} - min_{X(i)}} \right)$$

where:

$$max_{X(i)} \geq X(i) \geq min_{X(i)}$$

and:

$$0 \geq X_{tr}(i) \geq n_{int}$$

Using this kind of encoding the optimisation problem can be straight forward be faced using GA or Powell. While in the case of GA strings of integers are used, Powell algorithm uses reals providing in the end a *refining* technique more than a global optimiser.

20.4   MULTI POINT NUMERICAL EXPERIMENT

In the first three numerical experiments a single objective optimisation have been performed doing the pressure reconstruction problem for the reference airfoils and the design of a compromise shape with fixed weights. In details the three cases considered, referring to equation (1) are the following:

1. $W_1 = 1.0$ $W_2 = 0.0$
2. $W_1 = 0.0$ $W_2 = 1.0$
3. $W_1 = 0.5$ $W_2 = 0.5$

The Genetic Algorithm has been run for the three cases with the characteristics summarised in table 20.2 while the variables feasible range is given in the following table. The coordinate system where the variables are defined is a cartesian frame that have the origin at the leading edge of the airfoil where the mean line starts as shown in figure 20.1. The airfoil length is posed equal to one.

In all the cases three different optimisation tests have been made:

- one optimisation with GA
- one optimisation with Powell starting from the best obtained with GA
- one optimisation with Powell starting from a configuration with initial values equal to 1 for all the eighteen adimensionalised variables, i.e. with variables close to the minimum allowable value

Table 20.2   Single objective run

pop. size	alphabet base	n. of generation	cross-over prob.	mutation prob.
64	20	64	0.75	0.1

Table 20.3   Variables range

variable	min	max	variable	min	max
X1	0.00	X7	X10	0.00	0.475
X2	0.10	0.45	X11	0.00	0.475
X3	0.00	X2	X12	X18	0.00
X4	0.00	1-X2	X13	0.10	0.45
X5	X2+X4	1.00	X14	0.00	X13
X6	0.00	X7	X15	0.00	1-X13
X7	0.05	0.11	X16	X13+X14	1.00
X8	-11	16	X17	X18	0.00
X9	-11	16	X18	-0.11	-0.05

In this way a comparison of *robustness-accuracy-efficiency* can be made for each technique isolated or for a combination of both.

20.4.1   Numerical Results

Case 1.

The pressure reconstruction problem for this case can be seen as a relatively smooth optimisation problem. The solution is in fact unique and any multimodality of the objective function would be eventually due to the geometry parametrisation that might not allow a perfect reconstruction of the profile geometry.

In this case as it can be seen in figure 20.2 the reconstructed geometry obtained with a GA run is almost the same for all the computations even though the combined result is a little better. It is relevant to note that the computational effort, proportional to the number of calls to the CFD solver, of Powell and GA isolated are for this problem size comparable while a few iteration of Powell are needed after GA to converge. The reconstructed pressure distribution as shown in figure 20.3 is also very similar in the three cases.

Case 2.

The pressure reconstruction problem in the transonic regime is a much more difficult task. The shape of the transonic profile given as a reference is not smooth and presents a pressure distribution with a pronounced suction peek at the leading edge and a shock wave in the rear part.

In this case, as shown in figure 20.4, the geometry is well reconstructed with GA

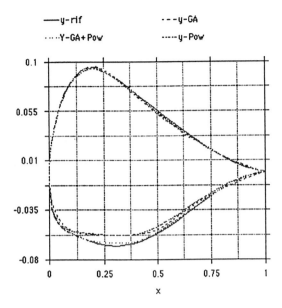

Figure 20.2   Case 1 Reconstructed geometries

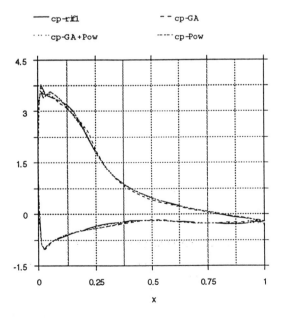

Figure 20.3   Case 1 Reconstructed pressure distribution

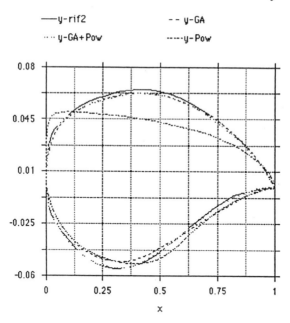

Figure 20.4   Case 2 Reconstructed geometries

and GA+Powell while the isolated Powell result is not acceptable. The strong non-linearity of the problem points out the weakness of a classical approach if the starting point of the optimisation is not closed to a good solution. Even more evident are these discrepancies in the pressure distribution as shown in figure 20.5.

The suction peek at the leading edge is not perfectly reconstructed by none of the algorithms but this is may be due to the smoothing of the shape given by the Bezier curves while the isolated Powell result is completely wrong.

Case 3.

In the case of the search for a compromise airfoil another important consideration must be made: uniqueness of the solution is not demonstrated and a complex multidimensional landscape can be foreseen for the objective function that in the end represent a trade-off between two conflicting objectives.

The results obtained in terms of objective function value by isolated GA and isolated Powell are close to each other and GA+Powell offer a small improvement. However if we compare as in figure 20.6 the geometries obtained, a relevant difference exist between the solution obtained with GA and a pure Powell. This can be explained by the observation that the objective function (1) equally weights two integrals of different absolute values for Cp that in the first design point do reach values close to 4. while in the second one reach only 1.4. As a consequence the geometry is probably unbalanced in favour of a high lift airfoil.

Figure 20.7 and 20.8 show the pressure distribution in the two operating point. In

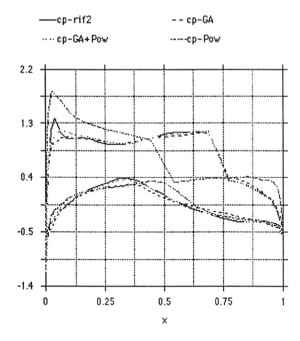

Figure 20.5   Case 2 Reconstructed pressure distribution

transonic flow the discrepancies, as it can be expected, are larger.

Convergence History

Some considerations can be made on the algorithm performances by looking at the convergence history of Powell and GA runs. It can be noted not only the computational cost of a Powell run but also the quality of the solution obtained is highly influenced by the starting point of the optimisation process. While a Powell run applied to a good solution obtained with a previous GA run converges quickly, a run from a bad starting point can be both more expensive than a GA as well as less accurate.

The hybrid approach results to be always better with a computational effort that is however less than two independent run with whatever algorithm. These results are shown in figure 20.9 and 20.10 where the objective function value adimensionalised with respect to the NACA4412 airfoil is plotted against the number of calls to the CFD solver.

20.5   MULTI OBJECTIVE NUMERICAL EXPERIMENT

As pointed out in the previous section, in the case of the search for a compromise solution that represents a trade-off between two or more conflicting objectives the common practise of fixing some weights in order to combine linearly the objectives

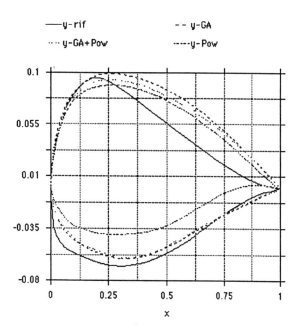

Figure 20.6   Case 3 Reconstructed geometries

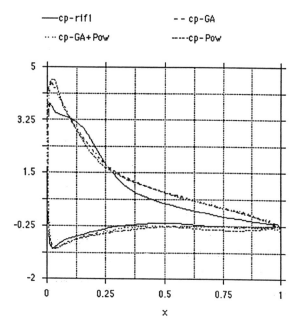

Figure 20.7   Case 3 Reconstructed pressure distribution -subsonic-

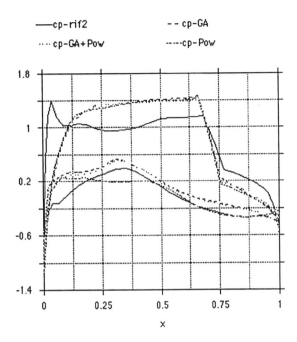

Figure 20.8   Case 3 Reconstructed pressure distribution -transonic-

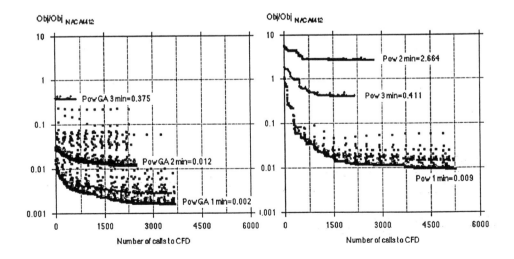

Figure 20.9   Convergence history for Powell runs

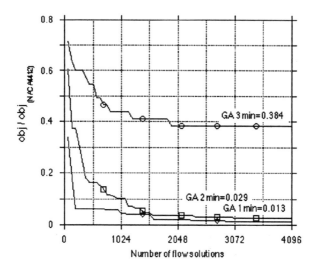

Figure 20.10   Convergence history for GA runs

is a questionable matter. A wrong weighting might drive any optimisation process toward wrong answers. Usually the weights tuning might be an expensive cut-and-try empirical procedure that make necessary a large number of optimisation runs before reaching satisfactory results. On the contrary the selection process described previously do offer a unique tool to avoid any weighting procedure giving a better understanding of the design problem by means of the so called "Pareto Set of efficient solutions".

In the case of the search for an airfoil that combine the pressure distribution of two other given profiles operating in two different conditions can be translated into the search for a *family of profiles* that have a smooth transition in shape from the first reference profile to the second one with a consequent satisfaction of different level of compromises between the two corresponding pressure distributions. In a classical way this correspond to a large number of shapes optimised with different weighted objective function.

It must be noted that while in the case of low subsonic speed the pressure distribution is in a linear relation with the profile geometry and therefore any linear combination of profile geometry gives a consequent linear combination in the pressure distribution, the transonic flow is highly non-linear ad therefore the combination of reference airfoils is no longer applicable.

In this case only a GA run has been made, considering an eventual further Powell runs to be done using a weighted multi-point approach with weightings taken from the analysis of the Pareto Set detected.

The GA was run with a local selection scheme with a random walk of 4 steps on a 1D and 2D grid of 16x8. Only the solution related to the 1D grid will be shown as better definition of the Pareto Set have been obtained in this case even though the differences found between the two where negligible. The setting parameters for the GA run are

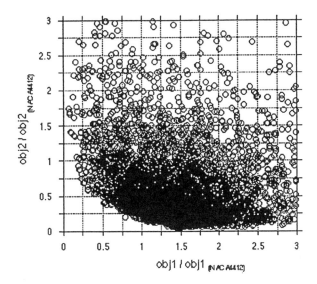

Figure 20.11   Configurations computed in the objectives plane

the same as for the previous examples with the exception of the population size that in the multi objective case have been doubled having 128 configurations evolving for 64 generations.

In this case a "convergence" history do not have any meaning and therefore a different representation of the computational results is needed. As usual in the field of multi objective optimisation the results are plotted in two dimensional sections of the multi dimensional space of objectives.

In this case being only a two objectives optimisation one plane is of course sufficient. In figure 20.11 all the computed configurations are shown in the two objective plane while in figure 20.12 the filtered Pareto Set is plotted. Even though the curve obtained is not smooth and could be probably be improved by further local refinement, a clear impression of *what can be done* looking for compromise airfoils.

A sample of airfoils that lie on the Pareto curve are shown in figure 20.13. Clearly a smooth transition in shape is detected from high lift to low drag airfoils.

20.6   COMPUTATIONAL RESULTS

All the computations have been made using the full-potential transonic solver [Jameson,A., 1986] with a 64x32 grid points running population size of 64 individuals for 64 generation for the single-objective cases and 128 individuals for 64 generations for the multi-objective case. For the two point airfoil design the tests have been done on a serial machine (the Cray XMP/116 of the University of Trieste) while only one design point pressure reconstruction problem (case 1.) was used to determine the performance of the code on a parallel machine, the 64 nodes Cray T3D of the Italian

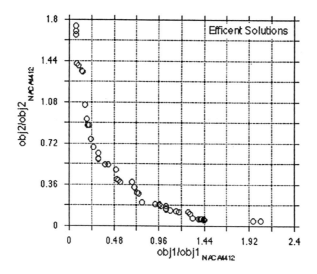

Figure 20.12   The filtered Pareto Set in the objectives plane

Table 20.4   Computational data

Computer	Problem Size	CPU [s]	CPU [s] 64x64 estimate
Cray XMP/116	64x128 (2 obj.)	15200	—
Cray XMP/116	64x5	244	3123
Cray C90	64x5	110	1408
Cray T3D 64PE	64x64	44.1	44.1
Cray T3D 32PE	64x64	83.9	83.9
Cray T3D 16PE	64x64	163.1	163.1

computing centre CINECA. The comparative results are summarised in the following table. A linear speed up can be seen. In the case of 64 processors the code was running at 3.5GFlops for the 30% of the time with a global speed of 1.5GFlops. These data have been estimated with the help of *Apprentice* library.

20.7   CONCLUSIONS

It has been shown that a comparison in terms of computational efficiency between GA and classical optimisation algorithms is not applicable for several reasons. In the aerodynamic problem shown GA not only has revealed a superior robustness in general but also, as pointed out in the case 1., if the Powell algorithm is not well initialised the computational costs are comparable. However a hybrid approach that uses Powell method as a local optimiser of good solutions obtained with GA always produced better results. If a multi objective optimisation is performed GA becomes far more

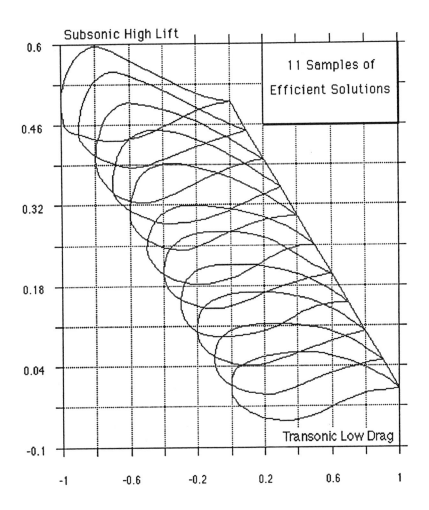

Figure 20.13 Smooth geometric transition from low drag to high lift airfoils
as obtained with a multi objective optimisation

efficient than a large number of differently weighted multi-point optimisation.

The GA code has also been parallelised and the results obtained proved that a high computational efficiency can be obtained on massively parallel computer even though it must be noted that this is strictly valid as far as the problem size is small enough to remain confined in one processor which might be not the case of large CFD problems.

Acknowledgements

The support of Prof. Giovanni Mosetti of the University of Trieste given to me has been determinant for the developement of all the fases of this research topic.

20.8 BIBLIOGRAPHY

1. Acton, F.S.(1990): *Numerical Methods That Work* , Mathematical Association of America, corrected edition (1990), 464-467
2. AGARD Report (1994): *Optimum Design Methods in Aerodynamics* , AGARD-R-803
3. Arcilla, A.S. et Al.(1991): *Numerical Grid Generation in Computational Fluid Dynamics and Related Fields* , Proceedings of the Third International Conference on Numerical Grid Generation in Computational Fluid Dynamics, North-Holland
4. Belegundu, A. et Al. (1994): *Multiobjective Optimization of Laminated Ceramic Composites Using Genetic Algorithms* , AIAA paper 94-4363-CP
5. Booker, L. (1987): *Improving Search in Genetic Algorithms* , in L. Davis (ed.), Genetic Algorithms and Simulated Annealing, Morgan Kaufmann Publishers, Los Altos, USA pp. 61-73.
6. Collins, R.J., Jefferson, D.R. (1991): *Selection in Massively Parallel Genetic Algorithms* , Proceedings of The Fourth International Conference on Genetic Algorithms, pp. 249-256, San Diego, USA.
7. Eschenauer,H., Koski, J. and Osyczka, A. (1990): *Multicriteria Design Optimization procedures and applications* , Springer-Verlag Berlin
8. Goldberg, D.E. (1989): *Genetic Algorithms in Search, Optimization and Machine Learning* , Addison-Wesley, Reading/Mass, USA.
9. Goldberg, D.E., Kelsey M. and Tidd, C. (1992): *Genetic Algorithm: A Bibliography* , (IlliGAL Report n. 92208), University of Illinois at Urbana-Champaign, Illinois Genetic Algorithm Laboratory, Urbana, USA
10. Hajela, P (1990): *Genetic Search-An Approach to the Nonconvex Optimization Problem* , AIAA Journal Vol.28, N. 7
11. Harik, G. (1994): *Finding Multiple Solutions in Problems of Bounded Difficulty* , (IlliGAL Report n. 94002), University of Illinois at Urbana-Champaign, Illinois Genetic Algorithm Laboratory, Urbana, USA
12. Horn, J. and Nafpliotis, N. (1993): *Multiobjective Optimization Using the Niched Pareto Genetic Algorithm* , (IlliGAL Report n. 93005), University of Illinois at Urbana-Champaign, Illinois Genetic Algorithm Laboratory,

Urbana, USA

13. Jameson, A. (1986): *A Vertex Based Multigrid Algorithm for Three Dimensional Compressible Flow Calculations* , In Proc. of the ASME Symp. on Numerical Methods for Compressible Flow, Anaheim.

14. Poloni, C. and G.Mosetti (1993): *Aerodynamic Shape Optimization by Means of a Genetic Algorithm* , 5th International Symposium on Computational Fluid Dynamics, Sendai, Japan.

15. Press, W.H. et Al.(1992): *Numerical Recipes* , Cambridge University Press, 2nd Editon

16. Quagliarella, D. and Della Cioppa, A. (1994): *Genetic Algorithms applied to the Aerodynamic Design of Transonic Airfoils* , 12th AIAA Applied Aerodynamics Conference, Colorado Springs, CO USA, AIAA-94-1896

17. Tong, S.S. (1991): *Integration of Symbolic and Numerical Methods for Optimising Complex Engineering Systems,in Proc. of Programming Environement for High Level Scientific Problem Solving* , Inter. Fed. for Information Processing, Germany

21

Genetic Algorithms Applications in Computational Fluid Dynamics

DOMENICO QUAGLIARELLA[1]

21.1 INTRODUCTION

Genetic Algorithms are a versatile tool for optimization, and are beginning to be used in aerodynamic design applications that range from airfoil drag minimization to multiobjective inverse airfoil design. The ability of evolutionary procedures in effectively searching solution spaces for optima of non linear problems has been proved, even if these spaces are not simply connected and characterized by a large number of variables. Furthermore, problems characterized by a mix of continuous and discrete variables can be successfully dealt with using such techniques. However their application to aerodynamic design presents some drawbacks that have to be taken into account. In particular, the evaluation of the objective function and of the constraints requires the determination of aerodynamic characteristics that have to be evaluated using numerical programs which may need a significant amount of computational resources.

Various approaches have been suggested to overcome the above mentioned limits. Among these the most successful uses approximated evaluations particularly when the fitness function requires a high degree of computation. Parallelization of genetic algorithms is also suitable when trying to improve the performance in difficult aerodynamic design tasks. However the effective application of evolutionary procedures to such problems requires also the parallelization of the solver in conjunction with massively parallel architectures.

In other words, given a population size suitable for the class of problem to be treated,

---

[1] Centro Italiano Ricerche Aerospaziali, Via Maiorise, I81043 Capua (CE), Italy

the goals that have to be achieved are the minimization of the population elements that require exact evaluation, the speeding up of the exact evaluation and the simultaneous evaluation of all the population elements belonging to the same generation.

This work concerns the application of genetic algorithms in computational fluid dynamics reporting results obtained using these techniques in the design of aerodynamic shapes and in the determination of the optimal computational load balancing in parallel multiblock flow field solvers. Aerodynamic shape design involves optimal computational load balancing when the evolutionary procedure is implemented on a massively parallel computer. In fact it can be implemented on a multiprocessor architecture, parallelizing both the generation evaluation phase and the flow field solver used for fitness evaluation.

The population size and the available number of processing units determines the balance between the parallelization of generation evaluation and the degree of parallelism in flow field solver. In fact, the efficiency obtainable in the parallelization of the evaluation loop of the genetic algorithm may be very high because the evaluation of each population element is independent of any other. On the other hand, the efficient parallelization of an aerodynamic code is more critical and depends on the data. It is, therefore, more difficult to obtain satisfactory speedup, especially when many processing units are used, because of the increased communications overhead.

Parallel generation evaluation, sequential flow solver

Sequential generation evaluation, parallel flow solver

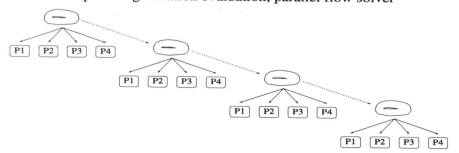

Parallel generation evaluation, parallel flow solver

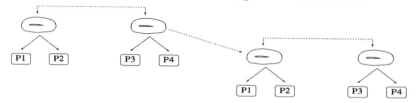

Figure 21.1   Levels of parallelism in a genetic optimization procedure

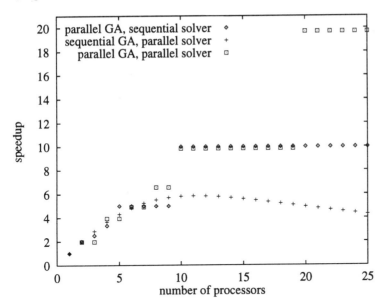

Figure 21.2 Ten member population: obtainable speedup using different
approaches

Optimum speedup may be obtained using a combination of the two approaches,
i. e. few processors, for example only two, for the fitness calculation and optimizing
the related load balancing (see fig. 21.1). Fig. 21.2 compares the obtainable speedups
when different approaches to parallelization are adopted for a population size of ten
elements. When one flow solution is attributed to each processing unit, the speedup
increases up to ten processors. Thereafter no increase of speedup is obtained although
more processors are available. Parallelizing the flow solver only demonstrates a strong
dependency on load balancing between the processors for the partitioned flow domain.
The use of parallel genetic algorithm and of parallel flow solver offers the best
settlement.

21.2 OPTIMUM AIRFOIL DESIGN

A wide variety of techniques have been developed to deal with the aerodynamic design
problem [Dul91], ranging from inverse methods [Slo84], to numerical optimization
[Van79, CH86], to control theory [Jam88] and to expert systems techniques [Ton85,
GQ92]. Evolutionary procedures are now emerging as an alternative optimization
technique [TPS89, MP93, QD94, PM95, QV95] that offers better robustness
characteristics than traditional "hill climbing" search techniques that can be easily
trapped into local optimum configurations. This makes such procedures more suitable
to deal with multipoint and multiobjective design problems.

In the following sections a computational procedure is presented that can be used for
the inverse and direct design of transonic airfoils. It can use either a genetic algorithm

or a traditional gradient-based method as a numerical optimization routine. Some application examples of this procedure to transonic airfoil design will be described.

21.2.1 The Optimization Problem

To solve the airfoil optimization problem, the values of the design variables x_i must be found in order to minimize the objective function $F(\mathbf{x})$, provided that the possible constraints $g_i(\mathbf{x}) \leq 0$ and $h_i(\mathbf{x}) = 0$ are satisfied. Constraints on lift coefficient and airfoil thickness have been considered adopting an explicit formulation. This means that, when the constraints are active, the actual airfoil thickness is scaled to the desired value after each geometry modification, and the aerodynamic solver searches for the particular angle of attack which provides the required lift coefficient. Both inverse and direct design problems have been addressed. In the first case, the objective function is computed as:

$$F(\mathbf{x}) = \int_{s_0}^{s_1} (c_p(\mathbf{x}, s) - c_p^t(s))^2 ds \tag{21.1}$$

where c_p^t is the target pressure distribution at design conditions (M, α), c_p the actual pressure distribution, and s is the fractional arc length measured along the airfoil contour. The direct design problem consists in the minimization of:

$$F(\mathbf{x}) = c_{d_w}/c_l^2 \tag{21.2}$$

at a given design condition specified in terms of Mach number and lift coefficient. Two point design problems can also be dealt with; in this case, the objective function is the sum of expression (21.2) evaluated at the two specified conditions. The aerodynamic solver adopted for calculations is a non-conservative full-potential one, with a 128×32 grid size.

21.2.2 Geometry Description

Two different techniques have been used for the airfoil geometry handling. A well known and widely used approach [HH78] consists in representing the airfoil upper and lower surfaces by means of a linear combination of a baseline shape and some given modification functions:

$$
\begin{aligned}
y^u &= y_b^u + \sum_{k=1}^{m} w_k f_k \\
y^l &= y_b^l + \sum_{k=m+1}^{n} w_k f_k
\end{aligned} \tag{21.3}
$$

where f_k is a given modification function and w_k the related weight. This ensures shape continuity and differentiability if base shape and modification functions are continuous and differentiable. Of course, the choice of the modification functions set is to some extent arbitrary, which constitutes a possible drawback of this kind of representation. In fig. 21.3 a typical set of modification functions is illustrated. An alternative technique is that of representing the airfoil contour by means of B-spline polynomials. Letting $\mathbf{p}(t)$ be the position vector along the curve as a function of the parameter t a B-spline is given by:

$$\mathbf{p}(t) = \sum_{i=1}^{n+1} \mathbf{b}_i \, N_i^k(t) \qquad t_{\min} \leq t \leq t_{\max}, \quad 2 \leq k \leq n+1 \qquad (21.4)$$

where the $\mathbf{b}_i$, are the position vectors of the $n+1$ defining polygon vertices and the N_i^k are the normalized B-spline basis functions [RA90]. For the ith normalized B-spline basis function of order k (degree $k-1$) the basis functions $N_i^k(t)$ are defined by the Cox-de Boor recursion formulas:

$$N_i^1(t) + \begin{cases} 1 & \text{if } x_i \leq t \leq x_{i+1} \\ 0 & \text{otherwise} \end{cases} \qquad (21.5)$$

and

$$N_i^k(t) = \frac{(t - x_i) N_i^{k-1}(t)}{x_{i+k-1} - x_i} + \frac{(x_{i+k} - t) N_{i+1}^{k-1}(t)}{x_{i+k} - x_{i+1}} \qquad (21.6)$$

The values of x_i are elements of a knot vector satisfying the relation $x_i \leq x_{i+1}$. The parameter t varies from $t_{\min}$ to $t_{\max}$ along the curve $P(t)$ and the convention $0/0 = 0$ is adopted.

The B-spline curve is as a polynomial spline function of order k (degree $k-1$) since it satisfies the following two conditions:

- The function $\mathbf{p}(t)$ is a polynomial of degree $k-1$ on each interval $x_i \leq t \leq x_{i+1}$, x_i being elements of a uniform knot vector satisfying $x_i \leq x_{i+1}$.
- $\mathbf{p}(t)$ and its derivatives to the order $k-2$ are continuous over the entire curve.

Each airfoil side is thus defined by assigning a starting *base polygon* $\{\mathbf{b}_1^0, \mathbf{b}_2^0, \ldots, \mathbf{b}_{n+1}^0\}$ and allowing, for each vertex, displacements within fixed boundaries. Fig. 21.4 shows a NACA 64A410 airfoil represented through a 6th order B-spline; the variation range for one of the control vertices is also illustrated. A displacement of one of the control points $\mathbf{b}_i$ introduces a modification of the curve in the interval between the points $\mathbf{b}_{i-k+1}$ and $\mathbf{b}_{i+k-1}$; the effect is then more "global" as the degree of the B-spline is increased. This technique for geometry handling allows a more general approach to the representation of the airfoil contour; the degrees of freedom can be adjusted by adding or deleting control points (or just freezing their positions), or by changing the order of the B-spline.

21.2.3 The Genetic Algorithm

Each design variable is assigned a variation range, and is then coded over the binary alphabet. The coded variables are then concatenated to form a single *chromosome*, containing all the *genetic* information about the corresponding geometry (fig. 21.5). An initial population of possible solutions is first generated by assigning random values to the design variables, or by perturbing a starting geometry in a deterministic or random fashion. This initial population then evolves under the action of selection, crossover and mutation mechanisms applied to the chromosomes. The fitness function drives

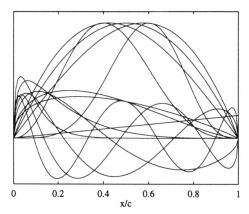

Figure 21.3   Set of modification functions

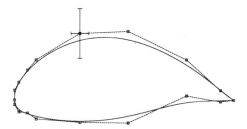

Figure 21.4   NACA 64A410 represented with a 6th order spline

the selection promoting the individuals with the best aerodynamic characteristics. The individuals suitable for reproduction are selected among the actual population by means of a biased roulette wheel, where each current string in the population has a roulette wheel slot sized in proportion to its fitness. After two parent strings have been selected, reproduction takes place by means of the *crossover* and *mutation* operators. Thus, the fundamental steps that characterize the algorithm can be summarized as follows:

1. a suitable starting population of *chromosome strings* is first created;
2. each population element is evaluated by computing its *fitness*;
3. a selection probability is assigned to each string on the base of its *fitness*;
4. new *chromosome strings* are generated in the reproduction phase by applying the *crossover* and *mutation* operators to the selected population members;
5. the evaluation, selection and reproduction phases are iterated and new populations are generated until a suitable solution is found.

Input to the procedure are the number of individuals per generation, and the probability of crossover and mutation to take place. The procedure outlined above can be regarded as the basic form of a genetic algorithm. Some techniques have been

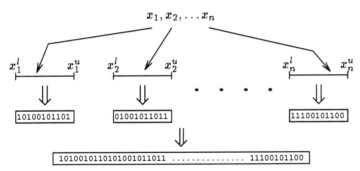

Figure 21.5 Codification of the design variables vector into a *chromosome* string

addded in order to improve the performance of the procedure.

In this work *generational reproduction* is adopted, which means that a new generation is formed by replacing the entire set of parents, selected among the actual generation, by their children. In this way, the best member of the population may fail to produce offsprings in the next generation, or the offsprings it produces may not inherit its favourable characteristics; the *elitist* strategy is used to fix this potential source of loss by copying the best member of each generation into the subsequent one.

A simple fitness scaling criterion is then used. At the start of a run it may be common to have a few extraordinary individuals in a population of mediocre colleagues, which could be an undesirable cause of premature convergence. On the other hand, later on during the run a different problem can arise, as there may still be significant diversity within the population, but the average fitness may be close to the best one. In this situation, average members and best members have nearly the same chance to reproduce, and improvements become more difficult to be found. In order to prevent this problems the fitness function can be conveniently scaled during the run; for the results presented here, this has been accomplished by evaluating the fitness of an individual as $1/obj^2$, obj being the quantity to be minimized.

A further aspect which may have an impact on the algorithm performance is the level at which mutation occurs; if mutation operates on the binary coding of the variables, small changes in the coded representation can correspond to large jumps of the variable itself (for example if the most significant bit undergoes a mutation). In the present procedure mutation can operate directly on the design variables of the physical problem as well. In this way, a better control is allowed on the changes that this operator produces on the variables.

21.2.4   The Gradient-Based Optimization Routine

In the applications reported here, constraints, when activated, have been formulated explicitly. In this way the numerical optimization routine faces an unconstrained problem. The first-order optimization method employed is the Broydon-Fletcher-Goldfarb-Shanno (BFGS) method. The gradient of the objective function F is evaluated by means of finite differences. The search direction at iteration q is defined as $s^q = -\mathbf{H}^{-1}\nabla F(\mathbf{x}^q)$, where $\mathbf{x}$ is the design variables vector. The matrix $\mathbf{H}$ is an approximation of the Hessian matrix, which is updated during the optimization [Van84]; for this reason, this method has convergence characteristics similar to second order methods. After the direction s has been found, the distance to move in direction s is calculated by means of a one-dimensional search technique. A few artifacts of the method need to be taken into account to obtain the best performances. For example, each design variable is normalized with respect to the pertinent variation range, so that all design variables have the same order of magnitude.

21.2.5 Applications

The first application presented is a direct optimization problem consisting in the minimization of the wave drag, c_{d_w}, of the RAE-2822 airfoil at mach $M = 0.73$ and lift coefficient $c_l = 0.87$, with fixed maximum thickness. For this case, a set of 30 modification functions was used. Results obtained using the genetic procedure are shown in fig. 21.6 while the results related to the gradient-based method are shown in fig. 21.7. Within the accuracy of the flow field discretization, the GA provides a shockless airfoil, whereas the BFGS algorithm converges more slowly to a solution with 2 drag counts. Also shown in the figure are the convergence history of the two methods, in terms of objective function value ($obj = 100\, c_{d_w}/c_l^2$). Each design variable is discretized through a seven bit string, yielding a chromosome length of 210 bits. Mutation is performed at *word* level, i.e. not on single bits but on the integer ranging from 0 to $2^7 - 1$. Crossover probability is set to 100%, as for all the other problems here described.

The second direct optimization problem is a two-point one with a fixed variation range for the airfoil maximum thickness t:

$$
\begin{cases}
\min \dfrac{c_{d_{w_1}}}{c_{l_1}^2} + \dfrac{c_{d_{w_2}}}{c_{l_2}^2} \\[2ex]
\text{subj. to:} \quad
\begin{aligned}
c_{l_1} &= 0.84 \\
M_1 &= 0.745 \\
c_{l_2} &= 0.063 \\
M_2 &= 0.846 \\
0.052 &\le t \le 0.055
\end{aligned}
\end{cases}
\tag{21.7}
$$

The optimization objective is the wave drag reduction in the two assigned operating conditions. Results reported in figure 21.9 were obtained after 30 generations and show a compromise performance between the two operating conditions. A 30 element population and a set of 24 modification functions have been used. Mutation is performed at *bit* level.

The first inverse problem reported is the recover of the Korn airfoil at $M = 0.75$ and angle of attack $\alpha = 0°$. The target pressure distribution is obtained by direct calculation of the flow around the Korn airfoil with the full-potential solver. No constraints are imposed to guarantee the well-posedness of the problem. The starting geometry for the BFGS algorithm is a NACA 64A410 airfoil; the same is used by the genetic algorithm (GA) to generate the starting population by random mutation. The airfoil is represented by means of two B-spline curves, with 10 control points each; sixth order curves have been used for the GA, whilst fifth order has been chosen for

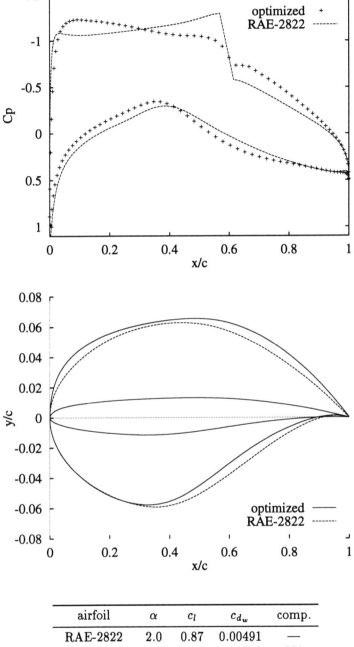

airfoil	α	c_l	c_{d_w}	comp.
RAE-2822	2.0	0.87	0.00491	—
OPTIMUM	1.15	0.88	0.00002	339

Figure 21.6 RAE-2822 airfoil wave drag minimization using the genetic
algorithm

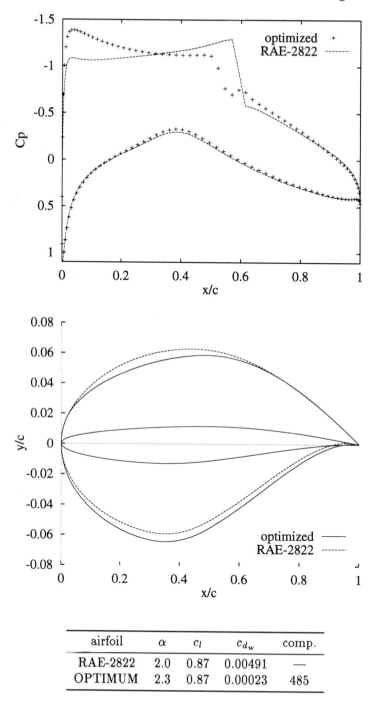

airfoil	α	c_l	c_{d_w}	comp.
RAE-2822	2.0	0.87	0.00491	—
OPTIMUM	2.3	0.87	0.00023	485

Figure 21.7 RAE-2822 airfoil wave drag minimization using the BFGS
algorithm

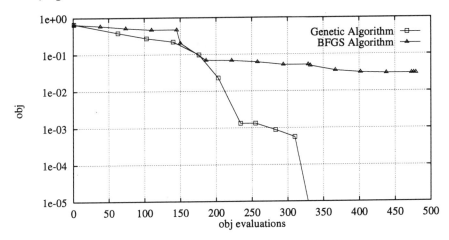

Figure 21.8   RAE-2822 airfoil wave drag minimization Genetic Algorithm
vs. BFGS convergence history

the BFGS. All control points are allowed to move except those at the leading and
trailing edges, for a total of 32 design variables. A number of elements per generation
equal to 50 has been chosen, and a total number of 80 generations per run. With
these parameter values, three successive GA runs have been performed, progressively
narrowing the design variables variation ranges and increasing at the same time the
probability of mutation to occur (1, 2, 4%). Mutation is performed at *word* level.
As a consequence of the high degree of freedom left to geometry at the start of the
run (where for generality vertices displacements of the order of 10% of the chord are
allowed), the genetic operators can generate shapes which are not compatible with the
flow solver. These shapes are assigned a very low fitness value, so that they are not
likely to be selected for reproduction. A routine for the correction of such incompatible
chromosomes has been introduced to avoid the reduction of the algorithm efficiency;
on the basis of the number of airfoil contour first derivative changes of sign, a least
square polynomial is fitted through the control points, and the control points are to
some extent "attracted" towards this curve, which results in a smoothing of the shape.

Fig. 21.10 shows the results obtained with the GA, while those obtained using
the BFGS algorithms are reported in fig. 21.11. The GA evolution history, shown
in fig. 21.12 illustrates the comparison among the results obtained with chromosome
correction and mutation at *word* level with those obtained without correction and
through mutation at *bit* level, respectively. The maximum fitness value for each
subsequent generation is shown; the curves are monotonically increasing by virtue
of the *elitist* technique. It is to be noted that the BFGS algorithm needed a few
restart procedures in order to obtain a satisfactory solution due to solution stagnation.
This was accomplished by shifting between two different numbers of points for
the evaluation of the integral (inverse); the corresponding slight changes in the
objective function were sufficient to obtain further improvements. In general, the BFGS
algorithm showed a remarkable sensitiveness on the initial position of the control points
as compared with the GA.

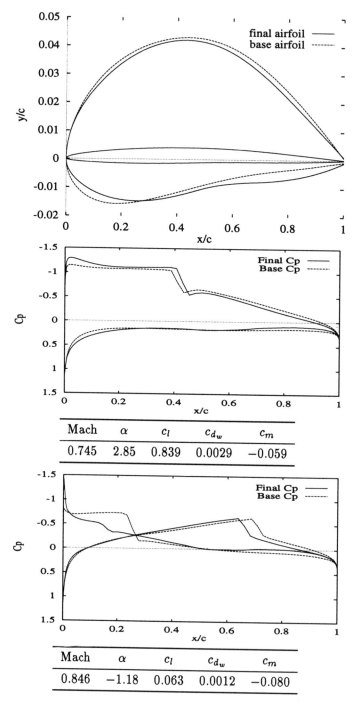

Mach	α	c_l	c_{d_w}	c_m
0.745	2.85	0.839	0.0029	−0.059

Mach	α	c_l	c_{d_w}	c_m
0.846	−1.18	0.063	0.0012	−0.080

Figure 21.9 Two-point design

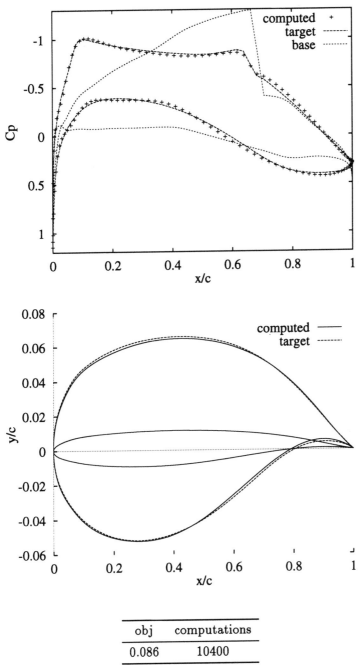

obj	computations
0.086	10400

Figure 21.10 Korn airfoil reconstruction problem; $M = 0.75, \alpha = 0°$; result obtained using the genetic algorithm

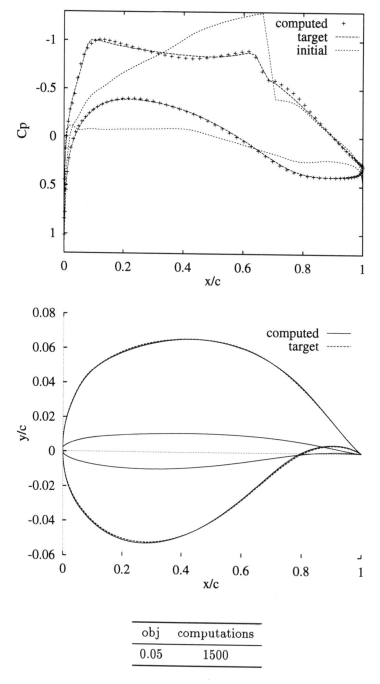

obj	computations
0.05	1500

Figure 21.11   Korn airfoil reconstruction problem; $M = 0.75, \alpha = 0°$; result
obtained using the BFGS algorithm

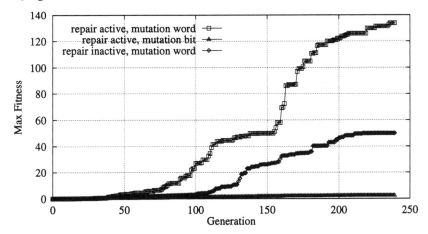

Figure 21.12   Korn airfoil reconstruction problem; Genetic Algorithm
evolution history

The second inverse problem consists in the reconstruction of a laminar airfoil at $M = 0.703$ and $\alpha = 1.05°$. The results are shown in fig. 21.13 and 21.14. It should be noted that the pressure distribution assigned as the target was evaluated with an Euler code; this justifies the differences between the geometry obtained by the calculations and the original one. Three GA runs have been performed using 32 design variables, 50 population elements and 80 generation per run.

21.3   OPTIMUM COMPUTATIONAL LOAD BALANCING OF PARALLEL FLOW-FIELD SOLVERS

Multiblock flow field solvers are well suited to parallel implementation on multiprocessor systems but the obtainable degree of parallelization is heavily data dependent so that optimal domain decomposition and efficient computational load balancing are necessary to obtain favourable speedup.

In some cases this problem is straightforward; in fact the domain decomposition can be easily tuned to optimize load balancing among processors of a parallel computer if constraints on the domain decomposition are not imposed, computational weight of each block is fixed and known, costs due to communication between processors can be neglected and each processor runs at the same speed.

On the other hand it is not easy to obtain good load balancing if:

1. the given domain decomposition is aimed at optimizing other characteristics of the flow field solver so that several computational blocks of different computational weights are present and it is not easy or convenient to split the heavier blocks;
2. it is important to take into account the communication time when processors exchange data;
3. the computational environment is heterogeneous, so that the speed of

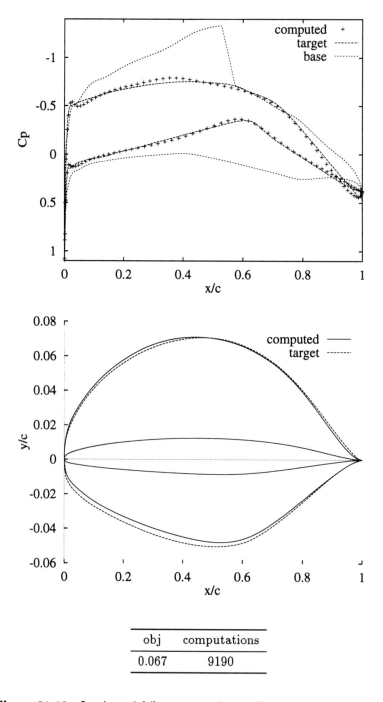

obj	computations
0.067	9190

Figure 21.13   Laminar airfoil reconstruction problem; $M = 0.703, \alpha = 1.05°$;
results obtained using the genetic algorithm

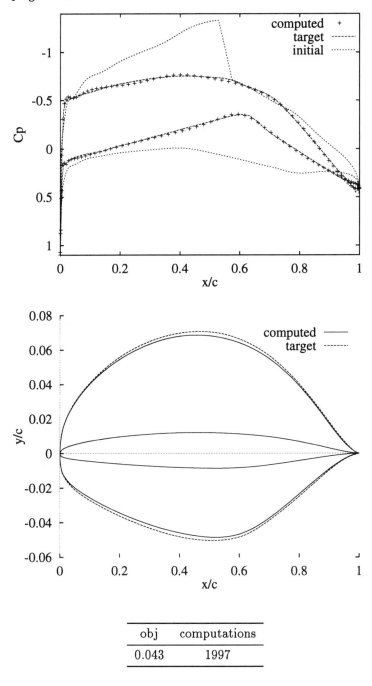

obj	computations
0.043	1997

Figure 21.14   Laminar airfoil reconstruction problem; $M = 0.703$, $\alpha = 1.05°$; results obtained using the BFGS algorithm

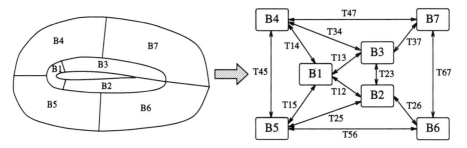

Figure 21.15 Computational domain topology

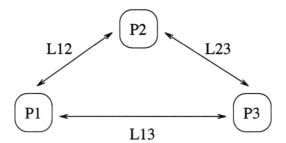

P_i = processor number and associated average processor speed
L_{ij} = link between processors i and j and associated cost

Figure 21.16 Parallel computing net configuration

each processor may be different.

When some of these problems are present, the load balancing problem becomes difficult to solve because it presents a heavy combinatorial aspect ($\mathcal{NP}$-hard).

A solution to this problem is described herein, based on a genetic algorithm. The technique developed can take into account heterogeneous computational environments with different processor speeds and data transfer rates among processors [Qua94].

21.3.1 Problem Statement

To correctly model the block-to-processor allocation problem data have to be available both on the computational domain topology of the particular CFD problem confronted and on the parallel computing net configuration adopted. In particular, the available data must allow an accurate evaluation of the computational resources related to a generic assignment of computational blocks to parallel computing net processors.

In the present application the cost function to be optimized takes into account the computational and communication weight of each block, the relative speed of each processor and the speed of the underlying communication network. For this reason, topological models have to be specified both for the computational domain and for the computing net. Modelling the computational domain topology requires the specification of the data exchange map among the computational domain blocks. Figure 21.15 shows an example of topologic specification of a multiblock grid around an airfoil. For each link T_{ij} a weight has to be assigned proportional to the amount of data that block B_i sends to block B_j at each iteration. Note that also T_{ji} has to be specified and that it can be different from T_{ij}. The cost function considered here is derived on the hypothesis that, for each block and for each iteration, the data transmission phase follows the calculation phase. These hypotheses are in good agreement with the actual behaviour of PRANSA, the parallel version of ENSOLV Flow Solver [AKB93, AMS94] in use at CIRA.

The network topology modelled herein is assumed to be very simple, as only the specification of the relative processor speed and the data transfer rate between each pair of processing elements are required.

The mathematical formulation of the optimization problem considered is as follows. Let

$$B \triangleq \{b_1, b_2, \ldots, b_n\} \tag{21.8}$$

be the ordered set of n blocks in which the computational domain is split, while:

$$P \triangleq \{p_1, p_2, \ldots, p_m\} \tag{21.9}$$

is the ordered set of the m parallel processing units. The function

$$w : B, P \longmapsto \Re \tag{21.10}$$

assigns a weight $w(b, p) \in \Re$ to each $b \in B$ when it is assigned to the processor $p \in P$ (here communication costs and processor speeds are taken into account). Let

$$f : B \longmapsto P \tag{21.11}$$

be a generic function that assigns a computational block to a processor, and

$$\mathcal{F}(P, B) \tag{21.12}$$

be the collection of functions defined in the above way. Let finally

$$C(p, f) = \sum_{\{b \in B: \, f(b) = p\}} w(b, f(b)) \tag{21.13}$$

be the computational cost associated to the processor $p \in P$.

The load balancing problem consists in finding a $f_o \in \mathcal{F}$ that optimizes the following cost function:

$$\min_{f \in \mathcal{F}} \left[\max_{p \in P} C(p, f) \right] \tag{21.14}$$

Introducing the integer variables vector

$$\mathbf{x} = (x_1, \ldots, x_n)$$

with $x_j \in \{1, \ldots, m\}$ so that if $x_j = i$, the block $b_j \in B$ has been assigned to the processor $p_i \in P$, the problem becomes:

$$\begin{cases} \min_{\mathbf{x}} \left[\max_{i=1,n} C(p_i, \mathbf{x}) \right] \\ \text{s.to:} \quad 1 \le x_j \le m, \quad \forall j \in \{1, \ldots, n\} \\ \quad\quad x_j \text{ integer} \end{cases} \tag{21.15}$$

21.3.2   Genetic Algorithm Based Solution

The genetic algorithm technique has been adopted to find *near-optimal* solutions to the problem described above.

Following this approach, a population of *chromosome strings*, each representing a possible solution to the problem, evolves under the action of fitness based selection, mutation and crossover mechanisms. The correspondence between $\mathbf{x}$ and a string s of n characters in the alphabet $\{1, \ldots, m\}$ is defined so that the assignment of the i-th block to the j-th processor corresponds to assigning the value "j" to the i-th character of the string. Then each character c of s is translated into a string of length $k = \lceil \log_2 m \rceil$ in the alphabet $\{0, 1\}$ that corresponds to the base 2 conversion of c.

21.3.3   Results

The first application reported is related to the computational domain topology of figure 21.15 mapped into the parallel computing net of figure 21.16. Because of its simplicity it does not justify the genetic algorithm application, nevertheless it is reported as an example.

Results have been compared to those obtained from a scheduling rule known as the *longest processing rule* (LPT) [HS76] that generates schedules very close to an optimal schedule when processors are identical and communications can be neglected. In fact, under these condition, each computational block can be assigned to an independent task and finding optimum load balancing on m identical processors is equivalent to obtaining minimum finish time schedule on the same processors.

An *LPT schedule* is a schedule that is the result of an algorithm which, whenever a processor becomes free, assigns to that processor a task whose time is the largest of those tasks not yet assigned. Ties are broken in an arbitrary manner.

Figure 21.17 is related to load balancing without taking into account communications. In this case the genetic algorithm performance is equal to the result of the application of the LPT rule. When communications are taken into account, the genetic algorithm performs better. In figure 21.18 the contribution of data transfer to the computational charge of each block is shown by the shaded zones. The genetic algorithm minimizes the global amount of data transferred between processors.

The second application presented is related to an Euler flow calculation around an RAE 2155 wing in a wind tunnel at $M = 0.744$ and $\alpha = 2.5°$. The computational domain produced by the domain modeller presents 29 blocks of different computational weight. The grid produced by the grid generator has 713216 cells (891272 points). The genetic algorithm is directly interfaced to the output of the domain modeller and of the grid generator from which it obtain the data about the domain decomposition and the computational weight of each block. At the moment no data are available about the data exchange between blocks. The output generated by the computational load balancing procedure is then directly read by the parallel flow field solver PRANSA.

The virtual machine was built using PVM 3.3.7 and was made up of by the following computers:

- Convex MetaSeries: 8 nodes HP PA RISC, $7 \times 64 + 96$ MB main memory, o.s. HP-UX 9.01 (6 nodes used for this application).
- INDIGO 2 SGI workstation: MIPS R4000, 96 MB memory, o.s. IRIX 5.3.

• IBM RS6000/550 workstation: POWER, 256 MB memory, o.s. AIX 2.3.

The relative processor speed was obtained by using the standard suite of benchmarks used at CIRA and is reported in the following table:

MetaSeries	RS6000	INDIGO 2
1.00	0.61	0.49

Fig. 21.19 reports the genetic algorithm convergence history and, finally, fig. 21.20 shows the resulting load balancing data compared with the actual CPU times after 600 iterations. Comparison with the actual elapsed time has not been made because no communication data were available, however it is 1.27 times slower than the worst CPU time.

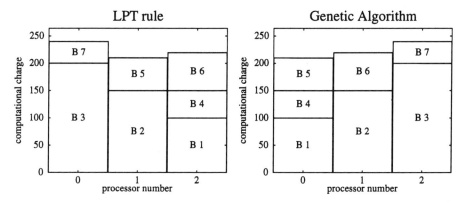

Figure 21.17   Load balancing without communications

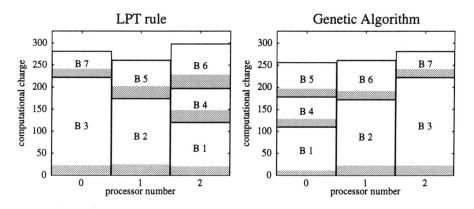

Figure 21.18   Load balancing taking into account communications

21.4 CONCLUSIONS

Two applications of evolutionary programming techniques to computational fluid dynamics have been described.

The airfoil design procedure shows how such techniques can be applied to problems that traditionally are treated using calculus based methods. The comparison with these methods has shown that, in many cases, evolutionary procedures require more computational resources to obtain comparable results. However the situation is expected to be different for optimization problems of greater size. On the other hand, if robustness characteristics are considered, even for the class of problems treated here, genetic algorithms exhibit a better behaviour. Besides, more complex problems, as multiobjective optimization, are beginning to be treated successfully and efficiently with these algorithms.

The computational load balancing procedure shows a completely different field of application of evolutionary programming that is, to some extent, more natural for such optimization approach. Remarkable is the capability of the procedure to match complex flow field topologies to heterogeneous computational environment. This can be exploited in the development of large aerodynamic design systems based on genetic algorithms and implemented on massively parallel architectures.

ACKNOWLEDGEMENTS

Thanks are due to A. Vicini for the assistance in the preparation of the airfoil optimization test cases, to M. Amato for the RAE 2155 wing flow field topology definition and grid generation and to R. Mella and A. Matrone for the valuable assistance in running the PRANSA flow field solver in an heterogeneous computing environment.

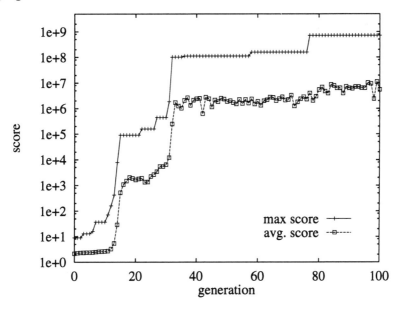

Figure 21.19 RAE 2155 wing genetic algorithm convergence history

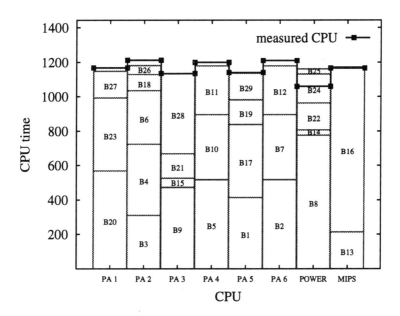

Figure 21.20 RAE 2155 wing computational load balancing

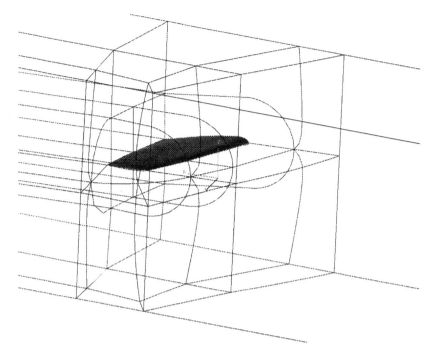

Figure 21.21   RAE 2155 wing; computational domain decomposition

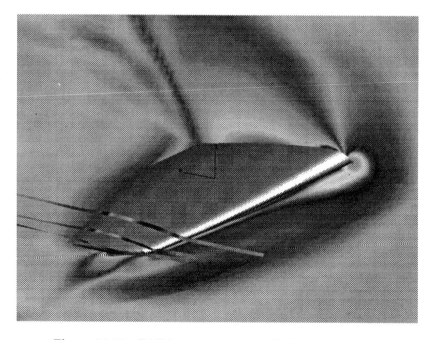

Figure 21.22   RAE 2155 wing; pressure field and streamlines

REFERENCES

[AKB93] Amato M., Kok J. C., and Bosse S. (1993) Software design of ENSOLV. A flow solver for 3d Euler/Navier-Stokes equations in arbitrary multiblock domains. Technical Report NLR-CR-93190-L, The National Aerospace Laboratory, The Nederlands.

[AMS94] Amato M., Matrone A., and Schiano P. (1994) Experience in parallelizing a large cfd code: the ENSOLV flow solver. In *Proceedings of HPCN Europe 1994 — The International Conference and Exhibition on High Performance Computing and Networking.*

[CH86] Cosentino G. B. and Holst T. L. (1986) Numerical optimization design of advanced transonic wing configurations. *Journal of Aircraft* 23(3).

[Dav91] Davis L. (1991) *Handbook of genetic Algorithms.* Van Nostrand Reinhold, New York.

[Dul91] Dulikravich G. S. (1991) Aerodynamic shape design and optimization. American Institute of Aeronautics and Astronautics (AIAA). AIAA Paper 91-0476.

[GQ92] Ghielmi L. and Quagliarella D. (1992) IDEA: An expert system as a support to the design of airfoils. In Houstis E. H., Rice J. R., and Vichnevetsky R. (eds) *Expert Systems for Scientific Computing*, pages 261–282. International Association for Mathematics and Computers in Simulation (IMACS), Elsevier Science Publishing B. V. (North-Holland), Amsterdam.

[HH78] Hicks R. and Henne P. A. (1978) Wing design by numerical optimization. *Journal of Aircraft* 15(7): 407–412.

[HS76] Horowitz E. and Sahni S. (1976) *Fundamentals of data structures.* Computer Science Press.

[Jam88] Jameson A. (1988) Aerodynamic design via control theory. Technical Report ICASE Report 88-64.

[MP93] Mosetti G. and Poloni C. (August–September 1993) Aerodynamic shape optimization by means of a genetic algorithm. In Daiguji H. (ed) *Proceedings of the 5th International Symposium on Computational Fluid Dynamics*, pages 279–284. Japan Society of Computational Fluid Dynamics, Tokyo, Japan.

[PM95] Poloni C. and Mosetti G. (July 1995) Aerodynamic shape optimization by means of hybrid genetic algorithm. In *Proceedings of the Third International Congress on Industrial and Applied Mathematics*. ICIAM 95, Hamburg, Germany.

[QD94] Quagliarella D. and Della Cioppa A. (June20–22 1994) Genetic algorithms applied to the aerodynamic design of transonic airfoils. In *12th AIAA Applied Aerodynamics Conference*, pages 686–693. American Institute of Aeronautics and Astronautics (AIAA), Colorado Springs, CO. AIAA Paper 94-1896 CP.

[Qua94] Quagliarella D. (1994) Optimal domain decomposition for paralel multiblock flow-field solvers using genetic algorithms. In *Proceedings of Parallel CFD'94 Conference*. Kyoto Research Park, Japan.

[QV95] Quagliarella D. and Vicini A. (September 11–15 1995) Transonic airfoil design by means of a genetic algorithm. In *Atti del XIII Congresso Nazionale*. Associazione Italiana di Aeronautica e Astronautica (AIDAA), Roma.

[RA90] Rogers D. F. and Adams J. A. (1990) *Mathematical Elements for Computer Graphics*. Mc Graw-Hill.

[Slo84] Sloof J. W. (1984) Computational methods for subsonic and transonic aerodynamic design. In *Proceedings of ICIDES-I*, pages 1–68. Austin, Texas.

[Ton85] Tong S. S. (1985) Design of aerodynamic bodies using artificial intelligence / expert system technique. American Institute of Aeronautics and Astronautics (AIAA). AIAA Paper 85-0112.

[TPS89] Tong S. S., Powell D., and Skolnick M. (June 1989) Engeneous: Domain independent, machine learning for design optimization. In Schaffer J. D. (ed) *Proceedings of the Third International Conference on Genetic Algorithms,* pages 151–159. M. Kaufmann Publishers inc., San Mateo, CA.

[Van79] Vanderplaats G. N. (December 1979) Efficient algorithm for numerical airfoil optimization. *Journal of Aircraft* 16(12): 842–847.

[Van84] Vanderplaats G. N. (1984) *Numerical Optimization Techniques for Engineering Design: with Applications.* Mc Graw–Hill.

22

Shape Representations for Evolutionary Optimization and Identification in Structural Mechanics

Marc Schoenauer [1]

22.1 Introduction

This paper presents an Evolutionary approach to problems of shape optimization and identification: the aim is to find a partition of a given *design domain* of the 2D-plane or the 3D-space into two subsets (e.g. material and void for the Optimal Design problems). The central issue of this paper is the representation of such repartition suitable for its evolutionary optimization, and its dependency with respect to the problem at hand.

Evolutionary Computation (EC) is a field that now regroups different stochastic optimization methods based on the same paradigm: *evolve* a *population* (set of fixed size) of *individuals* (points of the search space) using stochastic operators (e.g. *crossover* and *mutation*) in a crude parody of the Darwinian principle of *survival of the fittest*.

The main streams of EC are Genetic Algorithms (GAs), first described by [Hol75] and popularized by [Gol89a], Evolution Strategies (ESs), introduced by [Rec73, Sch81], and Evolutionary Programming (EP), imagined by [FOW66]. A good overview of EC can be found in the Proceedings of the first International IEEE Conference on the subject [ZMR94], or in [Fog95].

One of the main differences between these streams lies in the representation

---

[1] Centre de Mathématiques Appliquées, CNRS-URA756 - Ecole Polytechnique - 91128 Palaiseau - E-mail : Marc.Schoenauer@polytechnique.fr

issue: historically, GAs deal with bit-strings of fixed sizes, ESs with real vectors and EP with Finite State Automata. Another point of discussion,somewhat related to the representation issue, lies in the type of approach: *bottom-up* for GAs vs *top-down* for both ESs and EP: GAs basic assumption is that the optimal solution can be found by assembling *building blocks*, i.e. partial pieces of solutions, while ESs and EP assess for the environmental pressure to ensure the emergence of the best solutions. The most visible consequence of this debate deals with the recombination operator, viewed as essential for GAs, as potentially useful for ES, and as possibly harmful for EP.

The "modern" tendencies are more pragmatic, and put the discussions back to the representation issue: GAs users have turned to real number representations when dealing with real numbers, following experimental results by [JM91] and heuristic demonstrations by [Rad91]. ESs researchers have included recombination as a standard operator, as in [Sch81], and are designing specific operators for non real-valued problems (e.g. in [BS95]). And EP is being used on any representation, as for instance by [Ang93, FS95]. But the whole EC community now takes for granted that the definition of both the genotypic space (the space where the evolution operators apply) and the phenotypic space (the space on which the fitness is computed, or behavioral space) plays a central role in determining the success or failure of an EC optimization.

This paper introduces representations for 2-dimensional shapes, i.e. partitions of a compact set of $\mathbb{R}^2$ into two subsets, and discusses their properties in the light of their use for optimization by Genetic Algorithms. Section 22.2 first presents the "GA-natural" bit-oriented representation. Though standard bit-string crossover operators apply to that representation, the need to design specific crossover operators is emphasized. Such operators are introduced and experimental comparisons with the standard 1- and 2-points operators are performed. The Mechanical problem of Topological Optimal Design is then presented in section 22.3, and the design of the fitness function is carefully detailed, guided by some mechanical considerations. Section 22.4 then presents results obtained by GAs using this bit-oriented representation: in situations where standard methods are not applicable, a real breakthrough in Topological Shape Optimization is achieved. But these results also point out the limits of this bit-oriented representation, namely that it heavily depends on the a priori discretization of the design domain. So two other representations are introduced in section 22.5. Both representations overcome the dependency on a fixed discretization by allowing their complexity to evolve. They differ from the bit-array representation in the amount of *degeneracy* they contain, as defined in [Rad94]. Comparative results on the Optimal Design problem are given in section 22.6: both representation outperform the bit-oriented representation, and the representation in which genetic traits are strongly transmitted seems to give better results. Finally, section 22.7 discusses the pros and cons for EC in the field of Structural Mechanics, and presents on-going and further work.

22.2 Bit-array representation

[2] When the goal is to find an optimal shape, i.e. a partition of the design domain into two subsets, the most "natural" representation for using GAs is a bit-oriented representation based on a regular discretization of the design domain: First, GAs were initially designed to work on bit-strings; Second, the application problems are Structural Mechanics problems, usually involving Finite Elements Analyzes, based on some discretization of the domain at hand. And such a discretization induces some error on the Analysis. A way to minimize the influence of this error is to work with a constant mesh, without re-meshing every shape. And a straightforward representation for a shape is then, for a given mesh, the corresponding bit-string, with 0 and 1 representing each one of the two subsets of the desired partition.

Figure 1 gives an example of such a representation for shapes in a square domain design, discretized here into 15 × 15 square elements, and the corresponding partition of the design domain.

```
100000001110110
100100001000010
010011011111110
001000000000000
110000000011111
011000000010000
001111111010000
001000101010010
001011101110100
001111000100000
001101111100000
001010000000110
011001100000110
010000111110000
110000010000000
```

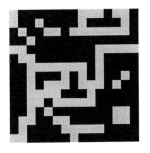

a - *The Bit-array genotype.*   b - *The phenotype.*

Figure 1 Shape representation: GA applies on the bit-array genotype (a), and the corresponding phenotype is the partition of the design domain (b).

22.2.1 *Bit-arrays are not bit-strings*

This section discusses the differences between linear bit-strings and 2-dimensional bit-arrays. Specific crossover operators are introduced to account for that difference.

---

[2] All results based on that representation were obtained by C. Kane, and are part of her PhD dissertation [Kan95].

Schema analysis

By analogy with the bit-string case, schema can be defined in bit-arrays by adding the "*" symbol to the binary alphabet. But the analogy fails when it comes to define the building blocks, or short low-order schemata with above average fitness ([Gol89a]). Figure 2 illustrates such situations. From a geometrical point of view, the schema of Figure 2-a should be considered as a potential building block. But its emergence (if we suppose it is related to highly fit individuals) is not likely to happen in the early generations, if only one-dimensional crossover operators are used: its one-dimensional order is 31. On the opposite, the schema of Figure 2-b is erroneously considered as a low order schema (of order 2), although it is hardly related to the fitness.

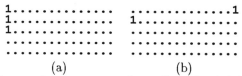

(a)   (b)

Figure 2 Differences between one- and two-dimensional defining lengths of schemata. Schema (a) has a two-dimensional length of 3, but a one-dimensional length of 31. Schema (b) has a two-dimensional length of 15, but a one-dimensional length of 2.

More generally, when bit-arrays are handled as bit-strings, one-dimensional cross-over operators induce a strong geometrical bias against vertical building blocks.

A quantification of this bias remains to be done, for instance by studying the variance of the fitness on schemas, as a function of the schema length (in the line of what is done in [Rad94] for evaluating various representations of the TSP problem).

Nevertheless, the expected inaccuracy of classical n-point crossover (confirmed by experimental results, section 22.2.2) leads to design specific genetic operators, suited to our two-dimensional problem.

22.2.2   Crossover operators

This section is devoted to the study of the crossover operators for the bit-array representation. All crossover operators presented in this section are symbolically drawn on Figure 3 and 4: two offsprings of a black parent and a grey parent are plotted. The color of the mesh elements of the children only tells which parent this element comes from, regardless of its actual value (0 or 1).

Standard crossover operators

In order to evaluate the bias induced by handling a bit-array as a bit-string, we first consider the standard crossover operators; One-point, two-points [Gol89a] and uniform [Sys89] crossovers are considered.

$\qquad$ a - *1 point standard* $\qquad\qquad\qquad$ b - *2 point standard*

Figure 3 Examples of offsprings from one dimensional crossover operators

As pointed out in section. 22.2.1, one- and two-points crossovers are geometrically biased: only horizontal bands of the parents get exchanged. Experimental results without mutation presented in next subsection witness for the bad influence of this bias. On the opposite, the uniform crossover (easy to imagine, though not presented in Figure 3) does not suffer from such a bias.

Specific two-dimensional crossover operators

Two specific two-dimensional operators are introduced and studied in this section.

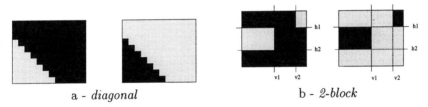

$\qquad\qquad$ a - *diagonal* $\qquad\qquad\qquad\qquad$ b - *2-block*

Figure 4 Examples of offsprings from 2-dimensional crossover operators

Diagonal crossover.
The basic idea of diagonal crossover is to generalize the popular one-point crossover to the two-dimensional case. As shown on Figure 4-a, a randomly selected straight line separates the rectangle in two parts which are exchanged between both parents. A generalization of this operator has been recently introduced in [KM95].

The block crossover
First introduced in [Jen92], the idea of the block crossover is to cut the whole two-dimensional domain by two horizontal lines and two vertical lines and to exchange some of the large blocks defined by these lines. The values v_1 and v_2 (respectively h_1 and h_2) are chosen uniformly, and the 2 (or 3) blocks to

be exchanged are selected randomly. Figure 4-b shows an example of 2-block crossover.

Comparison without mutation

These operators have been experimentally compared on an instance of the problem described in next section. Figure 5 shows the results obtained without any mutation, and with a crossover rate of 1. Of course, these results give only hints that something goes wrong with standard crossover operators: the absence of mutation forbids to draw any definite conclusion. Nevertheless, such result confirm the points discussed in section 22.2.1.

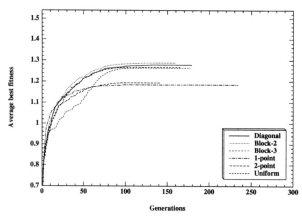

Figure 5 Performances of recombination operators without mutation on an Optimal Design Problem.

The first conclusion that can be drawn from Figure 5 is the ineffectiveness of the one-dimensional crossover operators, similarly outperformed by both uniform and two-dimensional crossovers.

The second remark is that the uniform operator performs poorly in the early stage of the runs, before catching up in the late generations. The explanation might be the following: in the beginning of a run, the uniform crossover equivalently disrupts schemas of any order, while all other crossovers (even one dimensional) preserve some large areas. These disruptive effects do not occur in the end of evolution, since neither kind of crossover does disturb regions where convergence already occurred: if the bits of the parents at a given position are the same, crossing over has no effect on those bits. In the meantime, the one-dimensional crossovers fail to beneficially exchange schema whose vertical precision is small compared to the horizontal one, like long vertical bars.

22.3   Topological Optimal Design

The shape representations presented in this paper can apply to different problems of Mechanical Engineering: Topological Optimal Design, where the optimization of a structure whose topology is not the continuous transformation of a given shape; Inclusion Identification, where a structure is made of an unknown repartition of two materials, and this unknown repartition is to be identified from a given set of experiments on that structure. This latter problem will not be addressed in this paper, though the first results obtained using the representations presented here are very good indeed (see [JS95]).

This section is devoted to the detailed presentation of the Problem of Topological Optimal Design, on the impact of this problem on the representation and on the design of the fitness function.

22.3.1   The mechanical problem

The goal is to find a design for a structure that minimizes some cost function (usually the weight of the structure, but it can include as well technological costs) while respecting some given constraints on mechanical behaviors (such as limit values for the displacement, the stress, and so forth) under some given loadings. The design is sought inside the design domain, a given subset of $\mathbb{R}^2$ or $\mathbb{R}^3$.

We shall restrict here to 2-dimensional cantilever problems, to plane strain model, and elastic or hyper-elastic materials. The design domain is a rectangular subset of $\mathbb{R}^2$, the plate is fixed on its left boundary, the loading is made of a single force applied at some point of the right boundary (see Figure 6), and a constraint is imposed on the displacement of that loading point.

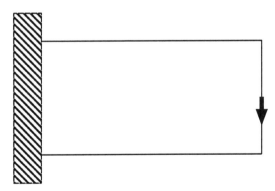

Figure 6 The 1×2 cantilever plate. Ω is the design domain, the plate is fixed on the $x = 0$ axis, and a force is applied at point M.

22.3.2   The fitness function

The problem here is a constrained problem. Many methods have been designed and used to take constraints into account in evolutionary algorithms (see [Mic95] for a survey of such methods). In particular, the method described in [SX93] has been successfully applied to another problem of structural mechanics, the optimization of truss structures. But our point here is to study the influence of representation, not the way to handle constraints. Therefore, the standard method of penalization has been chosen, as a non specific and quite robust way of handling constraints.

In summary, the first draft for the fitness function (to be minimized here) is computed from the weight of the structure (or its area in 2 dimensions) and the maximal value of the displacement, obtained through a Finite Element Analysis. It has the following expression:

$$\mathcal{F} = Area + \alpha (D_{Max} - D_{Lim})^+ \qquad (22.1)$$

where D_{Max} is the maximal displacement of the structure, D_{Lim} the limit value for the displacement and α is a positive user-supplied parameter (a^+ denotes the positive part of a). Some mechanical difficulties detailed in next sections will be addressed by modifying this simple expression for the fitness.

22.3.3   Deterministic methods

There is only one known deterministic method for topological shape optimization, namely the homogenization method. It is based on a relaxation of the problem considering not only 0/1 repartition of void/material, but any continuous "mixing" of void and material with density in $[0, 1]$. Initiated by [BK88], it has been studied theoretically by [AK93], and wonderful results have been obtained by [BFJ95], including 3-dimensional results. But the limits of that method are well-known.

- The homogenization method searches solutions in the space of composite materials having a density of material taking continuous values in $[0, 1]$. Hence, the solutions found have to be projected on feasible part (with density of material either 0 or 1). This projection step does not always succeed in giving feasible solutions.
- The only mechanical context in which the homogenization applies is that of linear elasticity;
- It gives only one solution, even when multiple optimal solutions exist;
- It is limited to a single loading;
- It cannot handle loadings on the unknown boundary.

A more extensive discussion between the homogenization method and the GA approach, together with some comparative results, can be found in [KJS95].

22.3.4   From genotype to phenotype

We have to reconsider the definition of the phenotype given in section 22.2 in the light of the mechanical problem at hand: a solution to that problem is a shape of the design domain *that link the loading point(s) to the fixed boundary*, otherwise the structure is statically unstable, that is not viable (any Finite Element Analysis on such structure will abort abruptly). The first difficulty for representing shapes of structures is that possible shapes (repartition void/material) do not correspond to viable structures. Moreover, only the material parts connected to both the fixed boundary and the loading point have something to do with the underlying mechanical problem. So two different shapes can actually represent the same structure.

The first difficulty can be overcome by giving null fitness to non viable structures. But there is a risk that almost all random structures are non viable (for long narrow shapes for instance). A better approach is to smoothly increase the cost function as the gap between the fixed boundary and the connected component containing the loading point increases, thus favoring almost connected structures.

But the second difficulty raises the problem of the *degeneracy* of the representation with respect to the problem at hand, as defined in [Rad94]: a representation is said to be degenerate when the application from the genotypic space into the phenotypic space is not injective. Figure 7 gives a new look at the example presented in Figure 1. Any genotype belonging to the genotypic schema represented in Figure 7-a (i.e. having any value replacing the "*" signs) correspond to the same structure (Figure 7-b). So even though the bit-array representation is not degenerate to represent partitions of the design domain into void and material, it becomes degenerate when it comes to represent actual structures, if no precaution is taken when designing the fitness.

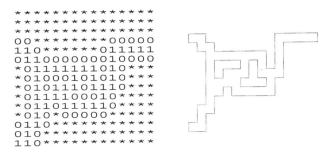

a - *The genotypic schemata.*   b - *The actual structure.*

Figure 7 An example of degeneracy: The Finite Element Analysis applies on structure (b), to compute the fitness of any genotype of schemata (a).

There are 3 ways to handle this problem-dependent degeneracy:

- Replace the genotype by that of the mechanical structure, removing all unconnected bits. But possibly valuable genetic material would then be lost, therefore resulting in bad performance of the genetic search.
- Ignore the degeneracy, using the fitness defined in section 22.3.2. But degeneracy is acknowledged to be generally bad for GA optimization.
- Account for the unconnected material in the fitness function, thus getting rid of the degeneracy.

This last option was chosen here, and the draft fitness function (22.1) is modified and becomes, for viable shapes:

$$\mathcal{F} = Area_{usefull} + \lambda Area_{useless} + \alpha(D_{Max} - D_{Lim})^+ \qquad (22.2)$$

Of course, the choice of λ is critical: too high, some actual optima can be lost. Too low, it has no effect. In the following, a value of 0.1, experimentally determined, was used for all experiments presented in the following.

22.4 Evolutionary Optimal Design with Bit-array Representation

The results presented in this section definitely witness for the power of the evolutionary approach on the Optimal Design problem: such results are a real breakthrough in Optimal Design. The price to pay is what was expected in terms of computational cost.

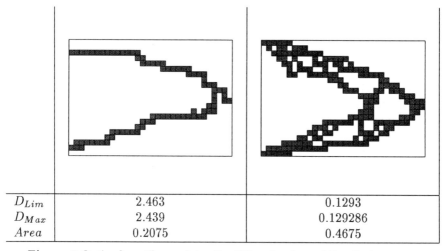

D_{Lim}	2.463	0.1293
D_{Max}	2.439	0.129286
$Area$	0.2075	0.4675

Figure 8 Optimal cantilever plates for different values of the displacement constraint D_{Lim}. D_{Max} is the actual maximum displacement of the structure.

22.4.1 Results in linear elasticity

In order to validate the evolutionary approach, the linear elasticity model is used on the standard 2×1 benchmark cantilever plate. The results presented in Figure 8 are the best results obtained out of 10 independent runs. The plate is discretized according to a 32×22 regular mesh. The population size for all runs is 125, and the number of generations arbitrarily fixed to 1000. One run thus requires about 100000 Finite Element Analyzes, taking approximately 6 hours of a powerful HP workstation for the 32×22 discretization. The block crossover is applied at a rate of 0.6, and the standard mutation (presenting no geometrical bias) at a rate of 0.01. All these parameters were adjusted after exhaustive tests (see [Kan95]).

These first results can be obtained more precisely and more rapidly (up to 2 orders of magnitude) by the homogenization method, and only demonstrate the validity of the approach.

Further results in linear elasticity deal with problems the homogenization method cannot handle (nor any other method, as far as we know): multiple optimal or quasi-optimal solutions can be found [KJS95], simultaneous multiple loadings can be considered, including uniform pressure on the unknown boundary [Kan95].

But the most important field that opens to optimization is that of nonlinear elasticity. From the GA point of view, it seems at first sight a straightforward adaptation of the linear case: replace the linear Finite Element Analyzes by Nonlinear Finite Element Analyzes. The following section demonstrates how such an approach might lead to erroneous results.

22.4.2 Nonlinear mechanical model

This section considers standard plane stress problems in the context of large displacements. The material still obeys a linear law (the extension to any other law is straightforward), but the nonlinear geometric effects due to the large displacement hypothesis are taken into account. A thorough description of the theoretical model together with the numerical algorithm can be found in [Cia88]. Details on the Finite Element Method and its implementation used here can be found in [Jou93].

The first experiments use the penalized fitness function 22.2. The idea was to use different loadings $\mathcal{F}$, with a fixed ratio $\mathcal{F}/D_{Lim}$ (D_{Lim} is the constraint on the displacement). In the purely linear case, all such problems are of course equivalent. The problem is the standard benchmark 1×2 cantilever problem described in section 22.3 (the plate is fixed on its left side only).

A typical disastrous result of the GA based optimization in the large displacement context is shown in Figure 9-b, together with the result on the same problem in the pure linear model (Figure 9-a). And different values of $\mathcal{F}$ and D_{Lim} do produce similar results. It is disastrous at first sight for any Mechanical engineer, as having a too large stress at the point where the structure becomes thin. A closer look at the maximum value of the stress of the solution σ_{Max} confirms that first diagnosis. The reasons for that are both: First, the large displacement model can have different solutions for the same loading, some with higher stress than others. Second, the stress field itself, on a rough domain like the structures obtained for the cantilever plate problem, and with such coarse discretization, presents some singularities. The value of σ_{Max} in Figure 9-b witnesses for that phenomenon, being more than an order of magnitude higher than those of the linear case (for the same loading).

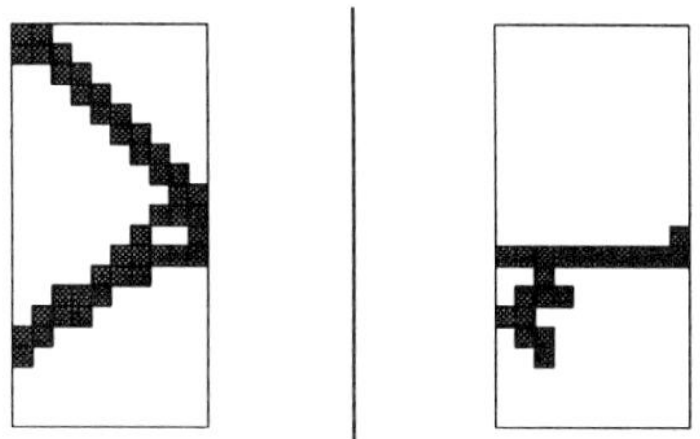

D_{Max}	0.022607	0.0199
σ_{Max}	0.076	0.77
$Area$	0.41	0.20
	(a): small displacements.	(b): large displacements.

Figure 9 Disastrous optimal designs found by GAs with displacement constraints. $\mathcal{F} = 0.009$ and $D_{Lim} = 0.02285$.

This suggests to incorporate a constraint on the maximal stress in the fitness function. The fitness 22.2 (to be minimized) becomes

$$Area_{usefull} + \lambda Area_{useless} + \alpha(D_{Max} - D_{Lim})^+ + \beta(\sigma_{Max} - \sigma_{Lim})^+ \quad (22.3)$$

where σ_{Lim} is the maximal stress allowed, σ_{Max} is the maximal stress (computed) of the structure, and β is some positive penalty parameter which has to be adjusted.

Figure 10 shows the optimal designs obtained with that modified fitness: Reasonable solutions are found, as long as the stress is imposed a strong constraint.

Moreover, the nonlinear effects are clear: all experiments of Figure 10 would have given the same result in the linear case, as the ratio $\mathcal{F}/D_{Lim}$ is constant and no multiple solution was ever observed on the simple 1×2 cantilever problem.

$\mathcal{F}$	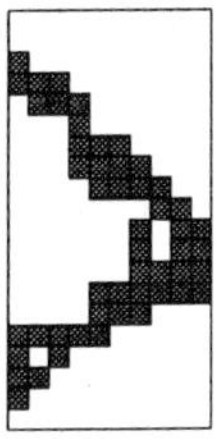 0.009	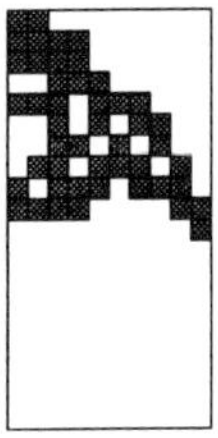 0.018	0.09
D_{Lim}	0.22856	0.457	2.2856
σ_{Lim}	0.53	1.0622	5.3
D_{Max}	0.2143	0.4504	1.687
σ_{Max}	0.550	0.9835	4.379
$Area$	0.21	0.47	0.1

Figure 10 Optimal designs for nonlinear elasticity with displacement and stress constraints, with fixed ratio $\mathcal{F}/D_{Lim}$.

22.4.3   The need for another representation

The preceding subsections have demonstrated the power of the evolutionary optimization of structures using the bit-array representation. Many dead-ends for the homogenization method have been put wide open:

- multiple solutions can be found when they exist
- any mechanical model can be used
- multiple loading, or loading on the unknown boundary, can be treated (though not presented here)

But the limits of the method also crudely appear, when it comes to refine the mesh (rather coarse in the results presented), or when we think of 3-dimensional problems: in both cases, the size of the individual will grow tremendously, and probably make the problem intractable for any evolutionary algorithm. In order to overcome this limitation, two other representations are now presented in the next section. They will then be compared to the bit-array representation, both by the cost of the resulting optimization process and by the accuracy of the solution.

22.5   Two new representations

22.5.1   The "holes" representation

Consider the design domain to be made of plain material, and imagine cutting holes into it. Neither the number of holes nor their size is fixed. The result of such cutting is a shape included in the design domain.

More precisely, we restrict ourselves to rectangular holes (merely for technical reasons), defined by their centers and side lengths. An individual is then a number R of such holes, and the description of the R holes. The corresponding shape is easily computed by superposing the holes in the design domain (see Figure 11).

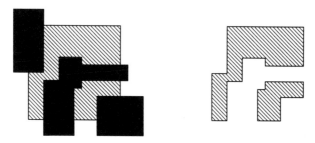

Figure 11 Five rectangular holes, and the corresponding structure.

The advantages of such a representation over the bit-array representation described in section 22.2 are quite clear.

- This representation is totally independent of any discretization of the design domain; Of course, the shape obtained from this representation will be discretized during the Finite Element Analysis. But the size of the genotypes is totally independent of this discretization;
- Large areas can be described by a few numbers of holes: the complexity of the representation is not fixed. Moreover, adaptative complexity can be achieved by including evolutionary operators that change the number of holes of an individual.

Genetic operators for the holes representation

The crossover operator is a straightforward adaptation of the *diagonal* crossover presented for the bit-array representation in section 22.2.1: a random line is drawn across the design domain, and the holes whose center are on one side of this line are exchanged between the parents (Figure 12).

There are 4 mutation operators:

- Add a hole
- Delete a hole
- Move a hole. The coordinates of the center of the hole are moved by addition of a Gaussian function, whose standard deviation depends on the fitness of the structure at hand, in an EP-like way [3].

---

[3] Adaptative mutation by having each hole carry its standard deviation, in an ES-way, is the following step for smart mutations

- Change the size of a hole. Here again both sides of the rectangular holes are changed according to a Gaussian law, with the same remarks than for the "move" mutation.

The probability of application of each mutation operator is governed by user-supplied weights (only one at the time is applied).

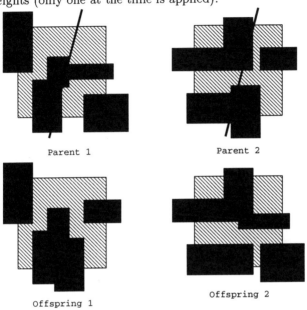

Figure 12 Example of crossover between two holes-represented structures.

Degeneracy of the holes representation

The holes representation clearly contains some degeneracy: some rectangular holes might be completely included in the union of other holes. Of course, another mutation operator could be used, aiming at deleting useless holes, i.e. holes that are themselves included in other unions of holes; but even without useless holes, the holes representation would remain highly degenerate, as different rectangles configurations can give the same shape without any useless hole.

Another argument go against the holes representation: diagonal shape boundaries require a lot of holes (but allowing triangular holes would probably fix that problem, and is on-going work).

Anyway, the Voronoï representation, introduced in next section, is an attempt to overcome these a priori weaknesses of the holes representation.

22.5.2   Voronoï-based representation

Voronoï diagrams provide nice way to describe any partition of the design domain into polygonal elements.

Voronoï diagrams

Given some points $S_1, \ldots, S_l$ of $\mathbb{R}^n$ called the *sites*, the Voronoï diagram defined by these sites is a partition of the space into convex polygons, or *cells*, each cell being defined as

$$Cell(S_i) = \{M \in \mathbb{R}^n; d(M, S_i) = min_{i=j,n} d(M, S_j)\}$$

To define a shape from a Voronoï diagram, it is sufficient to give each site (thus each cell) a color. Figure 13 gives an example of a shape, together with the associated Voronoï diagram.

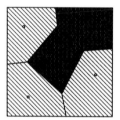 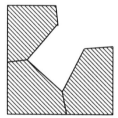

Figure 13 A Voronoï diagram with 5 sites, and the corresponding partition of the design domain.

Genetic operators

The genetic operators for the Voronoï representation look much like those for the holes representation: the crossover operator uses the same *diagonal* random line to exchange Voronoï sites between the parents. The mutation operators are here only three, namely the "add", "delete" and "move" operator, all acting on the Voronoï sites. Figure 14 shows an example of the crossover operator. An important remark is that each operator for the Voronoï representation is more global that the corresponding operator for the holes representation, due to the fact that each Voronoï cell actually depends on all sites surrounding it, where a hole is not modified whatever modification happen to other holes.

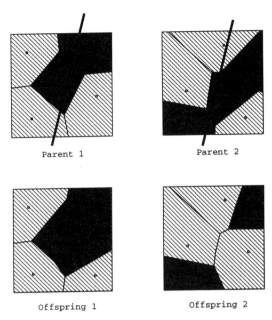

Figure 14 Example of crossover between two Voronoï diagrams.

Degeneracy of the Voronoï representation

The amount of degeneracy in the Voronoï representation is clearly not as obvious as for the holes representation. Nevertheless, it does happen that some sites are useless: they are surrounded by sites of the same color, and the resulting shape is not modified if they are removed. As in the holes representation, these sites could be removed, either systematically, or by some mutation operator, though at a somewhat high computational cost. And a Voronoï representation without useless sites is indeed a non degenerate representation.

The following section compares the new representations (the holes and the Voronoï representations) with the bit-array representation of section 22.2. The implementations of the holes and the Voronoï representations are straightforward; No attempt to remove any degeneracy is yet implemented.

22.6   First comparative results

A comprehensive experimental comparison of the three representations presented so far was made on the benchmark 2×1 cantilever plate using a 20×10 discretization, thus resulting in about 1 hour for 10000 evaluations.

Averages over 10 runs are considered, and the computational resources was limited to 7000 Finite Element Analyzes for all runs.

As expected, and in spite of its absence of intrinsic degeneracy, the bit-string representation was consistently outperformed by both other representations - and for a rather coarse discretization. But in every experiment, the holes representation performs significantly better than the Voronoï representation, both in average fitness and in best result. Figure 15 plots an example of such comparative experiments: For that particular problem, the solution is known (the one-element wide beam of weight 0.1), as the constraint on the displacement is very large. Out of 10 runs, it was found exactly 3 times by the holes representation, and never by the Voronoï representation. And similar results were observed, whatever D_{lim} and α.

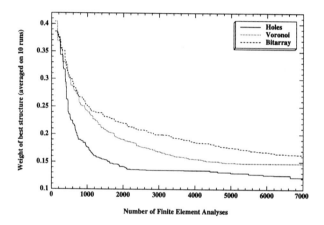

Figure 15 Comparison of the three representations on the benchmark cantilever problem. The weights are given in percentage of the full plate, $D_{Lim} = 371$, so that the perfect horizontal beam of weight 10% is <u>the</u> solution ($D_{Max} = 370.6$), and $\alpha = 1$.

22.6.1 Strong vs weak inheritance

An important feature that all works on problem instances of GAs try to capture is the amount of useful genetic traits that is transmitted from the parents to the offsprings during crossover. And a partial explanation of the reason why the holes representation seems to outperform the Voronoï representation can be derived by a close examination of that feature in both representations: when a hole is inherited by an offspring, the area covered by this hole is void, once and for all. But when a Voronoï site is transmitted to an offspring, the amount of what is really transmitted, in terms of void/material repartition, is totally dependent on the other sites present in the offspring. The holes representation implies *strong inheritance* of genetic traits (at least for the "void" trait), while the Voronoï representation achieves only *weak inheritance*, in which any genetic trait can only be determined by looking at the whole genetic characteristics of the individual.

More careful experiments are needed in order to make these statements precise, such as fitness variance analysis [Rad94] or fitness distance correlation [JF95]. And theoretical explorations certainly can help to understand these results, and more generally to a priori predict the behavior of a given representation on a given problem.

22.7 Conclusion: Evolutionary Computation and Structural Mechanics

Apart from the computational cost, which can hardly be denied, the lack of "sense of Mechanics" is generally the argument of many engineers and researchers in structural Mechanics against evolutionary methods. But such a position is generally the results of a "black box" use of GAs, resulting from the "find a bit-string representation for your problem and you're done" paradigm. We hope to have given arguments in that paper for a different approach to evolutionary optimization:

- at the level of the representation, first. It has been shown how a change of representation can increase the power of evolutionary optimization, while decreasing the complexity of the whole process. Both the Voronoï and the Holes representation outperform the bit-array representation, present a lower complexity for the representation of the same problem. Moreover, they are discretization-independent representations, allowing some iterated process: first find (rather quickly) a coarse solution, and then end with a few generations using a finer discretization, to refine the coarse solution. And finally, the same hierarchy should look even clearer in three dimensions: the bit-array representation clearly reached its limits with 2-dimensional problems, while Voronoï and holes representations hopefully can be used with success on 3-dimensional problems. But this remains to be experimentally demonstrated.
- At the level of the genetic operators, it should be clear from section 22.2.1 that these should be chosen according the problem at hand, and that the success can depend on that choice.
  But even more specific operators can be designed, depending on the problem, not only on the representation, as in section 22.2.1, and relying on that "sense of Mechanics": For the Optimal Design problem, the whole stress field provides some indications on where the structure at hand is the strongest and the weakest. Such information could be used by the

mutation operator, adding material with higher probability where the stress field is large, and more likely removing material where the stress field is small.

- Finally, at the fitness level, when it comes to understand the results. As was shown for the Optimal Design problem in the nonlinear case in section 22.4.2, only experts in the field can sometimes prevent the evolutionary process from giving non-viable answers.

Many more refinements of the evolutionary process can be imagined. They require knowledge in both Evolutionary Computation (e.g. to respect the fundamental principles of EC, as described in [Gol89b]) and the application domain, (e.g. to both give hints and respect here again the underlying basic principles of the domain). Further work tries to use such synergy to design intelligently biased genetic operators for both the Optimal Design problem and other Structural Mechanics problems (e.g. Inclusion Identification).

22.8   Summary

Three representations for 2D or 3D shapes have been presented, and tested on the problem of Topological Optimal Design of cantilever plates.

The first important issue is demonstrated by extensive experiments using the "natural" bit-array representation. Evolutionary Computation can bring a real breakthrough to the field of Structural Optimization, provided the design of the fitness function takes into account mechanical expertise: a modification of the expression of the fitness has been necessary, first to handle the problem-dependent degeneracy (section 22.3.4), and then after introducing the nonlinear mechanical model, to get mechanically reasonable solutions (section22.4.2).

But the bit-array representation is clearly limited, as its complexity depends on the underlying discretization of the design domain: refining this discretization, or optimizing 3D structures rapidly leads to intractable problems. The Holes and Voronoï representations both seem to overcome partially these difficulties, as their complexity is not fixed, and is evolved by the Evolutionary Algorithm. But they both exhibit a high level of degeneracy. Nevertheless, the first comparative experiments demonstrate they both outperform the non-degenerate bit-array representation, suggesting that complexity plays a greater role than degeneracy, for the problem at hand. Moreover, the Holes representation consistently performs best on the Optimal Design problem, both in quality of the results and in average computational cost. This may be due to the property of strong inheritance of some genetic traits that the Voronoï representation fails to possess. Nevertheless, theoretical studies and further experiments including different problems are needed in order to draw more affirmative conclusions.

REFERENCES

[AK93] G. Allaire and R. V. Kohn. Optimal design for minimum weight and compliance in plane stress using extremal microstructures. *European Journal of Mechanics, A/Solids*, 12(6):839–878, 1993.

[Ang93] P. J. Angeline. *Evolutionary Algorithms and Emergent Intelligence*. PhD thesis, Ohio State University, Colombus, 1993.

[BFJ95] E. Bonnetier, G. Francfort, and F. Jouve. Shape optimization by the homogenization method. *submitted to Nümerische Mathematik*, 1995.

[BK88] M. Bendsoe and N. Kikushi. Generating optimal topologies in structural design using a homogenization method. *Computer Methods in Applied Mechanics and Engineering*, 71:197–224, 1988.

[BS95] T. Bäck and M. Schütz. Evolution strategies for mixed-integer optimization of optical multilayer systems. In 5th *Annual Conference on Evolutionary Programming*, March 1995.

[Cia88] P. G. Ciarlet. *Mathematical Elasticity, Vol I : Three-Dimensional Elasticity*. North-Holland, Amsterdam, 1988.

[Fog95] D. B. Fogel. *Evolutionary Computation. Toward a New Philosophy of Machine Intelligence*. IEEE Press, 1995.

[FOW66] L. J. Fogel, A. J. Owens, and M. J. Walsh. *Artificial Intelligence through Simulated Evolution*. New York: John Wiley, 1966.

[FS95] A. Fadda and M. Schoenauer. Evolutionary chromatographic law identification by recurrent neural nets. In 5th *Annual Conference on Evolutionary Programming*, March 1995.

[Gol89a] D. E. Goldberg. *Genetic algorithms in search, optimization and machine learning*. Addison Wesley, 1989.

[Gol89b] D. E. Goldberg. Zen and the art of genetic algorithms. In *Proceedings of the 3rd International Conference on Genetic Algorithms*, pages 80–85, 1989.

[Hol75] J. Holland. *Adaptation in natural and artificial systems*. University of Michigan Press, Ann Arbor, 1975.

[Jen92] E. Jensen. *Topological Structural Design using Genetic Algorithms*. PhD thesis, Purdue University, November 1992.

[JF95] T. Jones and S. Forrest. Fitness distance correlations as a measure of problem difficulty for genetic algorithms. In *Proceedings of the 6th International Conference on Genetic Algorithms*, pages 184–192, 1995.

[JM91] C. Z. Janikow and Z. Michalewicz. An experimental comparison of binary and floating point representations in genetic algorithms. In R. K. Belew and L. B. Booker, editors, *Proceedings of 4th International Conference on Genetic Algorithms*, pages 31–36. Morgan Kaufmann, July 1991.

[Jou93] F. Jouve. *Modélisation mathématique de l'œil en élasticité non-linéaire*, volume RMA 26. Masson Paris, 1993.

[JS95] F. Jouve and M. Schoenauer. Evolutionary lamé coefficient identification. *In preparation*, 1995.

[Kan95] C. Kane. *Algorithmes génétiques et Optimisation topologique*. PhD thesis, Université de Paris VI, 1995. Soutenance prévue en cours d'année.

[KJS95] C. Kane, F. Jouve, and M. Schoenauer. Structural topology optimization in linear and nonlinear elasticity using genetic algorithms. In *Proceedings of the 21st Design Automation Conference*. ASME, Sept. 1995. To appear.

[KM95] A. B. Kahng and B. R. Moon. Toward more powerful recombinations. In *Proceedings of the 6th International Conference on Genetic Algorithms*, pages 96–103, 1995.

[Mic95] Z. Michalewicz. Genetic algorithms, numerical optimization and constraints. In *Proceedings of the 6th International Conference on Genetic Algorithms*, pages 151–158, 1995.

[Rad91] N. J. Radcliffe. Equivalence class analysis of genetic algorithms. *Complex Systems*, 5:183–20, 1991.

[Rad94] N. J. Radcliffe. Fitness variance of formae and performance prediction. In *FOGA3*, 1994.

[Rec73] I. Rechenberg. *Evolutionstrategie: Optimierung Technisher Systeme nach Prinzipien des Biologischen Evolution*. Stuttgart: Fromman-Holzboog Verlag, 1973.

[Sch81] H.-P. Schwefel. *Numerical Optimization of Computer Models*. John Wiley & Sons, New-York, 1981.

[SX93] M. Schoenauer and S. Xanthakis. Constrained ga optimization. In Forrest S., editor, *Proceedings of 4th International Conference on Genetic Algorithms*, pages 573–580. Morgan Kaufmann, 1993.

[Sys89] G. Syswerda. Uniform crossover in genetic algorithms. In R. K. Belew and L. B. Booker, editors, *Proceedings of the 4th International Conference on Genetic Algorithms*, pages 2–9. Morgan Kauffman, 1989.

[ZMR94] J. M. Zurada, R. J. Marks, and C. J. Robinson, editors. *Computational Intelligence Imitating Life*. IEEE Press, June 1994.